Azure SQL Hyperscale Revealed

High-performance Scalable Solutions for Critical Data Workloads

Zoran Barać
Daniel Scott-Raynsford

Apress®

Azure SQL Hyperscale Revealed: High-performance Scalable Solutions for Critical Data Workloads

Zoran Barać
Resolution Drive, Auckland, 0930, New Zealand

Daniel Scott-Raynsford
Victory Road, Auckland, 0604, New Zealand

ISBN-13 (pbk): 978-1-4842-9224-2
ISBN-13 (electronic): 978-1-4842-9225-9
https://doi.org/10.1007/978-1-4842-9225-9

Managing Director, Apress Media LLC: Welmoed Spahr
Acquisitions Editor: Jonathan Gennick
Development Editor: Laura Berendson
Editorial Assistant: Shaul Elson

Cover image by Felipe Portella from Unsplash

Distributed to the book trade worldwide by Springer Science+Business Media New York, 1 New York Plaza, Suite 4600, New York, NY 10004-1562, USA. Phone 1-800-SPRINGER, fax (201) 348-4505, e-mail orders-ny@springer-sbm.com, or visit www.springeronline.com. Apress Media, LLC is a California LLC and the sole member (owner) is Springer Science + Business Media Finance Inc (SSBM Finance Inc). SSBM Finance Inc is a **Delaware** corporation.

For information on translations, please e-mail booktranslations@springernature.com; for reprint, paperback, or audio rights, please e-mail bookpermissions@springernature.com.

Apress titles may be purchased in bulk for academic, corporate, or promotional use. eBook versions and licenses are also available for most titles. For more information, reference our Print and eBook Bulk Sales web page at www.apress.com/bulk-sales.

Any source code or other supplementary material referenced by the author in this book is available to readers on GitHub. For more detailed information, please visit www.apress.com/source-code.

Printed on acid-free paper

Table of Contents

About the Authors

Zoran Barać is a cloud architect and data specialist with more than 15 years of hands-on experience in data optimization, administration, and architecture. He is a Certified Microsoft Trainer (MCT) and Microsoft Certified Solutions Expert (MCSE) with a master's degree in information technology. Sharing knowledge and contributing to the SQL Server community are his passions. He is also an organizer of the Auckland SQL User Meetup Group, an active blogger, and a speaker at different SQL events such as Data Summits, SQL Saturdays, SQL Fridays, meetups, etc.

Daniel Scott-Raynsford is a partner technology strategist at Microsoft with 15 years of experience as a software developer and solution architect. He specializes in DevOps and continuous delivery practices. He was a Microsoft MVP in cloud and data center management for three years before joining Microsoft and is an active PowerShell open-source contributor and Microsoft DSC Community Committee member. He is also a contributor to the Azure Architecture Center on multitenant architecture practices.

About the Technical Reviewers

Jonathan Cowley is a cloud infrastructure engineer, DevOps practitioner, and technical documentation writer with more than 15 years of experience in various roles and environments, in both small and large corporations and public organizations. He has been Microsoft Azure and AWS certified. He enjoys connecting people and information and solving challenging problems.

Vitor Fava has earned the title of Most Valuable Professional (MVP AI and Data Platform) and the main certifications related to SQL Server, such as MCP, MCTS, MCITP, MCSA, and MCSE. A DBA with more than 20 years of experience in database and information technology, Vitor works on the development, implementation, maintenance, and support of large corporate database servers. Vitor graduated with a degree in information systems from Universidade Presbiteriana Mackenzie/SP, with an emphasis on high-performance and mission-critical database environment management.He has been a speaker at several technology events, such as SQLBITS, SQL Saturday, The Developers Conference (TDC), SQL Porto, InteropMix, and several discussion groups focused on database technology. Vitor also provides advanced training in SQL Server and Azure SQL Database, in addition to acting as an MCT in official Microsoft training. He is also the chapter leader of the PASS Chapter SQL Maniacs in São Paulo, Brazil.

Introduction

Azure SQL is a platform-as-a-service (PaaS) relational database management system (RDBMS) that is provided as part of the Azure cloud and based on the Microsoft SQL Server platform. This offering contains service tiers that make it suitable for a wide range of use cases, from small single-user applications to mission-critical lines of business workloads. However, there is only one service tier that provides the high level of scalability and resilience that is required by some of today's largest cloud-based workloads: Hyperscale.

In this book, we'll take you through the capabilities and basics of the Azure SQL Database Hyperscale service tier. You'll learn the basics of designing, deploying, and managing Azure SQL Hyperscale databases. We'll look in detail at the resilience (high availability and disaster recovery) and scalability of Azure SQL Hyperscale databases and how to tune them. We will also look at the different techniques you can use to deploy and configure Azure SQL Hyperscale.

Monitoring and securing Azure SQL Hyperscale databases will also be covered to ensure your workloads continue to perform and operate securely. This book will be your guide to deciding when Hyperscale is a good fit for migrating your existing workloads and what architectural considerations you should make in your application. Finally, you'll learn how to migrate existing workloads to Azure SQL Database Hyperscale.

This book is intended for data architects, software engineers, cloud engineers, and site reliability engineers and operators to learn everything they need to know to successfully design, deploy, manage, and operate Azure SQL Database Hyperscale.

You'll learn what makes Azure SQL Database Hyperscale architecture different from a more traditional database architecture and why it's important to understand the architecture.

To get the most out of this book, you should have a basic knowledge of public cloud principles and economics as well as a fundamental knowledge of Azure, Microsoft SQL Server, and RDBMSs. It is not required to have experience using Azure SQL Database. This book will provide you with everything you need to know to effectively build and operate the Hyperscale tier of Azure SQL Database without prior experience.

Although this book covers concepts that apply to Azure SQL Database in general, it is intended to provide everything you need to know to run an Azure SQL Database Hyperscale tier in Azure without requiring any other material. You may find content that is covered in other more general books on running SQL in Azure, but this book looks at these concepts through the Hyperscale lens. Hyperscale has special use cases that make it unique, which also require specific knowledge to get the most out of it.

Thank you for venturing into the world of Hyperscale database technology.

A note about the recently announced automatic compute scaling with serverless for Hyperscale in Azure SQL Database: At the time of writing this book, the possibility of an automatically scaling Hyperscale with a serverless compute tier was merely an item on the roadmap. We did not, therefore, cover it in detail, even though we were aware it was planned. However, serverless Hyperscale has just arrived in public preview. This development does not change the guidance provided in this book, but serverless should be considered for Hyperscale workloads that need high levels of elasticity. It is yet another tool in the amazing array of data storage services available in Microsoft Azure and should be on your list of considerations when determining if Hyperscale is right for you.

Serverless Hyperscale automatically adjusts the compute resources of a database based on its workload and usage patterns. It also allows different replicas of the database to scale independently for optimal performance and cost efficiency. Serverless Hyperscale is different from provisioned compute Hyperscale, which requires a fixed number of resources and manual rescaling.

For more information on the serverless Hyperscale announcement, see this page `https://techcommunity.microsoft.com/t5/azure-sql-blog/automatic-compute-scaling-with-serverless-for-hyperscale-in/ba-p/3624440`.

PART I

Architecture

CHAPTER 1

The Journey to Hyperscale Architecture in Azure SQL

Every day, many billions or even trillions of transactions are processed by the ever-growing number of Azure SQL databases deployed globally in the Microsoft Azure public and government clouds. The number of transactions and the size of these databases are continuously increasing. The accelerating adoption of public cloud technology has enabled new use cases for Azure SQL databases as well as the migration of larger line-of-business applications from on-premises SQL servers.

As the public cloud continues to mature, businesses are moving more of their mission-critical workloads to Azure. This results in the need to support larger and larger databases.

Accompanying this move to the cloud, the demand for IT departments and engineering teams to provide more application features with greater resilience and security has also increased. This pressure has caused organizations to look at how they can cut down on the operational costs of maintaining their infrastructure. The requirement to reduce operational expenditure has led toward running more databases as *platform-as-a-service* (PaaS) solutions—where the cloud provider takes care of the day-to-day operations of the infrastructure—freeing the organization to focus on higher-value tasks.

As well as the migration of enterprise-scale databases, an increase in the adoption of *software-as-a-service* (SaaS) solutions by consumers and organizations has led to SaaS providers needing larger databases, higher transaction throughput, and increasing resilience—while also being able to scale quickly.

Z. Barać and D. Scott-Raynsford, *Azure SQL Hyperscale Revealed*,
https://doi.org/10.1007/978-1-4842-9225-9_1

This explosive shift by organizations to adopting SaaS has also facilitated a need by SaaS providers to manage costs as they scale. This has often resulted in a change from a single database per tenant to a *multitenant approach*. A multitenant approach can help to reduce costs, but it also increases the need for larger capacities and throughput, while being able to assure isolation and avoid challenges such as *noisy neighbors*. For an in-depth explanation on that, see `https://learn.microsoft.com/azure/architecture/antipatterns/noisy-neighbor/noisy-neighbor`.

The growth trend toward SaaS is expected to continue as organizations and SaaS providers accommodate the expectations of their customers to gain greater value from their data with analytics and machine learning. This growth in demand for better analytics and machine learning abilities also facilitates the need to collect more data, such as usage telemetry and customer behavior data.

New types of application workloads and patterns, such as event streaming architectures and the Internet of Things (IoT), have also increased the demands on relational database management systems. Although alternative architectures exist to reduce the impact of these application workloads on the relational database, they don't completely mitigate the need for larger relational databases as a part of applications.

Customers are expecting solutions that can integrate with and accept data from other services and data sources. This is leading organizations and SaaS providers to build a greater level of integration into their software and also to be seen as platforms with an ecosystem of third-party applications and extensions springing up around them. This is further pushing the scale and elasticity required by the relational database management system (RDBMS) underlying these platforms.

Applications are also consuming data from more and more systems and providing their own combined insights and enhancements. For example, applications that integrate with Microsoft Teams might retrieve calling and messaging data and combine it with other data sources to provide additional functionality. As the volume of data increases within these systems, so does the need for more capacity in the application database that collects it.

Azure provides the Azure SQL Database service, a PaaS product, that includes several different tiers offering options for capacity and resilience while also enabling a high degree of scale.

The changing data landscape, driven by the cloud, SaaS, and newer workloads, has resulted in the need for a design change to the Azure SQL Database architecture to be pushed to the limit, requiring a completely new internal architectural approach: the Hyperscale architecture.

Hyperscale is the newest service tier of the Azure SQL Database and provides highly scalable compute and storage that enables it to significantly exceed the limits of the other tiers of service. This is achieved by substantial alterations to the underlying architecture. These changes are discussed in more detail in Chapter 2.

Tip Throughout this book we'll often refer to the Azure SQL Database Hyperscale service tier as just Hyperscale to make it easier on you, the reader.

At its core, Hyperscale helps us address these changing requirements, without the need to adopt radically different architectures or fundamentally change our applications—which in this new world of high demand for applications and features is a great thing.

SQL on Azure

It's been a long and progressive journey of more than 12 years from the beginnings of Azure SQL called CloudDB all the way up to the new Azure SQL Database Hyperscale service tier and Hyperscale architecture.

Before we can get into the details of Hyperscale, we will start by looking at the different ways that SQL workloads can be hosted in Azure. This will give us a view of where Hyperscale fits within the portfolio and when it is a good choice for a workload.

The Basics of Azure SQL

There are many benefits to hosting your data workloads in the cloud. Whether you want to utilize modern cloud offerings using Azure SQL Database, migrate an existing workload and retain your current SQL Server version running on an Azure virtual machine (VM), or simply adopt the newest SQL Server features, Azure SQL might be the right choice for you.

Fundamentally, there are two distinct ways SQL workloads can be deployed in Azure.

- As infrastructure as a service (IaaS) using SQL Server on Azure VMs
- As a platform as a service (PaaS) using Azure SQL Database (DB) or Azure SQL Managed Instance (MI)

Both Azure SQL deployment types (IaaS and PaaS) offer you redundant, secure, and consistent cloud database services. However, with IaaS you are responsible for more of the system, requiring a greater amount of effort from you in maintaining and ensuring the system is resilient and secure.

Tip Azure SQL MI is a PaaS database service with a higher level of isolation and compatibility with SQL Server on Azure VMs or on-premises SQL Servers.

Running SQL Server on Azure VMs provides you with the ability to configure every aspect of the underlying infrastructure, such as compute, memory, and storage as well as the operating system and SQL Server configuration. This high level of configurability comes with an increased operational cost and complexity. For example, operating system and SQL Server upgrades and patching are the user's responsibility.

By comparison, Azure SQL DB (and Azure SQL MI) is a managed database service where users do not need to worry about SQL Server maintenance tasks—such as upgrades, patching, or backups—or service availability. The Azure platform abstracts and manages these important aspects of your environment automatically, significantly reducing operational costs and complexity. This often results in a much lower total cost of ownership (TCO).

Generally, PaaS provides significant benefits that make it extremely attractive; however, in some cases, IaaS is still required. Some applications might need access to the underlying SQL instance or features of SQL Server that are not available on Azure SQL DB.

Tip For some workloads that need features that Azure SQL DB does not provide, Azure SQL MI might still be an option. This could still allow the workload to be moved to Azure, while leveraging the significant benefits of PaaS. For a comparison between Azure SQL DB and Azure SQL MI features, see `https://learn.microsoft.com/azure/azure-sql/database/features-comparison`.

Workloads that are being migrated from on-premises or a private cloud often require IaaS when they are first migrated into Azure but will be able to be re-platformed onto Azure SQL DB after some modernization. The benefits of this usually far outweigh the engineering costs of modernization. Another reason that IaaS might be needed is to gain a higher level of infrastructure isolation for regulatory compliance reasons.

Figure 1-1 compares the administration needs and costs of on-premises, private cloud machines, IaaS, and PaaS. We can see that PaaS has lower administration needs and costs than IaaS.

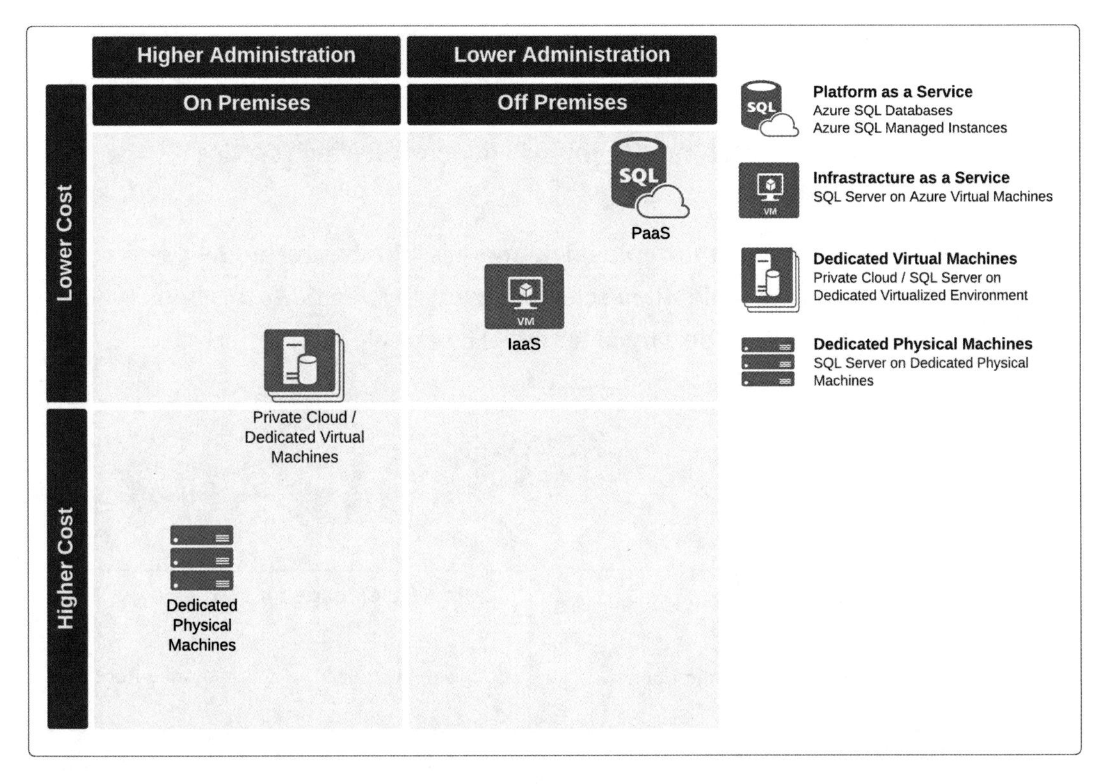

Figure 1-1. *Continuum of cost and administration of different hosting models*

We will be focusing primarily on Azure SQL DB, as this is the only deployment model that supports the newest service tier: Hyperscale.

Azure SQL Platform as a Service

When you select Azure SQL PaaS, you are getting a fully managed database service. Whichever deployment model you choose, you will always have some level of redundancy, starting with local redundancy and if available zone or geo-redundancy.

There are four key decisions you need to make when deploying a SQL workload on Azure.

1. Choose a deployment model.
2. Specify a purchasing model.
3. Specify a service tier.
4. Choose an availability model; this will depend on which service tiers you deployed.

Figure 1-2 shows a map of the deployment models, purchasing models, service tiers, and availability models available when selecting Azure SQL PaaS. As this book is all about Hyperscale, the most important area is the Hyperscale service tier.

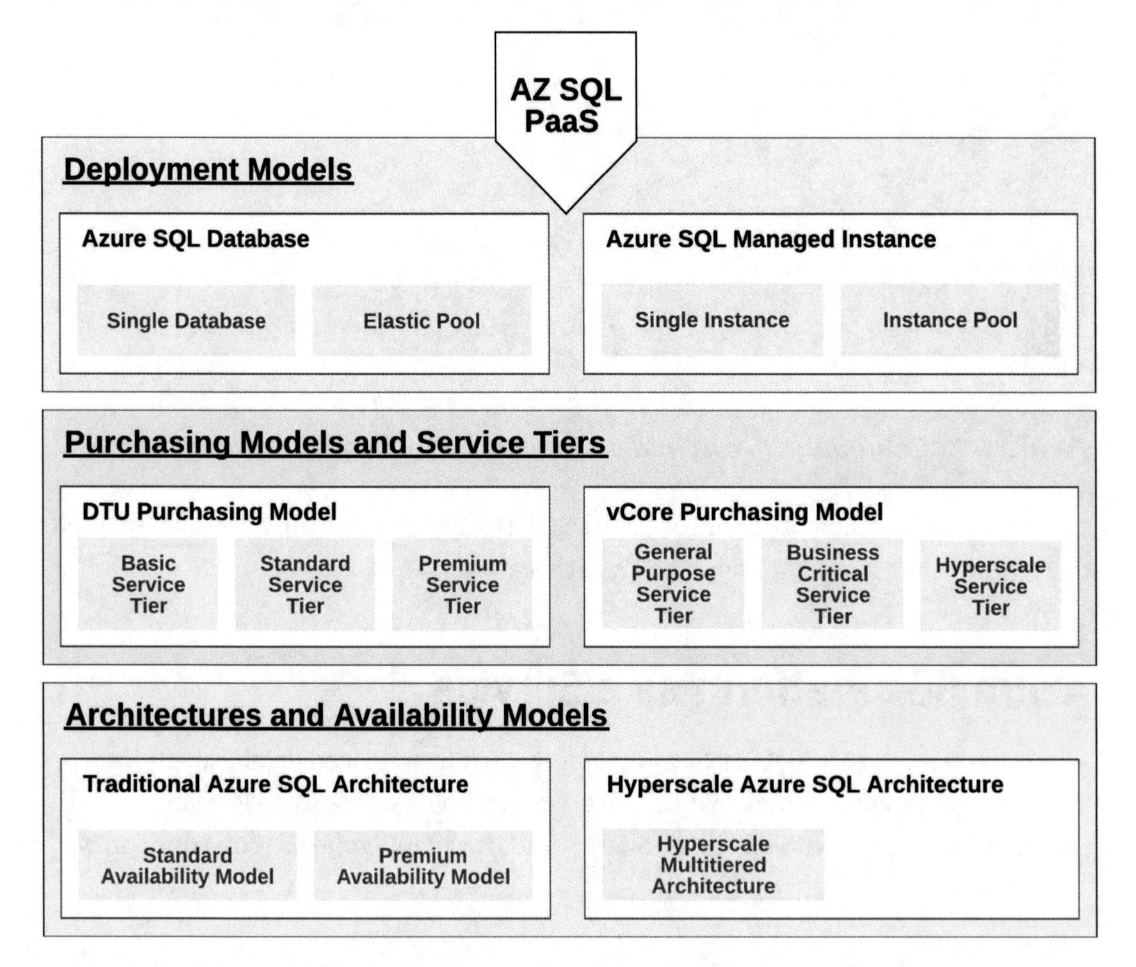

***Figure 1-2.** Azure SQL DB deployment models, purchasing models, service tiers, and availability models*

Deployment Models

The PaaS deployment models are divided into two types.

- Azure SQL Database
- Azure SQL Managed Instance

When you deploy an instance of Azure SQL DB, you have two deployment options: a single database or elastic pools. When using elastic pools, you can allow multiple databases to share a specified number of resources at a set price.

Tip Elastic pools are an excellent model for building SaaS applications, where each tenant has their own database; however, elastic pools don't currently support Hyperscale.

Azure SQL MI offers both single-instance and instance pool deployment options. Single instance supports only a single SQL Server instance, while instance pools allow multiple SQL Server instances.

Table 1-1 shows the deployment options available with Azure SQL PaaS, as well as the features and recommended use cases.

Table 1-1. Azure SQL PaaS Deployment Options

<table>
<tr><th colspan="2">Managed Instances</th><th colspan="2">Databases</th></tr>
<tr><th>Single Instance</th><th>Instance Pool</th><th>Single Database</th><th>Elastic Pool</th></tr>
<tr><td colspan="4">• Fully managed service.</td></tr>
<tr><td>• Minimum four vCores per instance.
• 16TB size storage for the General Purpose service tier.</td><td>• Hosting instance with smaller core number (two vCores).</td><td>• Hyperscale Service tier availability.
• Serverless compute tier for the General Purpose service tier.</td><td>• Price optimization—shared resources.
• Ability to limit necessary resources per database.</td></tr>
<tr><td colspan="2">• Supports most on-premises instance/database-level capabilities.
• High compatibility with SQL Server on-premises.
• At least 99.99 percent availability guaranteed.
• Built-in backups, patching, recovery.
• Latest stable database engine version.
• Easy migration from SQL Server.</td><td colspan="2">• The most used SQL Server features are supported.
• 99.99 percent to 99.995 percent availability guaranteed.
• Built-in backups, patching, recovery.
• Latest stable database engine version.</td></tr>
</table>

Tip You can use the Azure Portal to check the usage and properties for an existing MI pool, but not to deploy a pool. Instead, use a PowerShell session or the Azure CLI.

Purchasing Models and Service Tiers

Within Azure SQL all deployment offerings are provisioned on virtual infrastructure. Purchasing models and service tiers define the back-end hardware capabilities for specific deployments, such as machine instance and CPU generation, vCores, memory, and capacity. You need to consider the ideal hardware configuration based on your workload requirements. The exact offerings can vary depending on your deployment region.

Figure 1-3 shows the compute tier and hardware configuration for the General Purpose service tier for the Azure SQL Database deployment option.

Service and compute tier
Select from the available tiers based on the needs of your workload. The vCore model provides a wide range of configuration controls and offers Hyperscale and Serverless to automatically scale your database based on your workload needs. Alternately, the DTU model provides set price/performance packages to choose from for easy configuration. Learn more
Service tier
General Purpose (Scalable compute and storage options)
Compare service tiers
Compute tier
Provisioned - Compute resources are pre-allocated. Billed per hour based on vCores configured.
Serverless - Compute resources are auto-scaled. Billed per second based on vCores used.
Compute Hardware
Select the hardware configuration based on your workload requirements. Availability of compute optimized, memory optimized, and confidential computing hardware depends on the region, service tier, and compute tier.
Hardware Configuration
Standard-series (Gen5)
up to 128 vCores, up to 625 GB memory
Change configuration

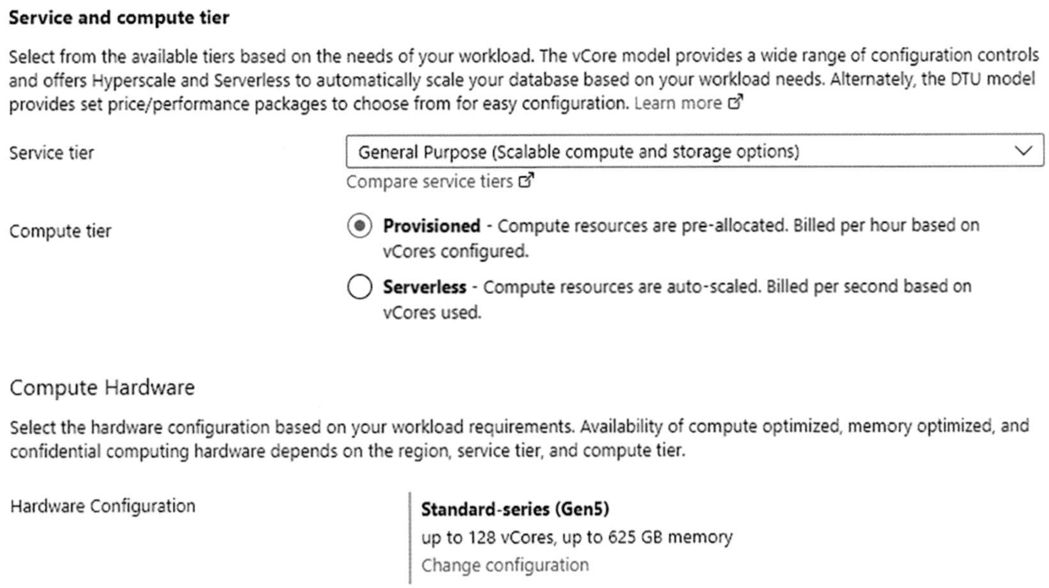

Figure 1-3. *Service and compute tier: Azure SQL DB General Purpose service tier*

There are two different purchasing models—vCore and DTU—each with multiple service tiers. The DTU purchasing model is based on data transaction units (DTUs). DTUs are a blended measurement of a machine's specifications—CPU, memory, reads, and writes. Figure 1-4 shows available purchasing models and service tiers for single Azure SQL Database deployment.

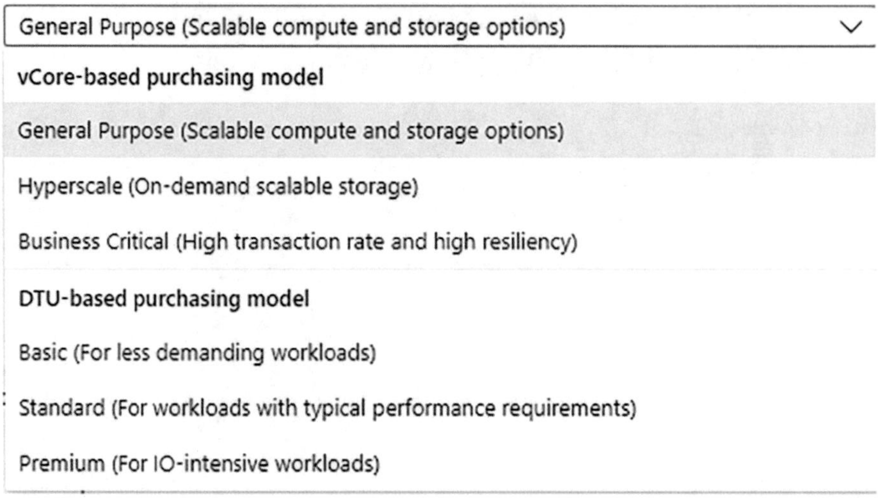

General Purpose (Scalable compute and storage options)

vCore-based purchasing model

General Purpose (Scalable compute and storage options)

Hyperscale (On-demand scalable storage)

Business Critical (High transaction rate and high resiliency)

DTU-based purchasing model

Basic (For less demanding workloads)

Standard (For workloads with typical performance requirements)

Premium (For IO-intensive workloads)

Figure 1-4. *Single Azure SQL DB purchasing models and service tiers available*

Tip For more information on DTU modeling, review this page: `https://learn.microsoft.com/azure/azure-sql/database/service-tiers-dtu`.

There are multiple service tier options within the DTU purchasing model:

- *Basic*: Use for less demanding workloads: up to 5 DTU, up to 2GB data, read latency of 5ms, and write latency of 10ms.
- *Standard*: Use for workloads with typical requirements: up to 3000 DTUs (S12), up to 1TB data, read latency of 5ms, and write latency of 10ms.
- *Premium*: Use for I/O-intensive workloads: up to 4000 DTUs, up to 4TB data, read and write latency of 2ms, read scale-out and zone-redundant options available.

The vCore purchasing model represents a choice of specific virtual hardware specifications. You select the generation of the hardware—starting with the CPU configuration, number of cores, maximum memory, and storage capacity. This model is transparent and cost efficient and gives you the flexibility to translate your on-premises workload infrastructure requirements to the cloud.

There are multiple service tier options within the vCore purchasing model.

- General Purpose
  - *Azure SQL DB*: 2–128 vCores, up to 625GB memory, and up to 4TB data.
  - *Azure SQL MI*: 4–80 vCores, Standard-series 5.1GB RAM/vCore, and up to 16TB data. Premium-series 7GB RAM/vCore, and up to 16TB data.
  - *Azure SQL DB Serverless*: 0.5–80 vCores, up to 240GB memory, and up to 4TB data.
- Business Critical
  - *Azure SQL DB*: 2–128 vCores, up to 625GB memory, and up to 4TB data.
  - *Azure SQL MI*: 4–80 vCores, Standard-series 5.1GB RAM/vCore and up to 4TB data. Premium-series 7GB RAM/vCore, and up to 16TB data.
- Hyperscale
  - *Standard-series (Gen5)*: 2–80 vCores, up to 415GB memory, and up to 100TB data.
  - *Premium-series*: 2–128 vCores, up to 625GB memory, and up to 100TB data.
  - *Premium-series memory-optimized*: 2–128 vCores, up to 830GB memory, and up to 100TB data.
  - *DC-series (enables confidential computing)*: 2–8 vCores, up to 36GB memory, and up to 100TB data.

You can choose the vCore purchasing model only for managed instances with the General Purpose or Business Critical service tier.

Availability Models and Redundancy

There are two main availability models in Azure SQL PaaS offerings.

- *Standard*: Separates compute and storage tiers. It relies on the high availability of the remote storage tier. This can result in some performance degradation during maintenance activities, so it is suitable for more budget-conscious users.
- *Premium*: Like the Always On availability option for VMs. It relies on a failover cluster of database engine processes. It is more reliable as there is always a quorum of available database engine nodes. This suits mission-critical applications with high I/O performance needs. It minimizes any performance impact to your workload during maintenance activities.

Depending on which service tiers you deployed, it may leverage the Standard availability architecture for the Basic, Standard, and General Purpose service tiers or the Premium availability architecture for the Premium and Business Critical service tiers.

There are three types of redundancy when running SQL workloads in Azure.

- *Local redundancy*: All Azure SQL PaaS deployments offer local redundancy. Local redundancy relies on the single data center being available.
- *Zone redundancy*: This replicates a database across multiple availability zones within the same Azure region. Azure availability zones are physically separate locations within each Azure region that are tolerant to local failures. To enable zone redundancy, you need to deploy your services within a region that supports availability zones. This protects your workload against failure of a single availability zone. For more information on Azure availability zones, see this page: `https://learn.microsoft.com/azure/availability-zones/az-overview`.
- *Geo-redundancy*: There are multiple ways to achieve geo-redundancy using failover groups (FOGs) or active geo-replication. Geo-replication is an Azure SQL Database business continuity and disaster recovery feature (asynchronous replication). This protects your workload against failure of an entire region.

Tip Zone-redundant configuration is available only for Azure SQL Database deployment and only for Gen5 hardware or newer. Using zone-redundant configuration with SQL Managed Instance is not currently supported.

Depending on which service tier you select, it may leverage either the Standard or Premium availability architecture model.

- *Standard availability model*: DTU purchasing model Basic and Standard service tiers and vCore purchasing model General Purpose service tier
- *Premium availability model*: DTU purchasing model Premium service tier and vCore purchasing model Business Critical service tiers

Standard Availability Model: Locally Redundant Availability

Basic, Standard, and General Purpose service tiers leverage the Standard availability model, which offers resilience to failure using locally redundant remote Azure Premium storage to store data and log files (`.mdf` and `.ldf` files). Backup files are stored on geo-redundant Azure Standard storage. High availability is accomplished in a similar way for the SQL Server Failover Cluster Instance (FCI), by using shared disk storage to store database files.

Figure 1-5 shows the Azure SQL PaaS Standard availability model architecture used by the Basic, Standard, and General Purpose service tiers.

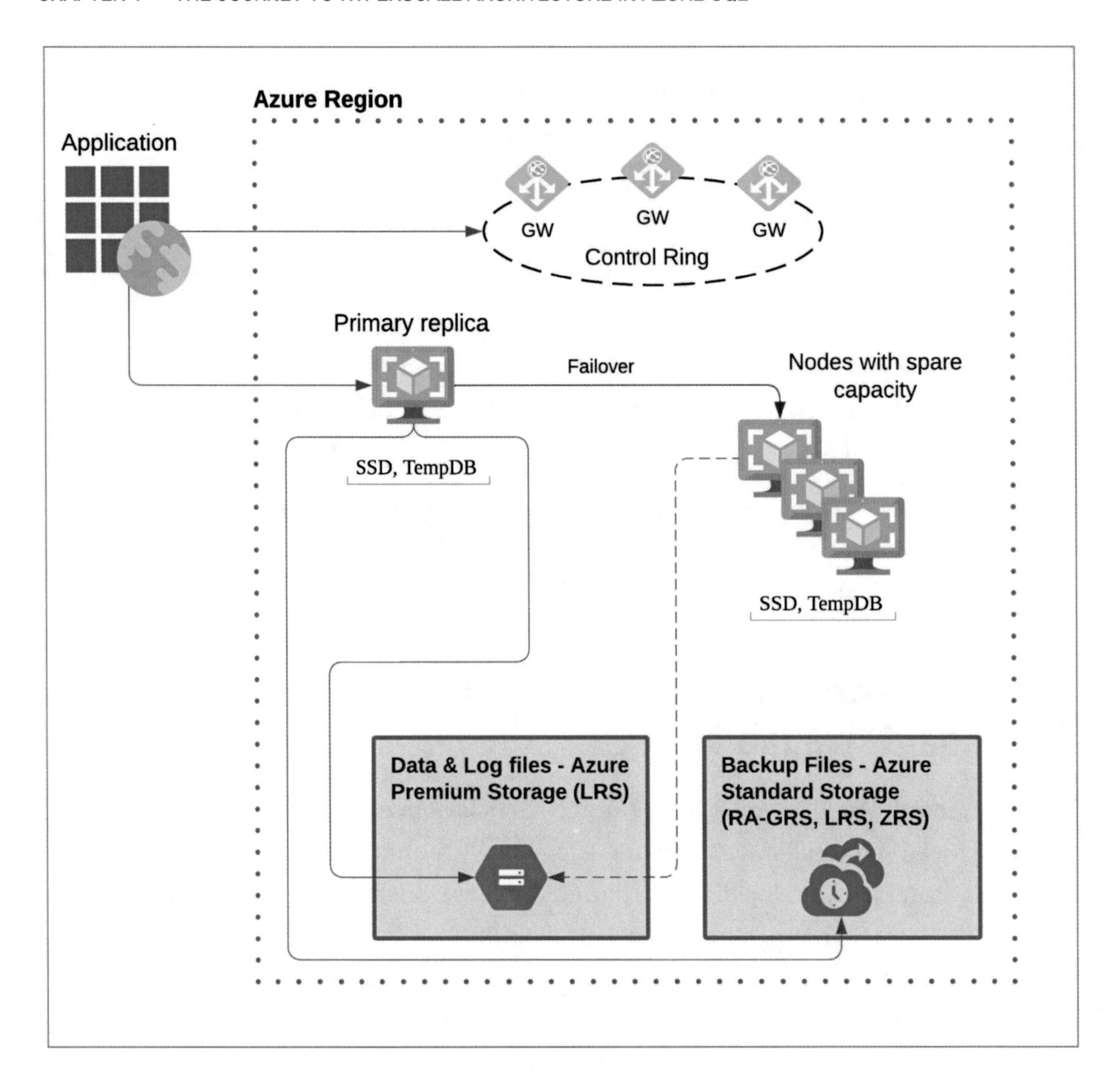

Figure 1-5. *Azure SQL PaaS Standard availability model architecture*

General Purpose Service Tier: Zone-Redundant Availability

Like Basic and Standard, the General Purpose service tiers leverage the Standard availability model. In addition to local redundancy, which is included with all Azure SQL PaaS deployment options, the General Purpose service tiers can offer zone redundancy as well.

This provides resilience to failure across different availability zones within the same region. Zone-redundant configuration for serverless and provisioned General Purpose tiers are available in only some regions.

Tip Be aware that with the General Purpose service tier, the zone-redundant option is available only when the Gen5 (or newer) hardware configuration is selected. Selecting zone redundancy incurs an additional cost.

Figure 1-6 shows the Azure SQL PaaS Standard availability model with zone redundancy used by the General Purpose service tier.

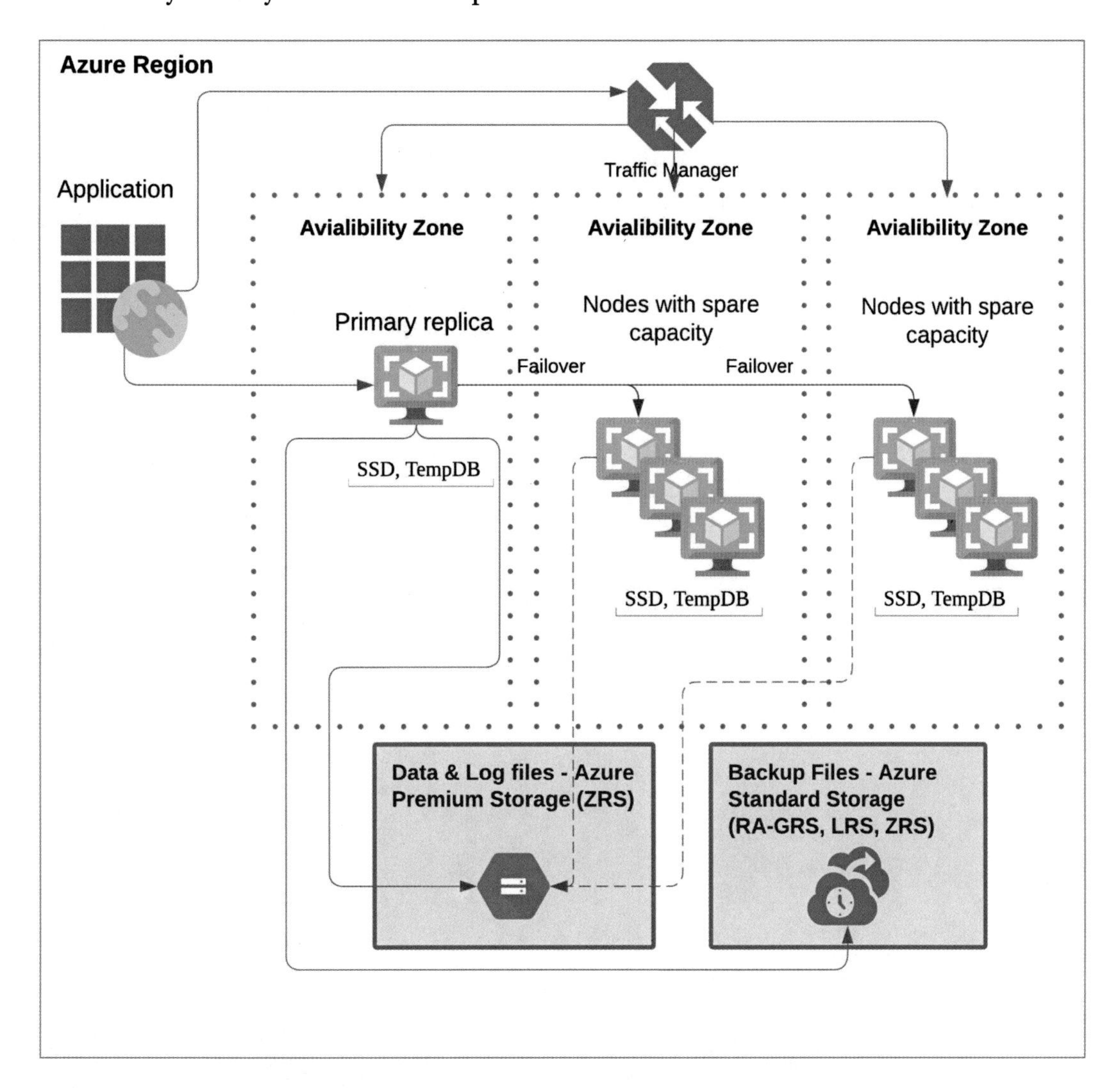

Figure 1-6. *Azure SQL PaaS Standard availability model with zone redundancy enabled*

Premium and Business Critical Service Tier: Locally Redundant Availability

The Premium and Business Critical service tiers leverage the Premium availability model, which offers resilience to failure using multiple replicas in synchronous data commit mode. Every replica has integrated compute and storage, which means that every node has an attached solid-state disk (SSD) with databases and log files on it (`.mdf` and `.ldf` files). High availability is accomplished in a similar way as with SQL Server Always On availability groups, by replicating nodes across the cluster using synchronous data commit mode.

Figure 1-7 shows the Azure SQL PaaS Premium availability model architecture used by the Premium and Business Critical service tiers.

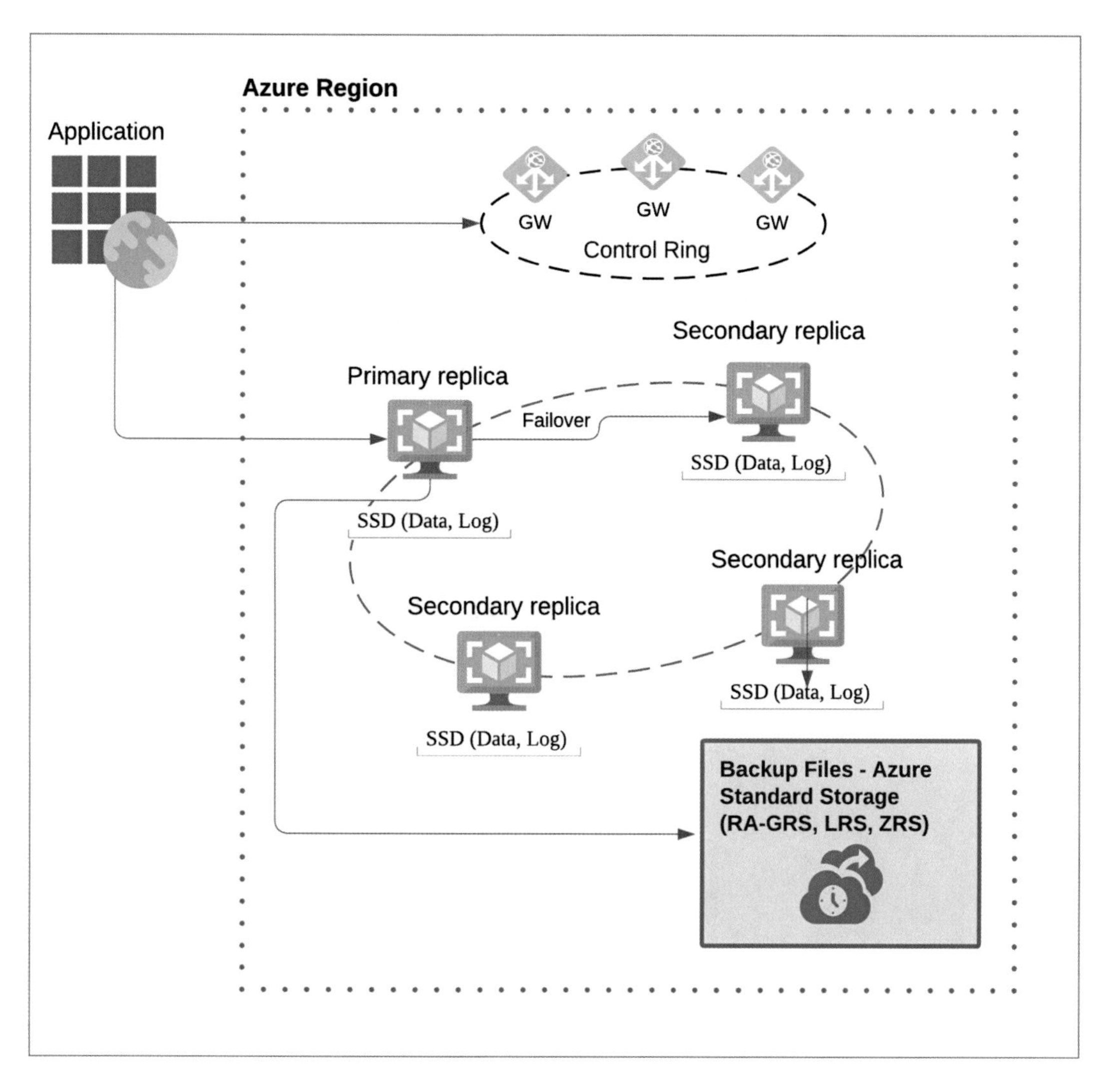

***Figure 1-7.** Azure SQL PaaS Premium availability model architecture*

Premium and Business Critical Service Tier: Zone-Redundant Availability

As mentioned earlier, the Premium and Business Critical service tiers leverage the Premium availability model, which offers resilience to failure using multiple replicas in synchronous data commit mode. Moreover, the zone-redundant deployment option provides additional resilience to an Azure availability zone (AZ) failure, if the failure is confined to a single AZ rather than an entire region. High availability is accomplished in a similar way for SQL Server Always On availability groups, replicating nodes across multiple availability zones in synchronous data commit mode.

Tip Azure SQL Database service tiers using the Standard availability model provides a greater than 99.99 percent uptime availability guarantee. Nevertheless, Premium and Business Critical service tiers—leveraging the Premium availability model with zone-redundant deployments—provide a greater than 99.995 percent uptime guarantee.

Figure 1-8 shows the Azure SQL PaaS Premium availability model with zone redundancy enabled for the Premium and Business Critical service tiers.

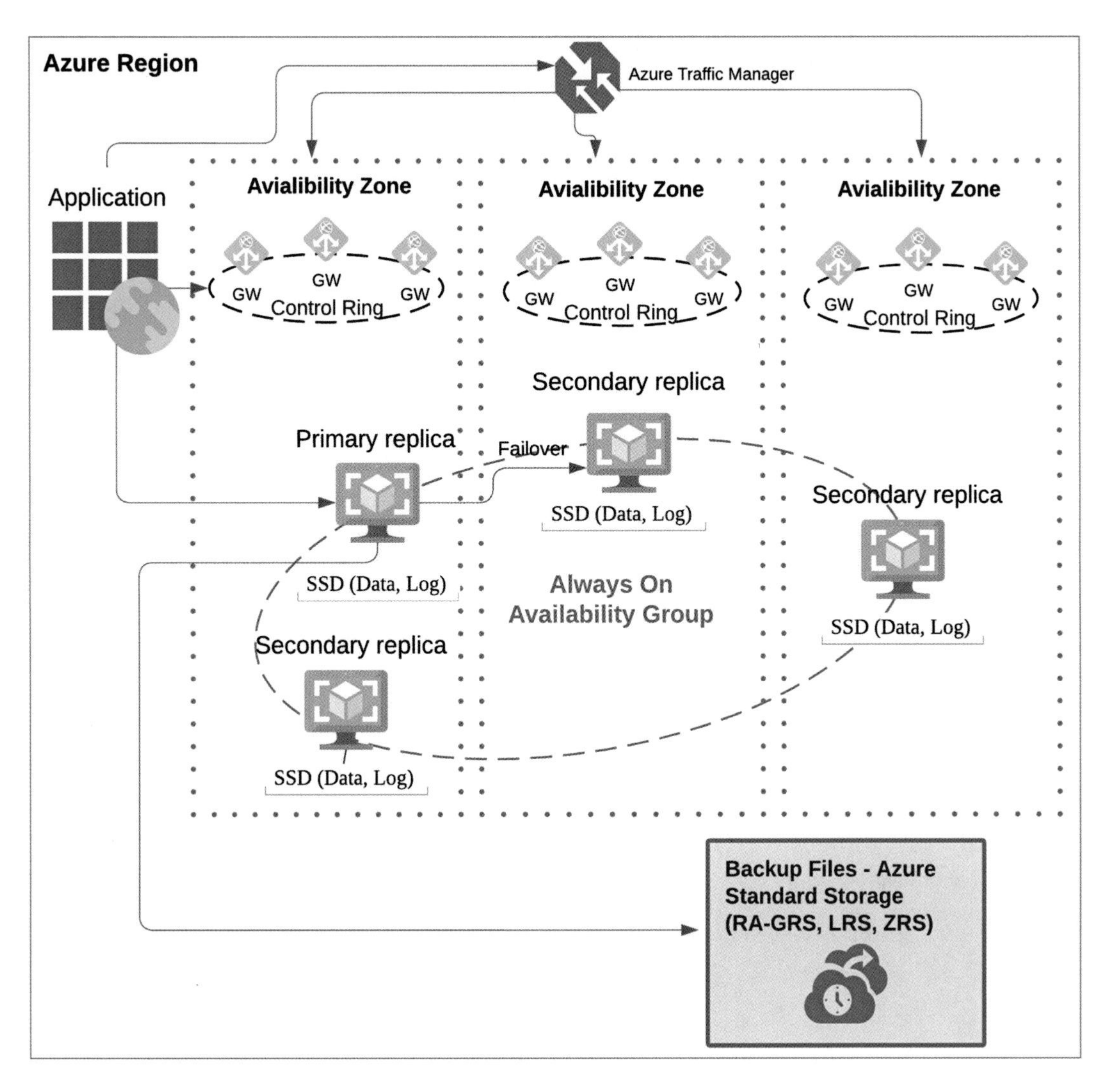

Figure 1-8. *Azure SQL PaaS Premium availability model with zone redundancy enabled*

Protecting Against Regional Outages Using Failover Groups with Geo-redundant Availability

To protect your SQL instance against failure of an entire Azure region and to achieve geo-redundancy with good disaster recovery capability, you can use failover groups. Failover groups enable asynchronous replication between multiple Azure regions (or within regions).

Figure 1-9 shows geo-replication using failover groups via asynchronous replication between multiple Azure regions.

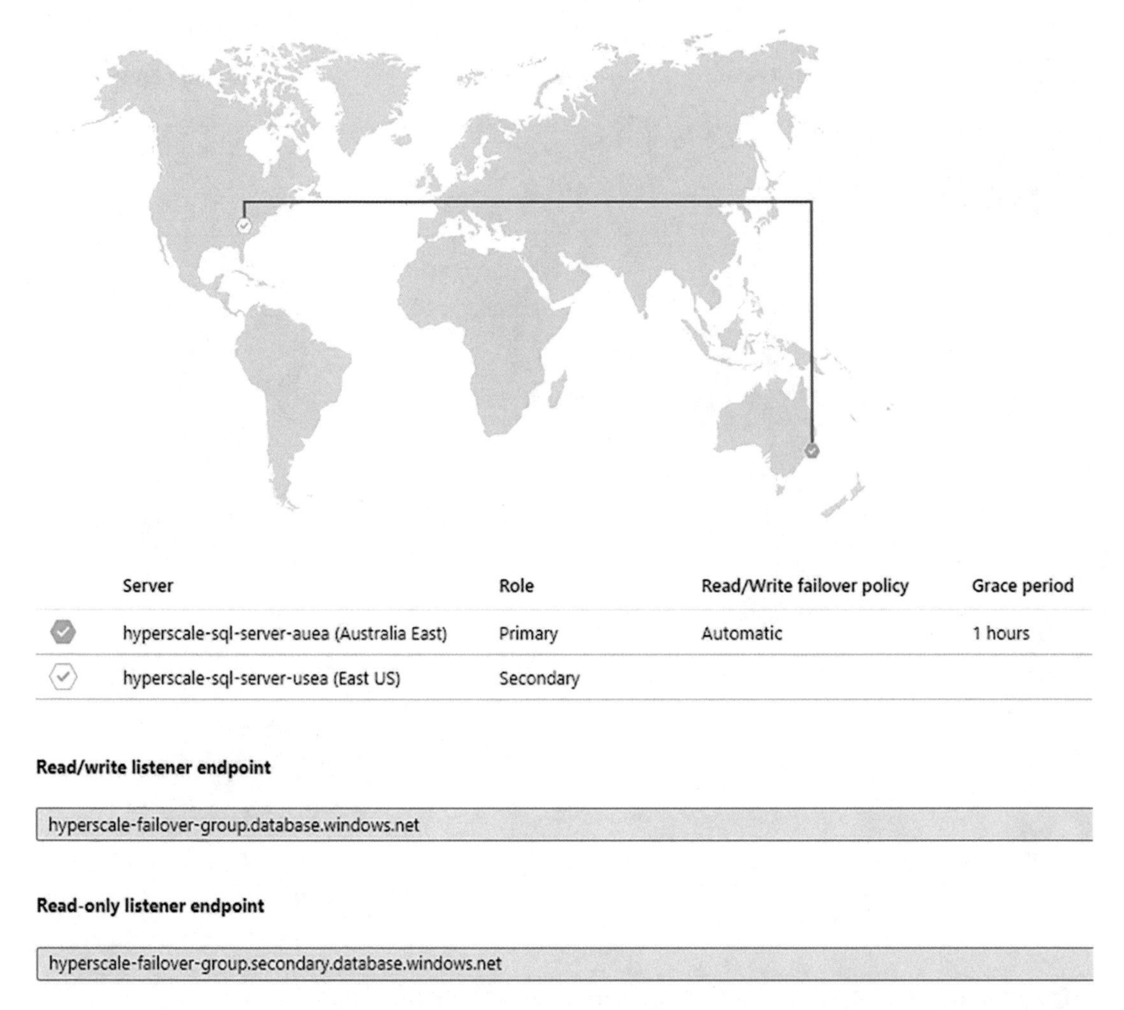

Figure 1-9. *Geo-replication between multiple Azure regions using failover groups*

The Hyperscale Service Tier

A traditional monolithic cloud architecture has its purposes but has several limitations, in particular relating to large database support and length of recovery time. For the past decade, Microsoft has been aiming to develop a new architecture model that can override these issues.

The goal of the new model has been to decouple compute, storage, and log services, while at the same time offering highly scalable storage and compute performance tiers.

At the time of writing, there is an Azure SQL database deployment model and the Hyperscale service tier under the vCore purchasing model, which leverages a new architecture, which in this book we will call the Hyperscale architecture.

Table 1-2 compares the traditional Azure SQL architecture and the Azure Hyperscale architecture.

Tip Hyperscale is available only under the Azure SQL Single Database deployment option; you cannot use the Hyperscale service tier with elastic pool deployment.

Table 1-2. *Traditional Azure SQL Architecture vs. Azure Hyperscale Architecture*

	Traditional Architecture	Hyperscale Architecture
Max Storage Size	Azure SQL DB: 4TB. Azure SQL MI Business Critical: 5.5TB. Azure SQL MI General Purpose: 16TB.	100TB.
CPU	Up to 128 vCores.	*Standard-series (Gen5)*: Up to 80 vCores. *Premium-series*: Up to 128 vCores.
Memory	*Standard-series (Gen5)*: up to 625 GB memory.	*Standard-series (Gen5)*: up to 415.23 GB memory. *Premium-series*: up to 625 GB memory. *Premium-series (memory optimized)*: up to 830.47 GB memory.
Latency	Premium availability model: 1–2ms. Standard availability model: 5-10ms.	Latency can vary and mostly depends on the workload and usage patterns. Hyperscale is a multitiered architecture with caching at multiple levels, data being accessed primarily via a cache on the compute node. In such cases, average latency will be similar as for Business Critical or Premium service tiers (1–2ms).

(*continued*)

Table 1-2. (*continued*)

	Traditional Architecture	Hyperscale Architecture
Log Throughput	Depending on vCores number. Up to 96 MB/s per node.	105MB/s regardless of the number of vCores.
Log Size	Up to 1TB log size.	Unlimited log size.
Read scale-out	1 read scale-out replica.	Up to 4 HA/Read scale-out replicas. Up to 30 named replicas.
Backup and Restore	Depending on database size.	Snapshot-based backup and restore operations do not depend on database size.

Tip The main question you should ask yourself is whether your data workload demands are exceeding the limits of your current Azure SQL service tier—in which case you might consider moving your data workload to a more suitable service tier such as Azure SQL Hyperscale, leveraging the Hyperscale architecture scale-out capability.

Hyperscale Architecture Overview

Azure SQL Hyperscale utilizes the four-tier architecture concept, which includes the remote durable storage service to achieve superior performance, scalability, and durability. Figure 1-10 shows the four-tier Hyperscale architecture concept of decoupling the compute layer from log services and scale-out storage.

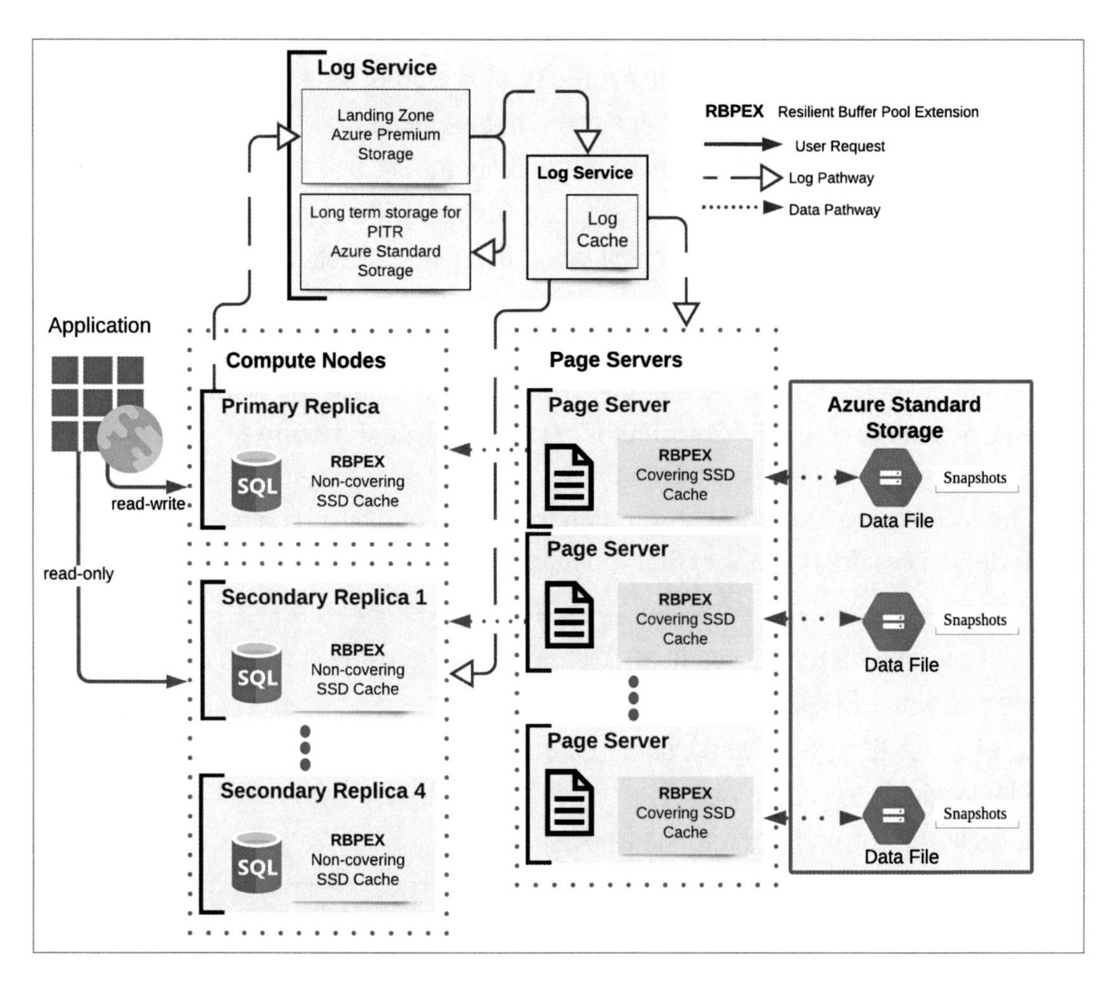

Figure 1-10. *Four-tier Hyperscale architecture concept*

Deploying Your First Hyperscale Database

Now that we have covered the basic options available when deploying a SQL workload in Azure, it's time to provision our very first Hyperscale database. For this, we're going to use a user interface–driven process in the Azure Portal.

Tip Deploying services using the Azure Portal is a great way to get started, but it is strongly recommended that as you mature in your use of Azure, you adopt infrastructure as code. To learn more about infrastructure as code, refer to `https://learn.microsoft.com/devops/deliver/what-is-infrastructure-as-code`. We will favor using infrastructure as code in the examples throughout this book.

You will need access to an Azure subscription that you have appropriate permissions to create new services in. The Contributor or Owner role in the Azure subscription will be adequate permissions. If you're not the subscription owner, you should first check with the account owner for the subscription if you may deploy a Hyperscale database as there will be a cost for every hour that you have the service deployed.

Tip If you don't have access to an Azure subscription, you can sign up for a free Azure one with USD $200 credit to use within 30 days; see `https://aka.ms/AzureFree`. After you have consumed the USD $200 credit, the free services will still be available to you for one year, but unfortunately the Hyperscale tier of Azure SQL DB is *not* currently among the free services.

The Azure Portal is under constant improvements, and it is quite possible that the user interface you see by the time this book is released will be slightly different. But the fundamental elements should remain. That is a reason why we will provide many examples as code.

In our example, we're going to deploy a zone-redundant Hyperscale database with two vCores running on Gen5 hardware in the Azure East US region with one high-availability secondary replica.

So, let's get started.

1. Open your browser and navigate to `https://portal.azure.com`.
2. Sign in using credentials that give you access to the Azure subscription you'll be deploying your Hyperscale database into.
3. Click the "Create a resource" button in the home dashboard.

4. Click Create under SQL Database on the "Create a resource" blade.

5. The Create a SQL Database blade will open showing the Basics configuration options of our database.

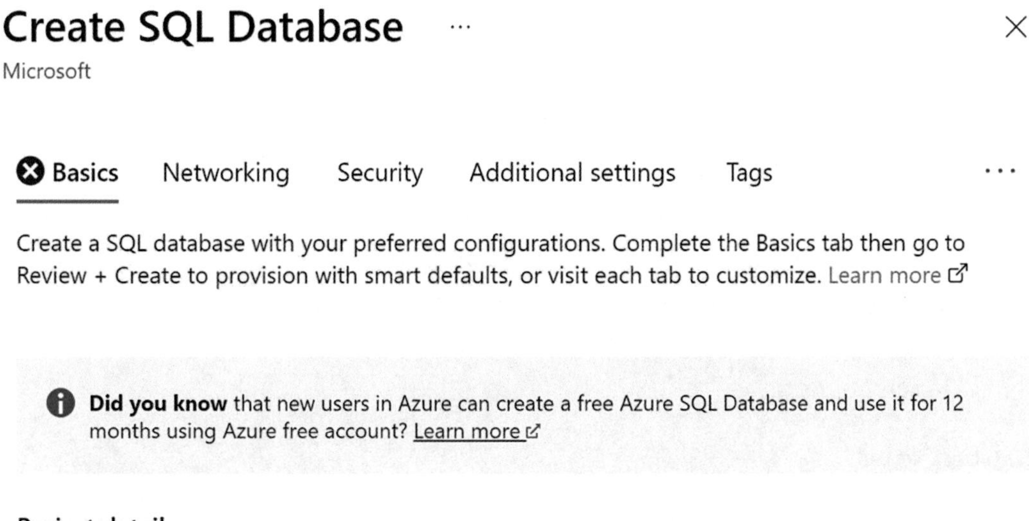

6. Make sure Subscription is set to the name of the subscription you have permission to deploy into.

7. Set the resource group to an existing resource group that you want to create your Hyperscale database into or click New to create a new resource group. In our example, we are going to create our hyperscale resources in an existing resource group called `sqlhyperscalerevealed-rg` that is in the Azure East US region.

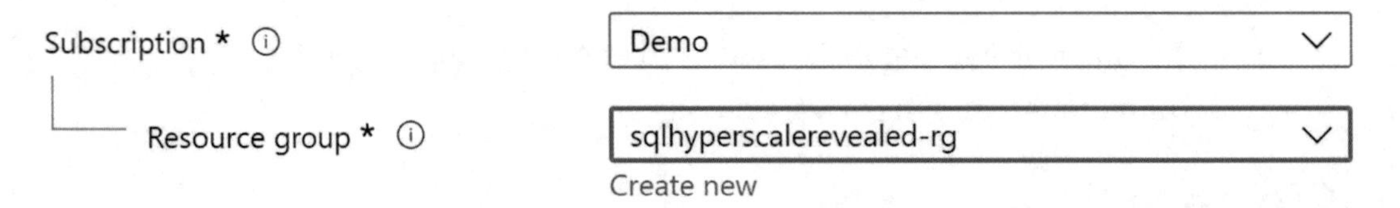
Project details
Select the subscription to manage deployed resources and costs. Use resource groups like folders to organize and manage all your resources.
Subscription *
Demo
Resource group *
sqlhyperscalerevealed-rg
Create new

8. Enter the name of the Hyperscale database to create in the "Database name" box. We are using `hyperscaledb` for this example.

9. Although Azure SQL Database is a PaaS service, there is still an underlying SQL Database server. We need to create a new logical instance of a SQL Database server by clicking the "Create new" button.

Database details

Enter required settings for this database, including picking a logical server and configuring the compute and storage resources

Database name * hyperscaledb

Server * Select a server

Create new

10. Every SQL Database server must have a unique name, not just within our environment, but globally. For this example, we are using `sqlhyperscalerevealed01`, but you should use a name that is likely to be globally unique.

11. Set the location to the Azure region in which you want to deploy the SQL Database server. This is the Azure region where your Hyperscale database will be stored and should usually be as geographically close to your application or users as possible.

Important Azure SQL Database Hyperscale is enabled in most Azure regions. However, if it is not enabled in the region you chose, select an alternate region, or see this page for more information: `https://learn.microsoft.com/azure/azure-sql/database/service-tier-hyperscale?view=azuresql#regions`.

---

12. Set the authentication method to "Use only Azure Active Directory (Azure AD) authentication."

Tip You can choose to use SQL authentication or to use both SQL and Azure AD authentication, but it is recommended you use Azure AD authentication to enable the central management of identities. However, some third-party applications may not support this. We will cover security best practices in detail in Chapter 15.

---

13. Click the "Set admin" button to select the Azure AD user or group that will be granted the admin role on this SQL Database server. But for this example, we will select our own Azure AD user account to be the admin.

Tip It is good practice to set the admin role to a separate Azure AD group that is created for this SQL Database server. Membership in this group should be carefully governed.

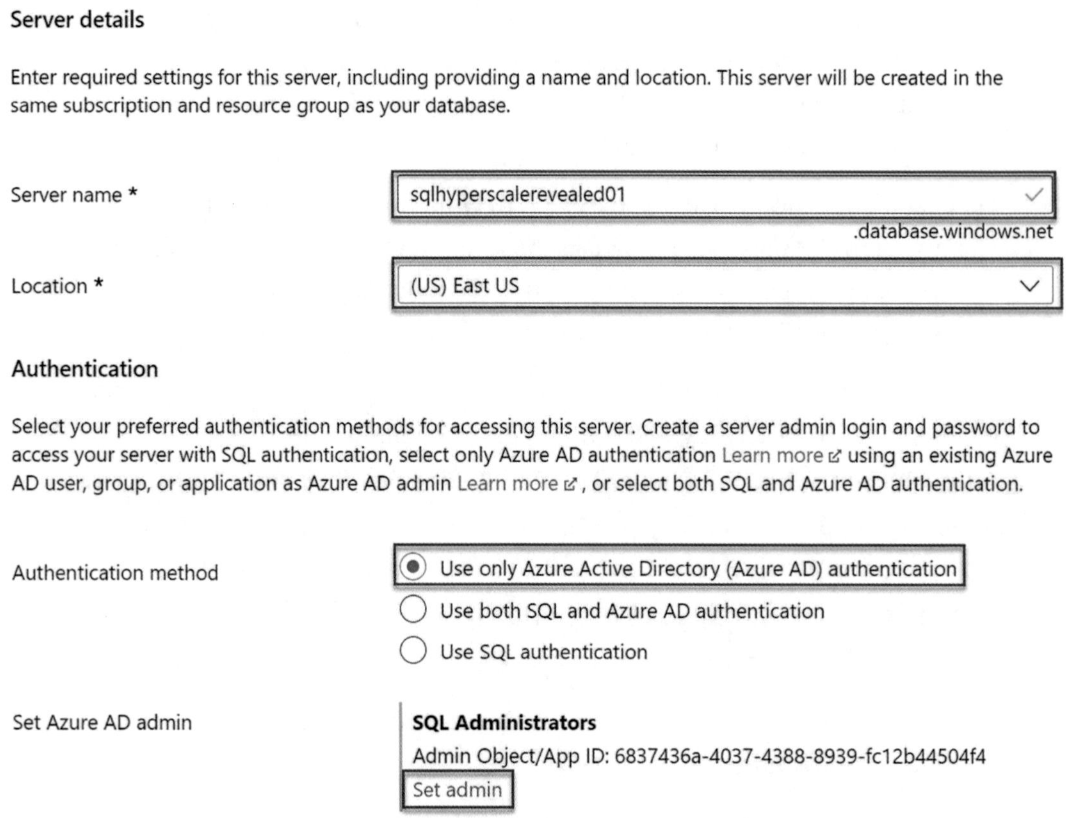

Server details

Enter required settings for this server, including providing a name and location. This server will be created in the same subscription and resource group as your database.

Server name * sqlhyperscalerevealed01 .database.windows.net

Location * (US) East US

Authentication

Select your preferred authentication methods for accessing this server. Create a server admin login and password to access your server with SQL authentication, select only Azure AD authentication Learn more using an existing Azure AD user, group, or application as Azure AD admin Learn more, or select both SQL and Azure AD authentication.

Authentication method
- Use only Azure Active Directory (Azure AD) authentication
- Use both SQL and Azure AD authentication
- Use SQL authentication

Set Azure AD admin **SQL Administrators**
Admin Object/App ID: 6837436a-4037-4388-8939-fc12b44504f4
Set admin

14. Click OK on the Create SQL Database Server blade.
15. We will return to the Create SQL Database blade to continue configuration.
16. As noted earlier in this chapter, Hyperscale does not support elastic pools, so specify No when asked if we want to use SQL elastic pools.
17. Select the Production environment. This affects the default compute and storage options that are suggested for the database. However, because we are going to specify Hyperscale, we can't accept the defaults anyway.
18. Click the "Configure database" button. This will open the Configure blade.
19. Set the service tier to "Hyperscale (On-demand scalable storage)."

20. The Hardware Configuration option should be set to "Standard-series (Gen5)."

21. We can leave the vCores option set to 2.

22. Set High-Availability Secondary Replicas to 1 Replica.

23. Specify Yes when asked if you would like to make this database zone redundant.

Service and compute tier

Select from the available tiers based on the needs of your workload. The vCore model provides a wide range of configuration controls and offers Hyperscale and Serverless to automatically scale your database based on your workload needs. Alternately, the DTU model provides set price/performance packages to choose from for easy configuration. Learn more

Service tier
Hyperscale (On-demand scalable storage)
Compare service tiers

Databases originally created in Hyperscale tier cannot be changed to other tiers.

Compute Hardware

Select the hardware configuration based on your workload requirements. Availability of compute optimized, memory optimized, and confidential computing hardware depends on the region, service tier, and compute tier.

Hardware Configuration
Standard-series (Gen5)
up to 80 vCores, up to 415.23 GB memory
Change configuration

vCores Compare vCore options
2

High-Availability Secondary Replicas

Increasing the number of High Availability replicas improves availability SLA. High Availability replicas can be used for simple read scale scenarios. Consider Named replicas for more complex read scale scenarios. Learn more

1 Replica

Would you like to make this database zone redundant?
Yes No

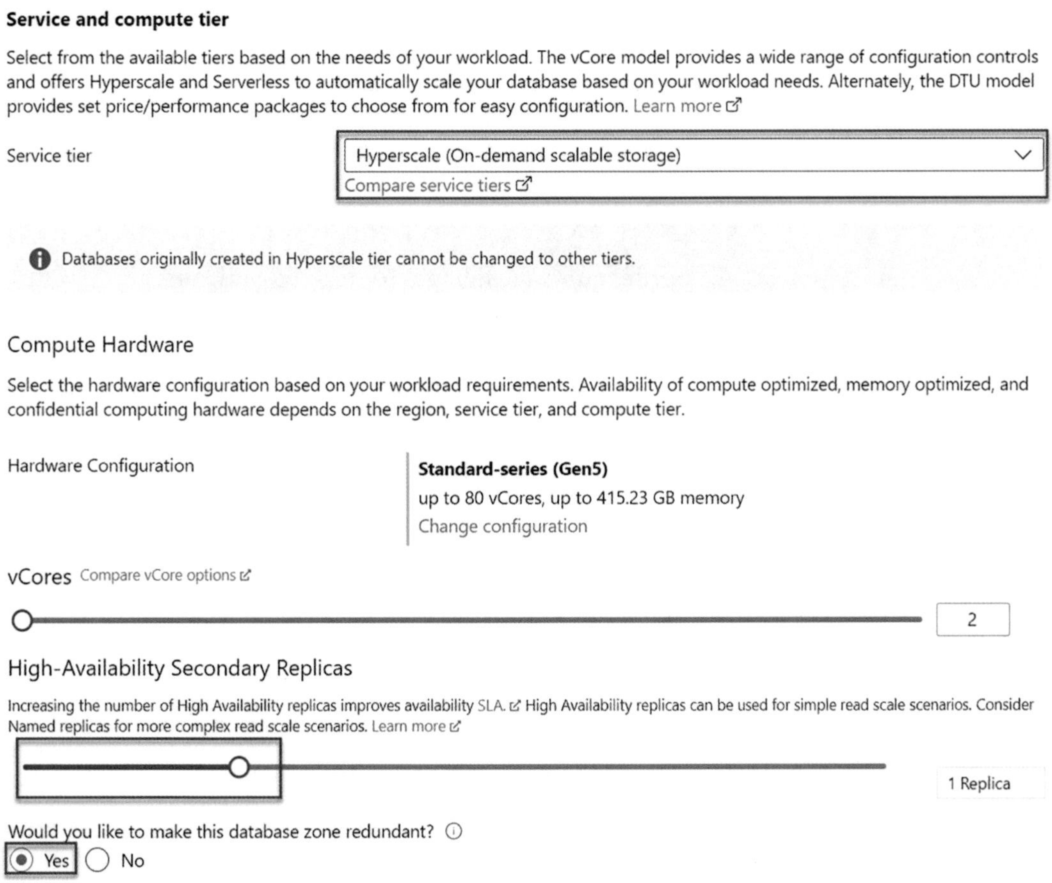

24. Click the Apply button.

25. We will return to the Create SQL Database blade again. We will not change any of the other deployment options, such as Networking and Security for this example, as we can change most of them after deployment.

26. Click the "Review + create" button.

27. Review the estimated cost per month and scan through the configuration to confirm we're configuring what we expected. There are ways to reduce the estimated cost per month (such as purchasing reserved capacity), but these will be discussed in Chapter 16.

Important The Azure SQL Database Hyperscale tier is billed hourly. So, once the resources are deployed, you'll begin to be billed. Deleting the database resource will cause you to stop being billed, but you will be charged for at least one hour.

---

28. Click the Create button.

29. After a few seconds, our resources will begin to be deployed, and a deployment progress window will be displayed.

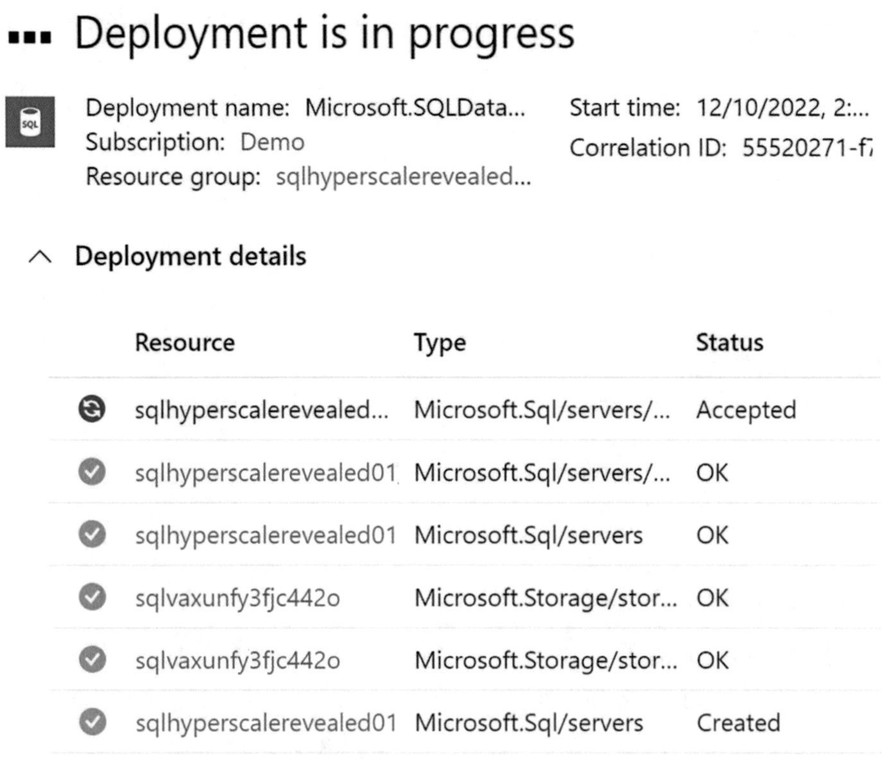

Deployment is in progress

Deployment name: Microsoft.SQLData... Start time: 12/10/2022, 2:...
Subscription: Demo Correlation ID: 55520271-f7
Resource group: sqlhyperscalerevealed...

Deployment details

Resource	Type	Status
sqlhyperscalerevealed...	Microsoft.Sql/servers/...	Accepted
sqlhyperscalerevealed01	Microsoft.Sql/servers/...	OK
sqlhyperscalerevealed01	Microsoft.Sql/servers	OK
sqlvaxunfy3fjc442o	Microsoft.Storage/stor...	OK
sqlvaxunfy3fjc442o	Microsoft.Storage/stor...	OK
sqlhyperscalerevealed01	Microsoft.Sql/servers	Created

30. Once the deployment has been completed, we can click the `sqlhyperscalerevealed01/hyperscaledb` resource (or whatever you called it in your subscription) to be taken directly to the database.

Resource	Type	Status
sqlhyperscalerevealed01/Default	Microsoft.Sql/servers/vulnerabilityAsses...	Created
sqlhyperscalerevealed01/Default	Microsoft.Sql/servers/advancedThreatPr...	Created
sqlhyperscalerevealed01/hyperscaledb	Microsoft.Sql/servers/databases	Created
sqlhyperscalerevealed01/Default	Microsoft.Sql/servers/connectionPolicies	OK

31. We will be taken to the Overview blade of our new Hyperscale database.

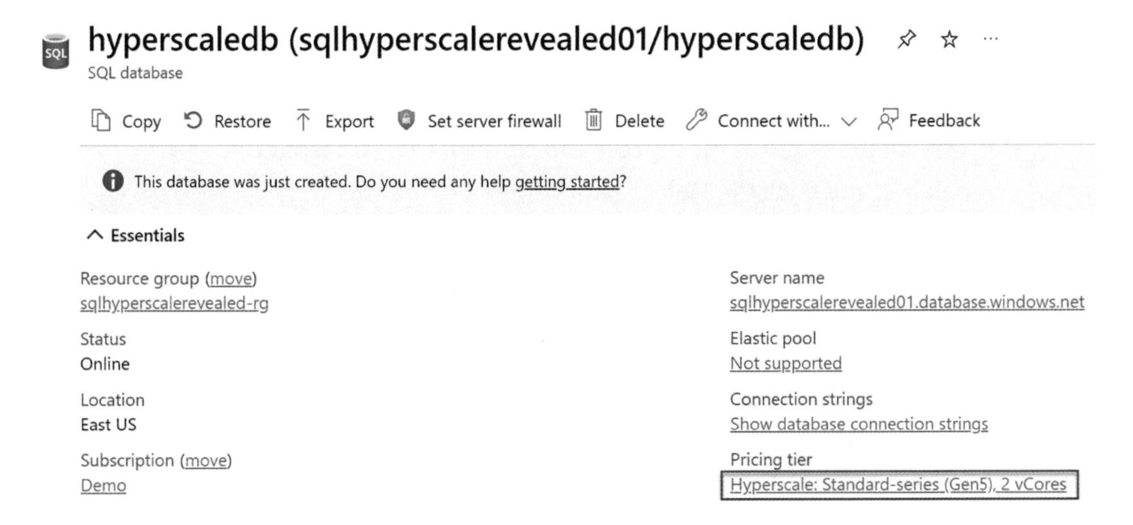

Congratulations! You have deployed your first Hyperscale database and can begin to use it. There are other administrative tasks we would typically do from here, such as granting access to the database and opening network access to our applications and users, but they will all be covered in later chapters. For now, this is enough to demonstrate just how easy it is to get a Hyperscale database up and running in Azure.

Cleaning Up

Like most services in Azure, you are charged only for what you consume. Therefore, cleaning up any unneeded resources is a good idea to manage your costs. Although we're going to use the Hyperscale database we've deployed here throughout this book, you can go ahead and delete it and then re-create it again later.

To delete the Hyperscale database using the Azure Portal, follow these steps:

1. Open your browser and navigate to `https://portal.azure.com`.
2. Navigate to the Overview blade of the Hyperscale database you deployed.
3. Click the Delete button.

4. Type in the name of the database to confirm that you want to delete it and click the Delete button.
5. You should also delete the SQL Database server that hosted the Hyperscale database. This is because it may also incur charges (such as Microsoft Defender for SQL licenses).
6. Navigate to the Overview tab of the SQL Server and click the Delete button.

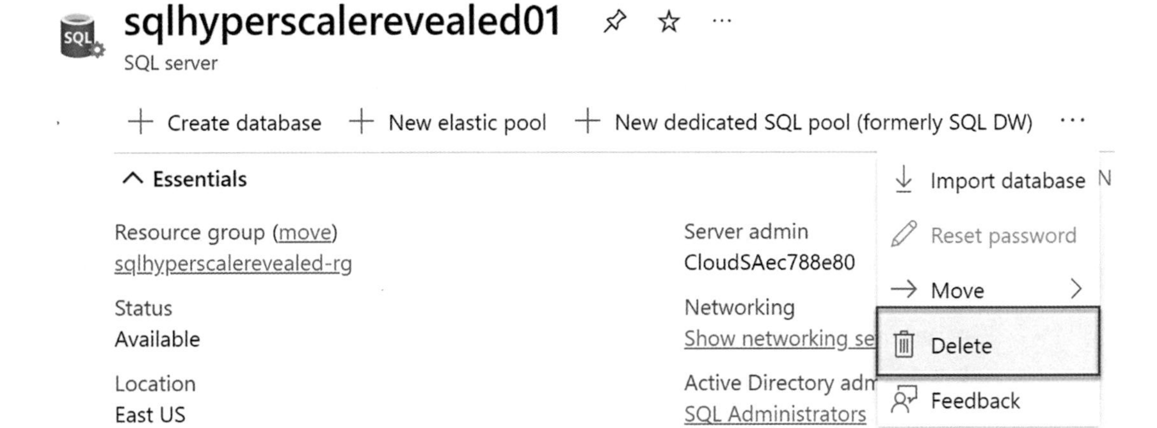

7. Type in the name of the SQL Database Server and click Delete.

8. Depending on other items that were configured during deployment, there may be additional resources that were also deployed into the `sqlhyperscalerevealed-rg` resource group. For example, when you choose to enable Microsoft Defender for SQL, an Azure Storage account is also created to store security assessment output. You can delete these resources to reduce costs and keep your environment clean.

Now that we have seen the different options for deploying SQL in Azure and have deployed our first Hyperscale database, in the next chapter we'll look at the Hyperscale architecture in more detail. This will help us understand how it differs from other Azure SQL PaaS architectures and why it can be used for workloads that need higher storage and scalability.

Summary

In this chapter, we took a high-level look at the basic options for running SQL workloads in Azure and introduced the basic concepts of the Azure SQL Database Hyperscale tier. We walked through some of the key concepts that should be considered when deploying SQL on Azure, including the following:

- *The PaaS deployment models*: Azure SQL Database and Azure SQL Managed Instances. The Hyperscale tier exists only in Azure SQL Database, so the remainder of this book will focus on this deployment model.
- *The purchasing models*: Whether your compute is provisioned or serverless. Hyperscale currently supports only the provisioned purchasing model.
- *The service tiers*: Provides you with additional control over the service-level objectives and features of the database. This also includes vCore and DTU capacity models. Hyperscale currently supports only vCore capacity.
- *The resilience and redundancy options*: The various redundancy approaches available in Azure SQL Database to keep your workload running in the case of a disaster or regular maintenance.

We wrapped up this chapter with an example deployment of the Azure SQL Database Hyperscale tier to show just how quick it is to get started using this advanced RDMS architecture.

In the next chapter, we will take a deeper dive into the architectural details of the Hyperscale technology to understand what makes it so different from previous technologies. You will learn what the key differences are in how compute and storage are managed by Hyperscale and how these differences enable it to be suitable for high-scale workloads that would be more difficult to implement with a traditional relational database management system.

CHAPTER 2

Azure SQL Hyperscale Architecture Concepts and Foundations

Over the years, many new features and improvements have been continuously developed as part of and alongside the traditional SQL database architecture. As cloud computing platforms began to take prominence in the technology sphere, a new form of cloud database architecture has progressively emerged.

This new form of architecture incorporates new methodologies involving multitiered architecture. A key characteristic is decoupling compute nodes from the storage layer and the database log service. It aims to incorporate many of the pre-existing and newly evolving features into the new architectural paradigm. This allows for the best combination of performance, scalability, and optimal cost.

Many of the characteristics and features that are the foundation of the Hyperscale architecture are adaptations providing significant improvements over their pre-existing implementations compared with a traditional SQL Server architecture.

Tip In this chapter there are some Transact SQL (T-SQL) and Azure CLI samples. These will all be available in the `ch2` directory in the files accompanying this book and in the GitHub repository: `https://github.com/Apress/Azure-SQL-Hyperscale-Revealed`.

Z. Barać and D. Scott-Raynsford, *Azure SQL Hyperscale Revealed*,
https://doi.org/10.1007/978-1-4842-9225-9_2

In this chapter, we will examine in more depth the architectural differences with Hyperscale that enable it to provide improvements in performance, scale, and storage when compared with other tiers of Azure SQL Database. The multitier architecture lies at the heart of this architectural transformation, so we will spend the majority of our time on it in this chapter.

Important The implementation details of the Hyperscale architecture are under regular revision and improvement by the team at Microsoft. You may find implementation differences or changes to the limits mentioned here. Therefore, it is worth testing Hyperscale with your specific workload. It is most important to understand the concepts and architecture rather than the implementation specifics, which do occasionally change. Learning how to determine these details will be covered in this chapter.

Hyperscale Azure SQL Scalability and Durability

As mentioned earlier, Azure SQL DB Hyperscale is designed in such a way as to achieve superior scalability, higher availability, and faster performance—all at a more efficient cost. It does this by utilizing a four-tier architecture concept (refer to Figure 1-10 in Chapter 1), which enables these fundamental improvements.

This fundamental shift in architecture results in several implementation differences. To be able to understand Hyperscale, it is useful to be aware of these differences. Table 2-1 shows the different availability methodologies utilized by traditional SQL and the new Hyperscale architecture in order to highlight some of the important differences.

***Table 2-1.** Summary of Traditional and Hyperscale Architecture Storage and Compute Differences*

Standard Availability Model: Basic, Standard, and General-Purpose Service Tier Availability
• *Compute*: Stateless, with `sqlserver.exe` running with locally attached SSD for transient data, such as `tempdb`, buffer, etc. • *Storage*: Stateful, with database files (`.mdf/.ldf`) stored in Azure Blob Storage. Premium locally redundant storage (LRS) is used by default. If the zone-redundant option of the General Purpose service tier is enabled, then zone-redundant storage (ZRS) is used. • *Zone redundancy*: Optionally available for the General Purpose service tier only, in which case secondary nodes will be created in different availability zones. • *Architecture*: Like the Failover Cluster Instance (FCI) approach to running Microsoft SQL Server with storage shared between all nodes in the cluster.
Premium Availability Model: Premium and Business Critical Service Tier Availability
• *Compute*: `Sqlserver.exe` running with locally attached SSD for transient data, such as `tempdb`, buffer, etc. Including SQL data and log files, they are as well stored on a locally attached SSD and automatically replicated to additional secondary nodes, creating a four-node cluster. • *Zone redundancy*: Available. Each deployment contains four nodes for achieving high availability. Secondary replicas are in sync commit and contain a full copy of the primary database. • *Architecture*: Like the Windows Failover Cluster (WFC) approach, which uses the Always On Availability Group configuration with all nodes in sync commit and hold full copy of database.

(*continued*)

Table 2-1. *(continued)*

Standard Availability Model: Basic, Standard, and General-Purpose Service Tier Availability
Hyperscale Availability Model
• *Compute*: Stateless, with `sqlserver.exe` running with a locally attached SSD for the RBPEX Data Cache (discussed later in this chapter) and transient data, such as `tempdb`, buffer, etc. One primary node with four secondary replicas. There is one primary node and up to four secondary high-availability replicas. You can have additional geo replicas for disaster recovery and up to 30 named secondary replicas. • *Storage* • *Page servers*: Stateless second cache layer in the Hyperscale architecture containing transient and cached data (covering RBPEX cache). Each page server also has a replica for redundancy and high-availability purposes, which means that we have two page servers in an active-active configuration. • *Log service*: Stateful completely separated tier containing transaction logs. • *Data storage*: Stateful, with database files (`.mdf/.ldf`) stored in Azure Blob Storage and updated by page servers. This uses the redundancy features of Azure Storage. • *Zone redundancy*: Available. Each stately compute replica can be optionally spread across availability zones. • Hyperscale utilizes the snapshot feature of Azure Standard Storage to make snapshot backups easy and fast without impacting user workloads.

Foundations of Azure SQL Hyperscale

The Hyperscale architecture is enabled by leveraging several foundational features. This section provides insight into these features as they are important components that enable the Hyperscale implementation. Understanding these concepts will enable you to get the most out of the database by tuning your workload and design to make better use of these features.

The Buffer Pool Extension

In SQL Server 2014, Microsoft introduced the Buffer Pool Extension (BPE) feature. The BPE feature was made available only in the Standard and Enterprise editions of SQL Server. The BPE feature expands the in-memory buffer pool content by extending it onto a local solid-state disk (SSD). This means that the buffer pool extension feature extends the memory buffer pool cache so that it can accommodate a larger number of database I/O operations. The buffer pool extension can significantly improve I/O throughput via offloading additional small random I/O operations from existing SQL configurations to a SSD.

The Resilient Buffer Pool Extension

The Hyperscale architecture utilizes a similar approach to the BPE but improves on it, by making the cache recoverable after a failure. This newer approach to the buffer pool extension is called the Resilient Buffer Pool Extension (RBPEX) data cache. Using RBPEX, the cache exists on both the compute replica and page servers. Like BPE, the RBPEX data cache contains the most frequently accessed data.

The size of the RBPEX cache is approximately three times the size of the memory allocated to the compute node. The larger the RBPEX cache, the more pages it can cache.

The compute replicas do not store the full database in the local cache. Compute replicas use a buffer pool in memory and the local resilient buffer pool extension (RBPEX) cache to store the data that is a partial noncovering cache of data pages. On the other hand, each page server has a covering RBPEX cache for the subset of data it manages.

Fundamentally, the larger RBPEX cache would reduce the frequency of compute nodes requesting data from remote page servers, which would lower the I/O latency of your database.

Investigating the Size of RBPEX

We will now look at some queries we can use to investigate the size of the RBPEX cache on our Hyperscale database.

Example 1: RBPEX on a Two-vCore Hyperscale Database

We deployed the Azure SQL Database Hyperscale service tier with two vCores for the primary compute node. To verify this, we can run the following query against the `sys.dm_user_db_resource_governance` DMV and look for the `slo_name` (service-level objective) and `cpu_limit` columns:

```
SELECT database_id as 'DB ID'
, slo_name as 'Service Level Objective'
, server_name as 'Server Name'
, cpu_limit as 'CPU Limit'
, database_name as 'DB Name'
, primary_max_log_rate/1024/1024 as 'Log Throughtput MB/s'
, max_db_max_size_in_mb/1024/1024 as 'MAX DB Size in TB'
, max_db_max_size_in_mb as 'MAX DB Size in MB'
FROM sys.dm_user_db_resource_governance
```

This DMV returns the configuration and capacity settings for the current database. Table 2-2 shows an example of these results on a Hyperscale database in Azure with two vCores assigned.

Tip For more information on the `sys.dm_user_db_resource_governance` DMV, see this page: `https://learn.microsoft.com/sql/relational-databases/system-dynamic-management-views/sys-dm-user-db-resource-governor-azure-sql-database`.

Table 2-2. *Query Result from the sys.dm_user_db_resource_governance DMV to Get the Configuration and Capacity Settings for a Two-vCore Hyperscale Database*

Service-Level Objective	CPU Limit	DB Name	Log Throughput MB/s	Max DB Size TB
SQLDB_HS_GEN5_2_SQLG5	2	hyperscale_db01	105	100

With the Hyperscale Standard-series (Gen-5) hardware configuration, scaling up is possible up to a maximum of 80 vCores. From a memory perspective, there will be 5.1GB of memory allocated per vCore, so for two vCores we will have approximately 10GB of memory.

To verify this, we can run a query against the `sys.dm_os_job_object` DMV and look for the `memory_limit_mb` column. That will show the maximum amount of committed memory that the underlying SQL Server can use.

```
SELECT memory_limit_mb as "MAX Memory MB" FROM sys.dm_os_job_object
```

This DMV returns a single row describing the configuration of the job object, including certain resource consumption statistics. In our example, this returns the `memory_limit_mb` column, which presents the maximum amount of committed memory for the SQL Server compute node. Table 2-3 shows the result of running this query on a new Hyperscale database with two vCores. It shows approximately 10GB of memory, which is what we expected based on the earlier calculation.

Tip For more information on the `sys.dm_os_job_object` DMV, see `https://learn.microsoft.com/en-us/sql/relational-databases/system-dynamic-management-views/sys-dm-os-job-object-transact-sql`.

Table 2-3. *Query Result from the sys.dm_os_ job_ object DMV to Get the Memory Available to a Two-vCore Hyperscale Database*

MAX Memory MB
10630

Finally, let's obtain the RBPEX cache size in storage. To do this we are going to query the `sys.dm_io_virtual_file_stats` DMV and look for the number of bytes used on the disk for this file (`size_on_disk_bytes`), as shown here:

SELECT database_id, file_id, size_on_disk_bytes/1024/1024/1024 as Size_on_disk_GB FROM sys.dm_io_virtual_file_stats(DB_ID(), NULL)

This dynamic management function returns I/O statistics for data and log files. The `database_id` and `file_id` columns will have a value of 0 for the Resilient Buffer Pool Extension. Table 2-4 shows the query results from querying this DMV on a Hyperscale database with two vCores assigned.

Tip For more information on the `sys.dm_io_virtual_file_stats` DMV, see `https://learn.microsoft.com/sql/relational-databases/system-dynamic-management-views/sys-dm-io-virtual-file-stats-transact-sql`.

Table 2-4. *Query Result from the sys.dm_io_virtual_file_stats DMV to Get the Size of the RBPEX Cache on Disk for a Two-vCore Hyperscale Database*

database_id	file_id	Size_on_disk_GB
0	0	32

As we can see, the RBPEX cache is 32GB, which is approximately three times the memory allocated to the compute node, which is 10GB.

Example 2: RBPEX on a Four-vCore Hyperscale Database

If we add an additional two vCores to the same hyperscale database, the RBPEX cache size will still be three times bigger than the memory provided. We expect a compute node with four vCores assigned to have approximately 20GB of memory and an RBPEX cache size of 60GB. We will verify this by running the same queries we ran in Example 1.

Table 2-5. *Query Result from the sys.dm_user_db_resource_governance DMV to Get the Configuration and Capacity Settings for a Four-vCore Hyperscale Database*

Service-Level Objective	CPU Limit	DB Name	Log Throughput MB/s	Max DB Size TB
SQLDB_HS_GEN5_4_SQLG5	4	hyperscale_db01	105	100

Table 2-6. Query Result from the sys.dm_os_job_object DMV to Get the Memory Available to a Four-vCore Hyperscale Database

MAX Memory MB
21260

Table 2-7. Query Result from the sys.dm_io_virtual_file_stats DMV to Get the Size of the RBPEX Cache on Disk for a Four-vCore Hyperscale Database

database_id	file_id	Size_on_disk_GB
0	0	63

This again demonstrates that the amount of storage assigned to the RBPEX cache is three times the amount of memory available to the compute node. In the case of our four-vCore Hyperscale database, it is 20GB memory with 60GB assigned to the RBPEX cache.

Row Versioning–Based Isolation Level

Read Committed Snapshot Isolation (RCSI) allows a read transaction to get to the initial versions of the committed value of the row stored in `tempdb` without any blocking. This is to increase concurrency and reduce locking and blocking. This ensures that write transactions do not block read transactions and that read transactions do not block write transactions.

For Microsoft SQL Server instances or SQL Server on Azure VM (IaaS), `READ_COMMITTED_SNAPSHOT` is set to `OFF` by default. However, on Azure SQL (PaaS) databases, including the Hyperscale service tier, it has `READ_COMMITTED_SNAPSHOT` set to `ON` by default.

Hyperscale databases have the following isolation levels set by default:

- *Primary replica*: Read committed snapshot
- *Read scale-out secondary replicas*: Snapshot

Tip To turn the `READ_COMMITTED_SNAPSHOT` settings on or off, only the command `ALTER DATABASE dbname SET READ_COMMITTED_SNAPSHOT ON` should be running. No other open connections are allowed until this command is completed.

Table 2-8 shows the relationship between isolation levels, reads, and locks.

Table 2-8. *Snapshot and Read Committed Snapshot Transaction Isolation Levels on a Hyperscale Database*

Isolation Level	Dirty Reads	Nonrepeatable Reads	Phantom Reads
Snapshot	No	No	No
Read Committed Snapshot	No	Yes	Yes

RCSI provides statement-level consistency and can produce nonrepeatable reads. This means you may get different results for the same SQL statement within the same transaction if another transaction started prior to committing the data. On the other hand, snapshot isolation provides transaction-level read consistency, which means that no reads made in a transaction will be changed, and there will be repeatable reads.

How does this work?

When single or multiple transactions try to modify a row, the values prior to the modification will be stored in `tempdb` in the version store, which presents a collection of data pages. This means that row versioning in all SQL Server PaaS deployment options, including the Azure SQL Hyperscale service tier, uses `tempdb` to store the row versions of previously committed values. Figure 2-1 shows a diagram of the read committed snapshot isolation behavior.

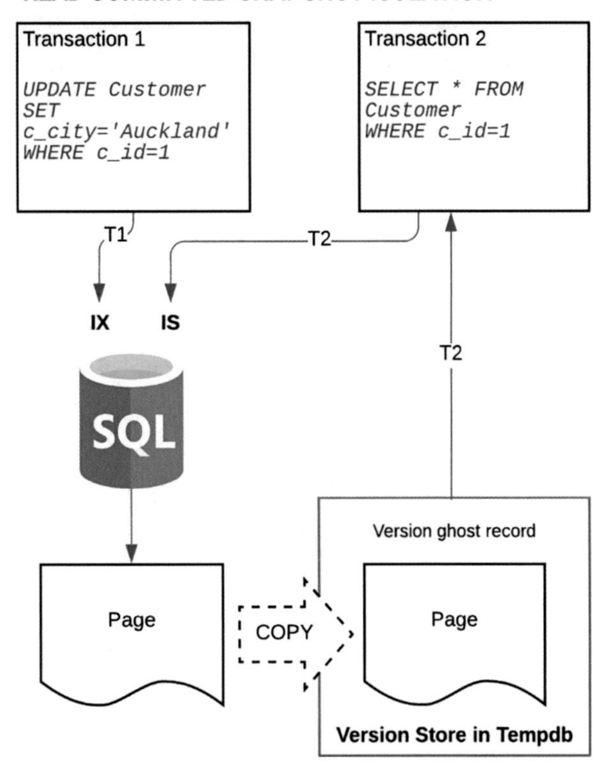

Figure 2-1. *Read committed snapshot isolation*

Note The copy data process will occur prior to any change being made. Copied data (`version_ghost_record`) will be stored in the version store within `tempdb` and will contain latest committed record before any new transaction changes the data.

What does this mean for the Hyperscale architecture? The main consistency behavior difference between traditional Azure SQL Database service tiers and the Hyperscale service tier is that with the traditional Azure SQL service tiers, versioning occurs in the local temporary storage. This means that row versions are available only locally to the primary replica. The Hyperscale architecture extends the shared disk ability in such a way that compute nodes must make row versions available in a shared storage tier to create a shared and persistent version store.

Tip `READ_COMMITTED_SNAPSHOT` is an additional database option that enables different behavior to the already existing `READ_COMMITTED` isolation level. RCSI introduces row versioning to increase concurrency and reduce blocking.

Accelerated Database Recovery

The idea of accelerated database recovery (ADR) in different forms is available from SQL Server 2019 onward. The main goal is to increase database availability during the recovery process of long-running transactions. In Azure SQL (PaaS) deployments, ADR is available by default; in addition, in SQL Server on Azure VMs (IaaS) from SQL Server 2019 on, you can manage and configure ADR according to your needs.

Before ADR, SQL Server used the ARIES recovery model, which included three phases.

- *Analysis*: The database engine checks the state of each transaction in the transaction log, from the latest successful checkpoint up to the final entry before the system crashed.
- *Redo*: The database engine checks the state of each transaction in the transaction log, from the earliest transaction not yet persisted on the disk.
- *Undo*: The database engine undoes all transactions that were active at the time of the crash.

Before ADR, the usual time necessary for a database engine to recover from an interrupted transaction, including the rollback of all incomplete transactions, was a lot higher—often close to the same amount of time originally needed for the execution. Long-running transactions were particularly affected by such a recovery model. Moreover, this could cause potential log issues since the database engine cannot truncate the transaction log until the transaction is committed or rolled back.

Tip Bear in mind that rollback is mostly single threaded. Even if your transaction used multiple cores to run, the rollback is going to use only one, so it will take more time for the rollback to finish.

The combination of the ADR feature and the persistent version store within Azure SQL offerings improved this recovery process significantly, by eliminating the need for the undo phase.

The Hyperscale architecture goes one step further by utilizing the shared drive concept and by making a shared persistent version store available to all page servers—instead of storing it on the local `tempdb` as per the traditional Azure SQL architecture. There are some ADR improvements in regard to the persistent version store publicly available in SQL Server 2022 such as the following:

- User transaction cleanup. Fixed lock conflict.
- ADR Cleaner improvements.
- Transaction-level persisted version store. Clean up versions belonging to committed transactions independently of any aborted transactions.
- Multithreaded version cleanup—available at the database level with the trace flag.

Why is this important for Azure SQL PaaS offerings? Azure SQL PaaS offerings are usually ahead of the on-premises version of SQL Server when compared to the SQL Server versions that are being used. This includes the latest security fixes and latest features available in the latest SQL Server versions.

Multitier Architecture Concepts in Hyperscale

To utilize Azure Storage's ability to hold very large databases without degrading the I/O performance, Microsoft built the Hyperscale service tier as an additional facet of the multitiered architecture. In this section, we will break down the four tiers and review them in greater detail.

- Compute nodes
- Log service
- Page servers
- Azure Standard Storage

Figure 2-2 shows a diagram of the four tiers making up the Hyperscale architecture.

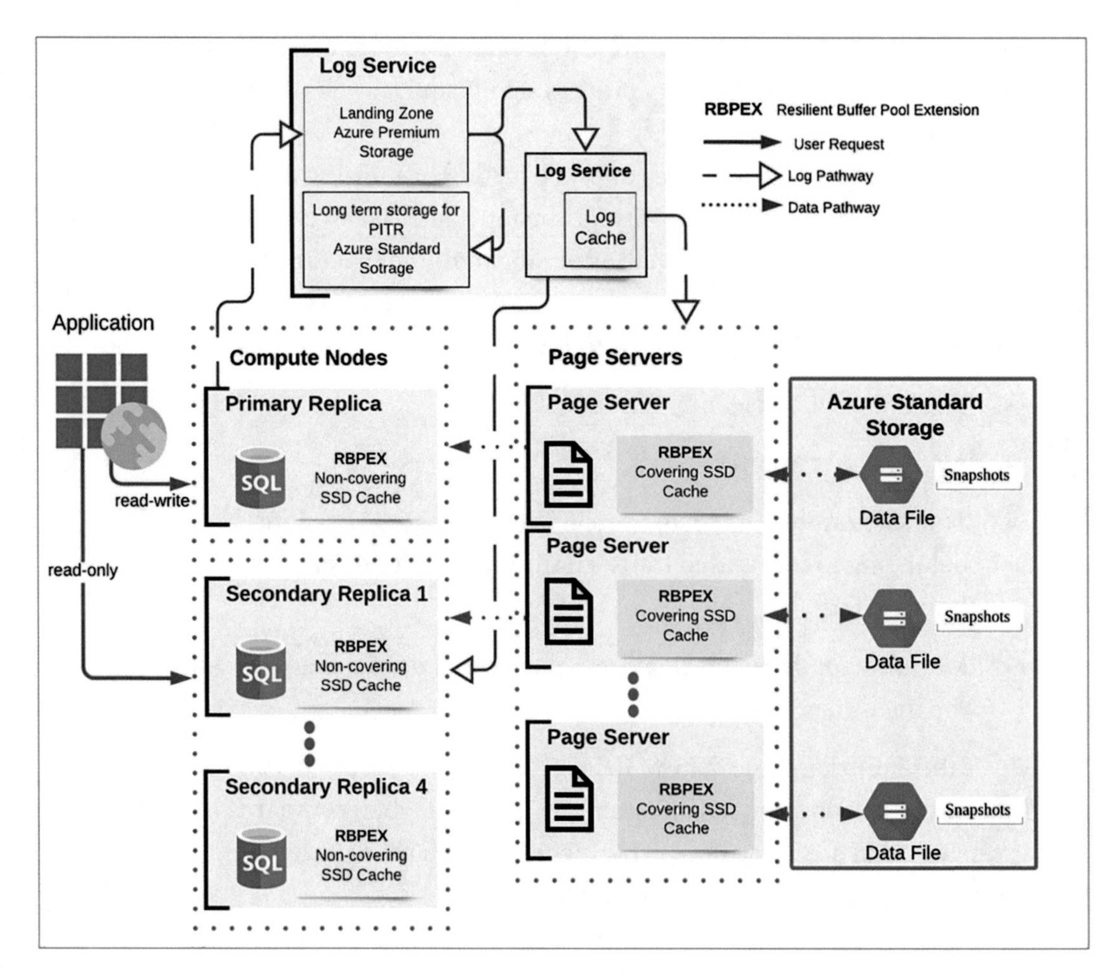

Figure 2-2. *The four-tier architecture of Hyperscale*

Compute Nodes

The Hyperscale architecture provides rapid compute scalability to support workload demands. The Azure SQL Database Hyperscale service tier has two types of compute nodes, as follows:

- *Primary compute node*: Serves all read-write operations.
- *Secondary nodes*: Provides read scale-out and geo-replication ability, which can simultaneously provide high availability and serve as a hot standby node.

Figure 2-3 shows the pane for creating a compute replica in the Azure Portal.

Service tier
Hyperscale (On-demand scalable storage)
Compare service tiers
Databases originally created in Hyperscale tier cannot be changed to other tiers.
Compute Hardware
Select the hardware configuration based on your workload requirements. Availability of compute optimized, memory optimized, and confidential computing hardware depends on the region, service tier, and compute tier.
Hardware Configuration
Standard-series (Gen5)
up to 80 vCores, up to 415.23 GB memory
Change configuration
vCores Compare vCore options
2
High-Availability Secondary Replicas
Increasing the number of High Availability replicas improves availability SLA. High Availability replicas can be used for simple read scale scenarios. Consider Named replicas for more complex read scale scenarios. Learn more
4 Replicas
Would you like to make this database zone redundant?
Yes
No

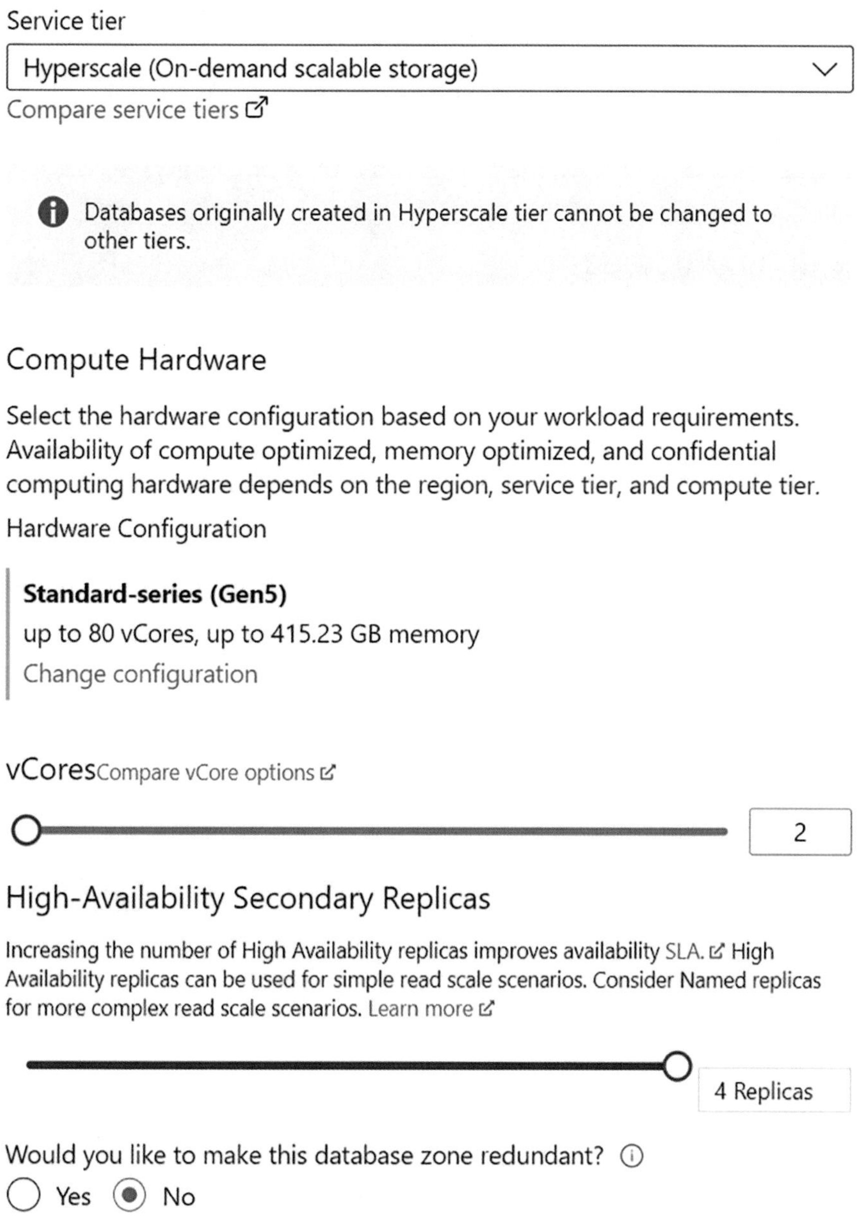

Figure 2-3. *Creating compute primary and secondary replicas via the Azure Portal*

One of the biggest advantages of the Hyperscale architecture is its rapid compute scalability—the ability to rapidly scale up or down your compute resources to support your current workload as needed. It also allows you to rapidly scale out by adding more read-only replicas to offload your read workloads if required.

Tip Bear in mind you cannot pause compute nodes to reduce the cost of the Hyperscale service tier; however, you can automate the downscaling of your compute node when there is no heavy workload.

Primary Compute Node

The compute node role within the Hyperscale architecture seems much simpler compared to the traditional SQL Server architecture, though in the end they fulfill the same function: to process read and write transactions.

In the traditional Azure SQL architecture, for service tiers leveraging the premium availability model, primary replicas use synchronous commit mode, and they must always be synchronized. Primary and failover replicas must communicate to each other to acknowledge that they have hardened the transaction log to the drive and that the transaction is committed. In the Hyperscale architecture, secondary replicas do not communicate directly with the primary compute node. Secondary replicas, like page servers, keep their interactions with the log service tier instead.

In the Hyperscale architecture, compute nodes present a first data caching layer. For a caching purpose, primary and secondary replicas utilize a locally attached SSD for storing the buffer pool and RBPEX data cache.

The most common configuration for the primary compute node for the Hyperscale service tier is Standard-series (Gen5) with a maximum of 80 vCores and 408GB of memory, which is approximately 5.1GB per deployed vCore.

Secondary Compute Node

There are three types of secondary replicas for the Azure SQL Database Hyperscale service tier.

- High-availability replicas
- Named replicas
- Geo replicas

Figure 2-4 shows the named and geo-replicas configuration options available when creating a new replica. The high-availability replica is part of the initial deployment process of the Hyperscale database.

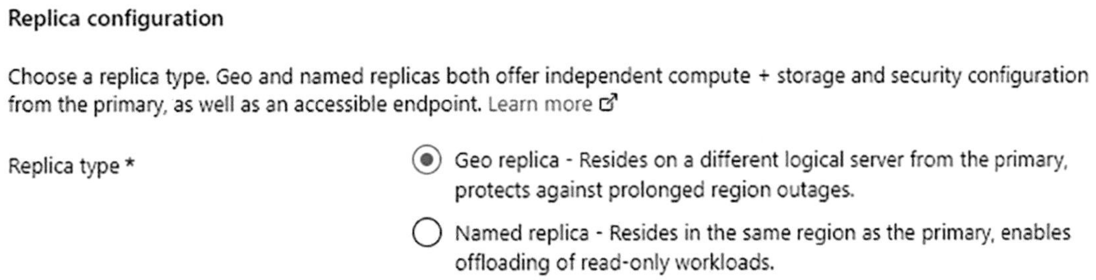

Replica configuration

Choose a replica type. Geo and named replicas both offer independent compute + storage and security configuration from the primary, as well as an accessible endpoint. Learn more

Replica type *

(•) Geo replica - Resides on a different logical server from the primary, protects against prolonged region outages.

( ) Named replica - Resides in the same region as the primary, enables offloading of read-only workloads.

Figure 2-4. *Named and geo replicas configuration*

Note Both named replicas and geo replicas can be deployed with their own high-availability replicas. This is especially important for using geo replicas in disaster recovery scenarios as we would usually want the same level of resilience in our failover region.

High-Availability Replicas

High-availability (HA) replicas are part of the Azure SQL Database Hyperscale deployment process. In a Hyperscale architecture, HA secondary replicas always have the same configuration and number of vCores as the primary node. HA replicas are a key part of high availability, and if a failover happens, they need to be able to provide the same performance to support the workload in place of the primary node. To improve availability, you can simply increase the number of HA replicas to a maximum of 4. Figure 2-5 shows the HA replicas configuration option.

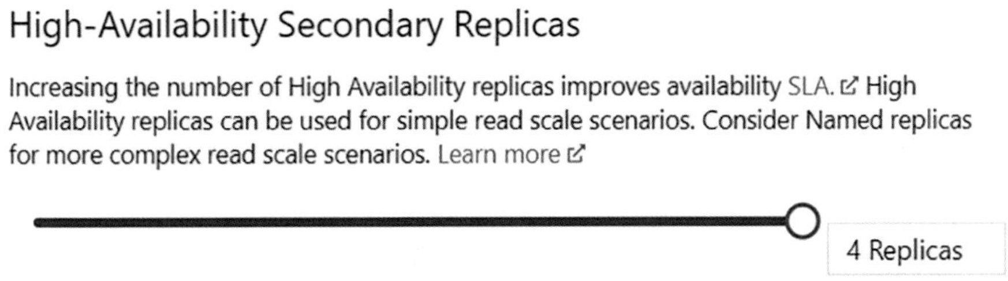

High-Availability Secondary Replicas

Increasing the number of High Availability replicas improves availability SLA. High Availability replicas can be used for simple read scale scenarios. Consider Named replicas for more complex read scale scenarios. Learn more

4 Replicas

Figure 2-5. *The HA replicas configuration when creating or updating a Hyperscale database*

In a nutshell, high availability replicas have two important roles.

- Failover
- Read-only replicas

Bear in mind, even if you did not choose any high availability replicas during deployment, you will still have local redundancy within the Hyperscale service tier. Data files are stored in Azure Standard Storage. If failover should occur, the first available node with spare capacity will take the role of the primary compute replica. It will then start feeding the data from the page servers and Azure storage. Failover will be somewhat slower than would be the case if you had a secondary replica in place.

HA replicas have several benefits.

- Easy deployment
- Extremely fast automatic failover
- Exceptionally reliable SLA (greater than 99.99 percent uptime with two HA replicas)

However, there are some considerations you should be aware of.

- You do not have full control over the HA replicas configuration. You will not have visibility of them in the Azure Portal.
- There is no dedicated connection string that can be modified for direct access to the HA replica(s); e.g., you may redirect read-only traffic by adding the parameter (`ApplicationIntent=ReadOnly`).
- If you have deployed multiple HA replicas, you cannot specify which one you are accessing.
- There are a limited number of replicas that can be used (a maximum of four).

Named Replicas

A named replica (read-only) is an independent node, and it uses the same page servers as the primary compute node. The Hyperscale service tier supports up to 30 named replicas, and it allows you to choose your desired compute tier for these replicas, independent from your primary compute nodes.

These replicas can be viewed and managed via the Azure Portal, Azure CLI, or PowerShell, and you will have full control over them. There are several advantages to these replicas.

- Dedicated connection string.
- Independent security configuration.
- Extremely easy to deploy.
- You can deploy up to 30 named replicas.

Some of the common use case scenarios would be the need for near real-time analytics, reporting, or other online analytical processing (OLAP) workloads. Moreover, it can apply anywhere there is a need for a massive online transaction processing (OLTP) read scale-out, such as for example API, ecommerce, or online gaming.

Additionally, you can utilize these named replicas based on your workload and limit or isolate the access per workload type (complex analytics, for example). You can also dedicate them to a particular service usage (PowerBI, for example), or you can route traffic to specific replicas based on traffic source (mobile apps, for example).

Tip Bear in mind that named replicas cannot be your choice for failover scenarios. For that purpose, Hyperscale uses high-availability replicas.

There are multiple ways of deploying named replicas to an existing Hyperscale database, such as via the Azure Portal, T-SQL, Azure PowerShell, Azure CLI, and Azure Bicep. Figure 2-6 shows the pane for creating a replica in the Azure Portal.

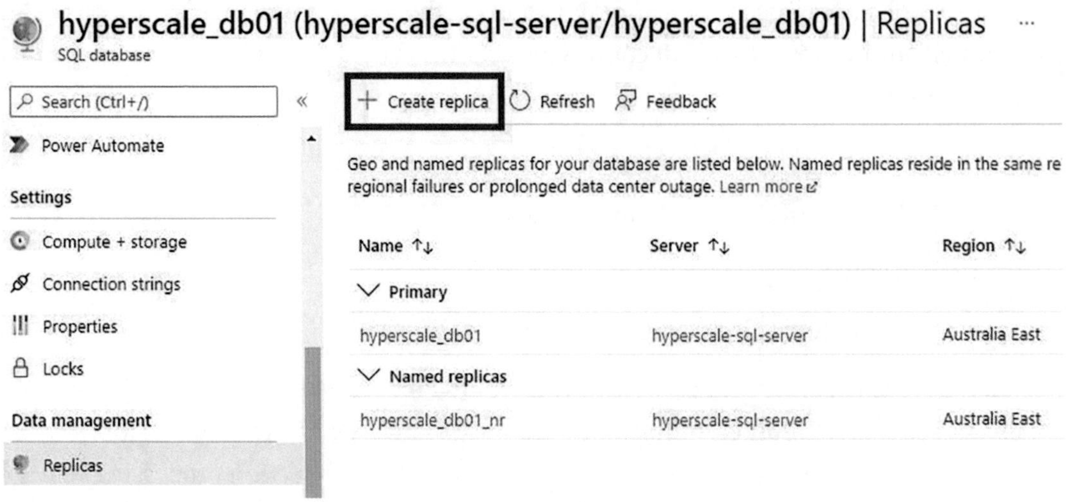

***Figure 2-6.** "Create replica" feature in the Azure Portal*

To create a new Azure SQL logical server and add a named replica to it using the Azure CLI, use the following:

```
# Create logical server
az sql server create -g resourcegroupname -n replicaervername -l region
--admin-user adminaccount --admin-password <enter-your-password-here>

# Create named replica for primary database
az sql db replica create -g resourcegroupname -n primarydatabasename
-s primaryservername --secondary-type named --partner-database
namereplicadatabasename --partner-server replicaervername
```

To create a named replica using T-SQL, use the following:

```
-- run on master database
ALTER DATABASE [primarydatabasename]
ADD SECONDASRY SERVER [replicaervername]
WITH (SECONDARY_TYPE=Named, DATABASE_NAME=[namereplicadatabasename])
GO
```

For more secure use-case scenarios, you have an option to create named replicas on a different logical SQL server.

Important A logical SQL Server needs to be in the same Azure region as the logical SQL server hosting the primary compute replica.

Geo Replicas

Geo replicas can cover disaster recovery scenarios in the case of a regional outage and must be on a different logical SQL Server instance from the primary replica host. Figure 2-7 shows the pane for creating a geo replica in the Azure Portal.

Create SQL Database - Replica ...

Microsoft

Primary database tpcc_hs

Region australiaeast

Replica configuration

Choose a replica type. Geo and named replicas both offer independent compute + storage and security configuration from the primary, as well as an accessible endpoint. Learn more

Replica type *

(●) Geo replica - Resides on a different logical server from the primary, protects against prolonged region outages.

( ) Named replica - Resides in the same region as the primary, enables offloading of read-only workloads.

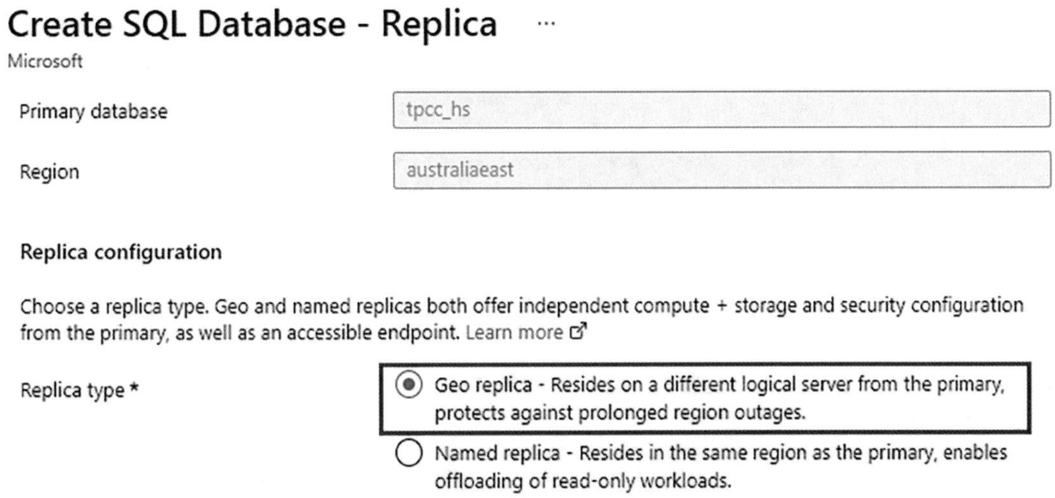

Figure 2-7. *Creating a geo replica in the Azure Portal*

The database name of a geo replica database will always remain the same as the database name of the primary node. Nevertheless, you can have different compute and storage configuration for primary and secondary named or geo replicas. You can also specify a different ratio of secondary replicas compared to what you have on the primary compute node configuration. A geo replica can also be configured with its own high-availability replicas.

Tip Keep in mind that latency may occur on any secondary replica, either HA, Named, or GEO. Moreover, if you deployed multiple secondary replicas, you may notice different data latency for each replica.

Figure 2-8 shows the ability to change the compute and storage configuration for your secondary replicas in the Azure Portal.

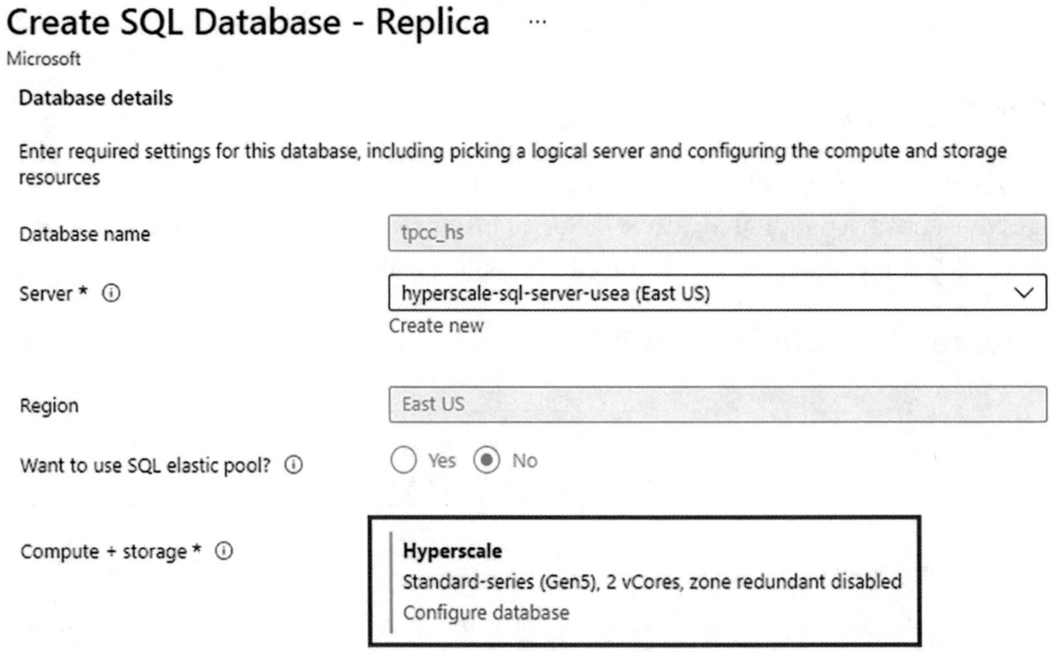

Figure 2-8. *Compute and storage configuration options for secondary named or geo replicas*

Log Service

One of the main advantages of the Hyperscale architecture compared to traditional architecture is that the transaction log service is separated from the primary compute node. This separation concept was introduced by Microsoft to help achieve lower commit latencies. The general concept of this would be to receive changes from the source, which is the primary compute node, and send those changes to multiple page servers and secondary compute replicas.

Tip The Log service process does not run `sqlserver.exe`. It has its own executable called `xlogsrv.exe`.

The Log service in Hyperscale is a multitier design and contains the following:

- Log service node with staging area, log broker, SSD cache
- Landing zone (Azure Premium Storage)
- Long-term storage for PITR (Azure Standard Storage)

How the Log Service Works in Hyperscale

As mentioned earlier, only the primary node writes all insert and update transactions to the log. The primary compute node writes log blocks synchronously and directly to achieve the lowest commit latency possible. As opposed to this, other Hyperscale architecture tiers, such as secondary replicas and page servers, receive transaction logs asynchronously.

When the insertion or update of transactions through the primary compute node occurs, the primary node will, using the Remote Block I/O (RBIO) protocol, write these log blocks directly to a landing zone, which is located on fast and durable Azure Premium Storage as well as to the XLOG process in parallel.

Initially the XLOG will store these log blocks in the staging area and wait for confirmation that those blocks have been made durable in the landing zone. Once the primary node confirms that a specific log block is hardened, Log Broker takes over that log block and prepares it for destaging. It will then move the log block to the SSD cache for fast access and to the long-term retention storage for the PITR. The reason for this additional processing is to avoid the log block arriving at the secondary replica before it has been made durable in the landing zone. Figure 2-9 shows the log service multitiered concept.

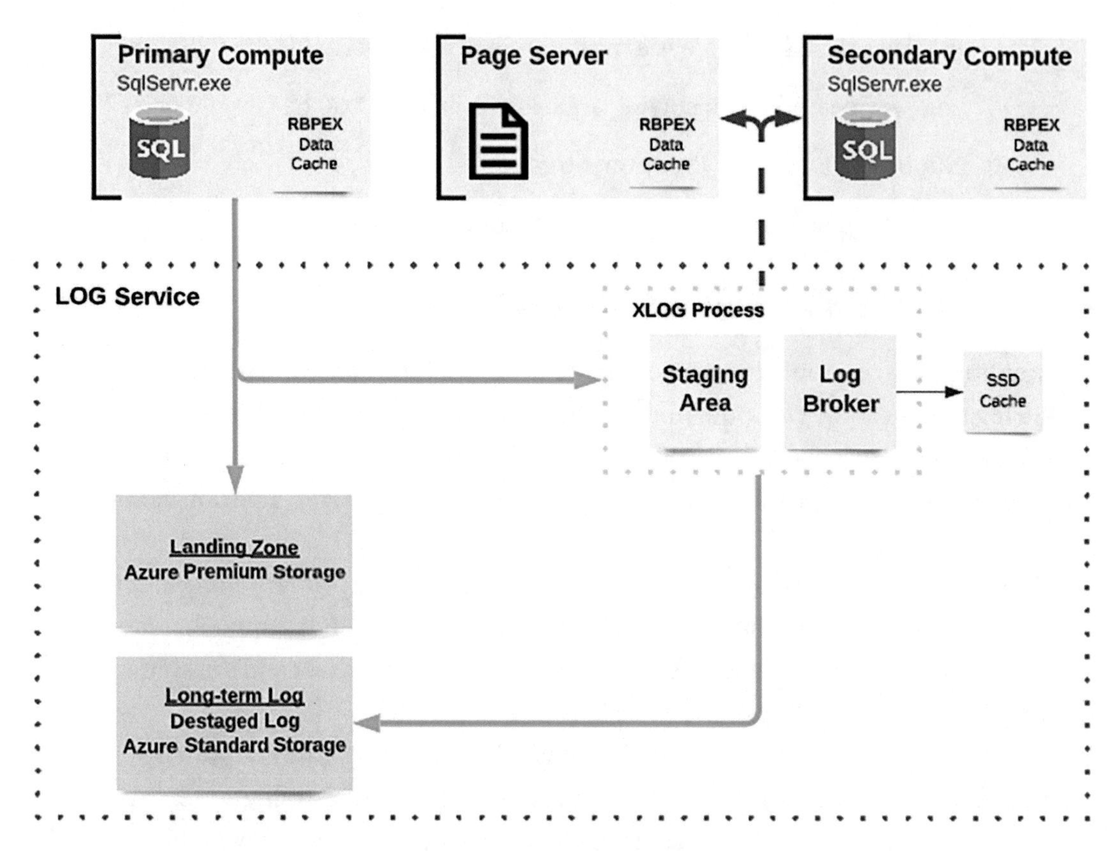

Figure 2-9. *Hyperscale Log service and XLOG process*

Log records are available for up to 35 days for point-in-time restore and disaster recovery in Azure Standard Storage. For the PITR backup retention period, any options longer than seven days were in preview at the time of writing.

Tip Differential backups are not supported for the Hyperscale service tier, unlike for other service tiers in traditional architecture.

Figure 2-10 shows the backup retention policy configuration pane in the Azure Portal.

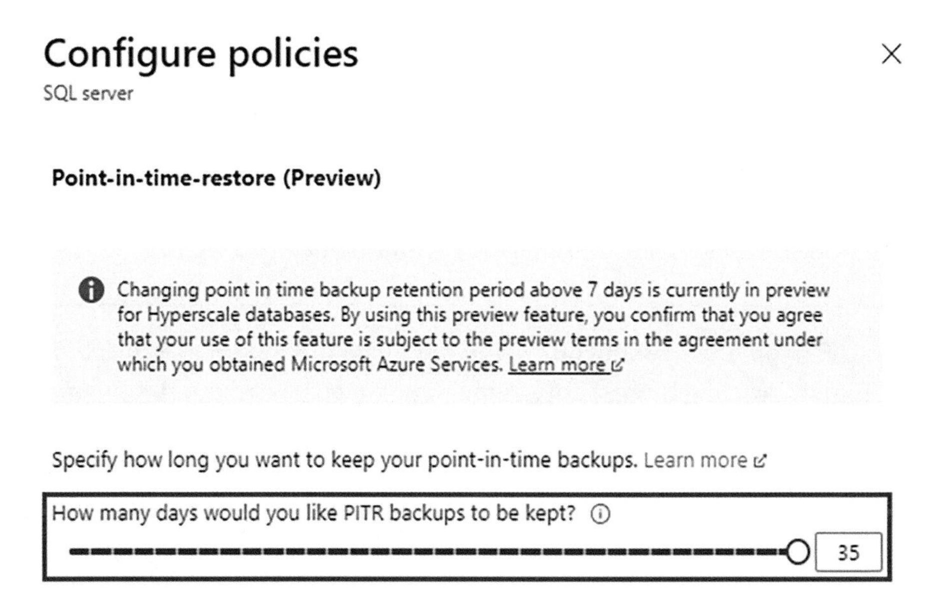

Figure 2-10. *Backup retention policy configuration*

Log Service Rate Throttling

The Azure SQL Database Hyperscale service tier has enforced log rate limits via log rate governance. It is currently set to a maximum of 105MB/sec irrespective of the compute size. This is on the Log service side of the Hyperscale architecture.

In Hyperscale, there is an additional process called *throttling* on the primary compute replica that may further limit the log throughput below the governance rate limit. The reason for this is to maintain the promised recoverability SLAs. There is a threshold that can come into play, whereby if any page server or secondary replica is more than threshold value behind the primary compute node, the throttling process will start. The length of throttling will depend on how far the page servers or secondaries are behind the primary compute node.

Page Servers

Page servers are the second cache layer in the Hyperscale architecture. They present a part of the storage layer that contains transient and cached data. To accomplish this caching objective, page servers rely on the RBPEX cache using the attached SSD and cached data pages in memory.

One of the roles of the page servers is to serve data pages to the compute nodes when required. The page servers themselves are updated by transaction log records from the log service. Each page server can contain between 128GB per file and approximately 1TB per file. As soon as the file size reaches a certain percentage of its maximum size, an additional page server node will be added automatically. The percentage for when this page server node increase occurs is determined automatically by Azure.

Note In the Hyperscale architecture, the data files are stored in Azure Standard Storage. Regardless, each page server maintains a subset of the data pages and has them fully cached on local SSD storage.

To obtain detailed information on the page servers, run this query:

```
SELECT file_id
, name AS file_name
, type_desc
, size * 8 / 1024 / 1024 AS file_size_GB
, max_size * 8 / 1024 / 1024 AS max_file_size_GB
, growth * 8 / 1024 / 1024 AS growth_size_GB
FROM sys.database_files WHERE type_desc IN ('ROWS', 'LOG')
```

This `sys.database_files` DMV returns a row per file of a database. In our example, this returns the current size of the file, in 8KB pages, as well as the maximum file size, in 8KB pages. Table 2-9 shows the results of running this query on an empty Hyperscale database.

Table 2-9. *The Page Servers in an Empty Hyperscale Database*

file_id	file_name	type_desc	file_size_GB	max_file_size_GB	growth_size_GB
1	data_0	ROWS	10	128	10
2	log	LOG	1022	1024	2

Tip For more information on the `sys.database_files` DMV, see this page: `https://learn.microsoft.com/sql/relational-databases/system-catalog-views/sys-database-files-transact-sql`.

After inserting dummy data into the database, in this case 64GB, we reach the threshold at which a new page server is added. Hyperscale automatically adds an additional page server with the max file size of 128GB.

If we run same query as earlier, we get the results shown in Table 2-10.

Table 2-10. *The Page Servers After Adding 64GB of Data*

file_id	file_name	type_desc	file_size_GB	max_file_size_GB	growth_size_GB
1	data_0	ROWS	60	128	10
3	data_1	ROWS	10	128	10
2	log	LOG	1022	1024	2

After inserting another 64GB data to the database (making a total of 128GB data inserted), an additional page server is created, and now we have three page servers, as shown in Table 2-11.

Table 2-11. *The Page Servers After Adding Another 64GB of Data*

file_id	file_name	type_desc	file_size_GB	max_file_size_GB	growth_size_GB
1	data_0	ROWS	110	128	10
3	data_1	ROWS	60	128	10
4	data_2	ROWS	10	128	10
2	log	LOG	1022	1024	2

Since the growth size per data file is in 10GB increments, this 180GB total size is not necessarily used space; it is rather allocated space for this example.

Important Although in the previous example each page server had a maximum file size of 128GB, we are charged only for the storage used. That is the `file_size_GB` value.

As mentioned earlier, each page server can contain between 128GB and 1TB total data per file. In all our previous examples, we were presenting page servers with a maximum of 128GB per file. In the following example, we will show the resulting page servers after we migrate an Azure SQL Database instance from General Purpose to Hyperscale if it contains more data than the initial maximum page server size of 128GB. If the initial size of the data file is larger than 128GB, the page server maximum data file size will be adjusted to accommodate the larger data file size. Figure 2-11 shows the used and allocated storage space of an Azure SQL Database General Purpose service tier in the Azure Portal before being migrated to the Hyperscale tier.

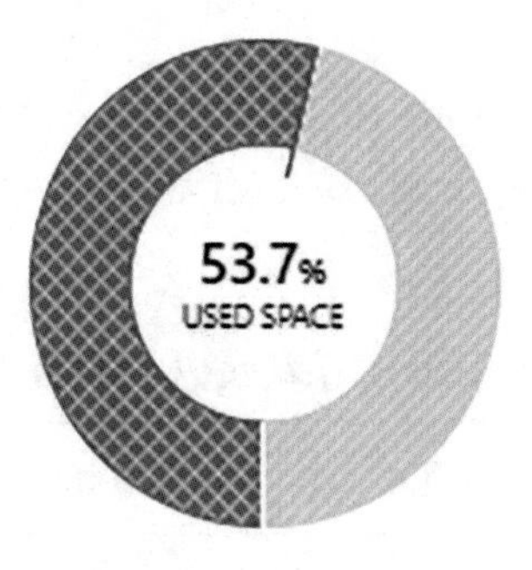

Figure 2-11. *Database data storage, used and allocated storage space in the Azure Portal*

After the migration from the General Purpose to the Hyperscale service tier is finished, we can run the same query as we did in the previous examples using the `sys.database_files` DMV. As we can see, the maximum file size has been changed and is no longer 128GB; it has increased to 384GB, three times larger than in the first example.

Table 2-12. *The Page Servers After Migrating from a Database with 134GB of Storage*

file_id	file_name	type_desc	file_size_GB	max_file_size_GB	growth_size_GB
1	data_0	ROWS	137	384	10
2	log	LOG	1022	1024	2

Note Bear in mind each page server can handle up to approximately 1TB data file size. If you are migrating your database to Hyperscale with a data file size bigger than 1TB, your data file is going to be split into smaller multiple files after migration is finished.

The next example shows the behavior of the page servers when we migrate a database that contains more than 1TB of data, which is larger than the maximum page server file size. This demonstrates that the maximum size of the data file per page server can differ within a database. For example, when we migrate an Azure SQL database from the Business Critical to the Hyperscale service tier with a database size of 1.12TB, we will have multiple page servers with a maximum data file size per each page server determined by Azure.

Figure 2-12 shows the used and allocated storage space of an Azure SQL Database Business Critical service tier with 1.12TB of data in the Azure Portal before being migrated to the Hyperscale tier.

Database data storage

Review the below metrics and monitor your applications and infrastructure.

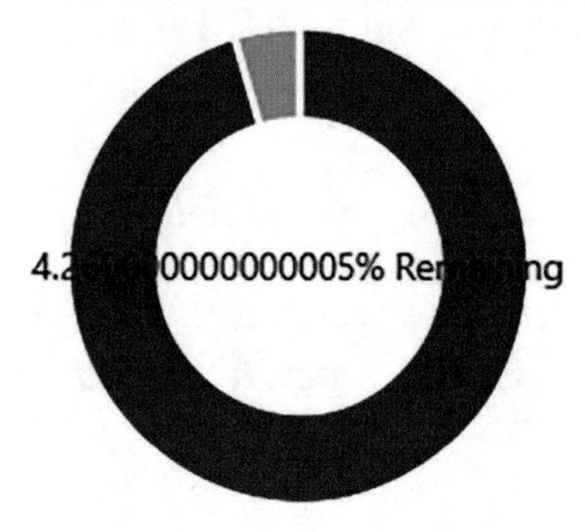

Used space 1.12 TB | Remaining space 51.09 GB | Allocated space 1.12 TB | Max storage 1.17 TB

Figure 2-12. *Azure SQL Database Business Critical service tier with more than 1TB*

After the migration to the Hyperscale service tier is complete, the Azure Portal database storage pane will look like Figure 2-13.

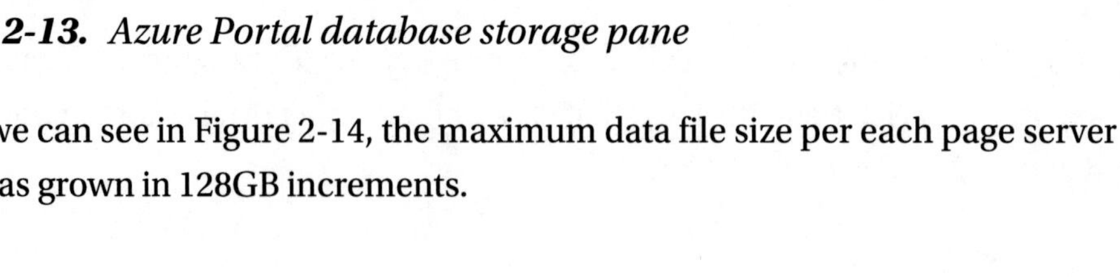

Database data storage

Review the below metrics and monitor your applications and infrastructure.

Used space 1.12 TB | Allocated space 1.12 TB

Figure 2-13. *Azure Portal database storage pane*

As we can see in Figure 2-14, the maximum data file size per each page server varies, and it has grown in 128GB increments.

Results | Messages

	file_id	file_name	type_desc	file_size_GB	max_file_size_GB	growth_size_GB
1	1	data_0	ROWS	400	512	10
2	3	data_1	ROWS	400	512	10
3	4	dfa_data_4	ROWS	348	384	10

Figure 2-14. *Page server distribution after migration to Hyperscale*

Azure determines what the maximum data file size for each page server should be, and once that size is set, it cannot be changed. Figure 2-15 shows the page servers after inserting more data into the previous Hyperscale database. It shows that each newly added page server can contain 128GB of data.

Results | Messages

	file_id	file_name	type_desc	file_size_GB	max_file_size_GB	growth_size_GB
1	1	data_0	ROWS	512	512	10
2	3	data_1	ROWS	512	512	10
3	4	dfa_data_4	ROWS	384	384	10
4	5	data_3	ROWS	128	128	10
5	6	data_4	ROWS	128	128	10
6	7	data_5	ROWS	128	128	10
7	8	data_6	ROWS	90	128	10
8	9	data_7	ROWS	40	128	10

Figure 2-15. *Page server distribution after inserting more data*

Each page server always has a replica, in active-active configuration, for the purposes of redundancy and high availability. This means both page servers are fully independent of each other and are both able to serve requests at the same time. Each page server has its own file data subset on Azure Standard Storage. The primary compute node can communicate with either of these page servers in an active-active mode. Figure 2-16 shows page servers in active-active configuration.

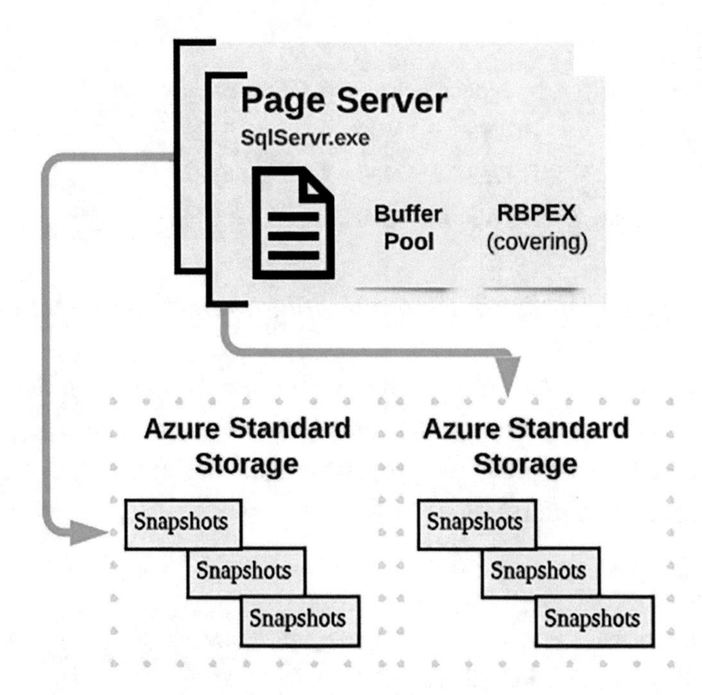

Figure 2-16. *The redundancy of the page servers*

The page servers are part of a scaled-out storage engine. This is because one of the roles of the page servers is to maintain and update the data files that are stored in Azure Standard Storage.

Note In the Hyperscale architecture, all data files are managed by page servers.

Azure Standard Storage

This tier provides the persistent storage used to hold both the data files (`.mdf`) and the log files (`.ldf`). Azure Standard Storage is used for this purpose. The multitier architecture, combined with the snapshot feature of Azure Standard Storage, enables significant improvements in backup and restore performance and behavior in Hyperscale. This includes the following:

- Extremely fast snapshot backups, regardless of database size
- A constant time for the PITR process to complete

- No compute impact for performing a snapshot backup

Having both the data files and log files on Azure Standard Storage makes this snapshot backup easy and fast.

Tip The snapshot backup file is in fact not a full backup, only a small differential copy from a base file with a timestamp on it.

This snapshot backup process on Hyperscale has almost no impact on your workload. At the time when the snapshot backup is taken, a brief freeze I/O operation occurs on the page servers to allow the snapshot to be taken. There is no compute impact on the snapshot process.

Because Hyperscale uses page servers that use Azure Storage, this is also where snapshots are kept. The Log service's long-term storage also keeps our logs for point-in-time restores in Azure Standard Storage. This ensures the constant time of PITR with the help of the ADR feature. This operation runs impressively quickly regardless of the database size.

When you initiate the PITR process, the following operations happen:

1. It finds and copies the latest snapshots to the Azure Storage allocated space for this new database.
2. It finds the LSN of the oldest transaction from the time of the latest snapshot from the long-term storage, which holds the transaction logs.
3. The new page servers that are created will be synchronized from the latest snapshots.
4. The new log service synchronizes its landing zone using premium storage from the available transaction logs in long-term storage.
5. Once the new primary compute node is online, it will pick up logs from the new landing zone. All this happens quickly due to the ADR feature.

Figure 2-17 shows a diagram of how the point-in-time restore process works.

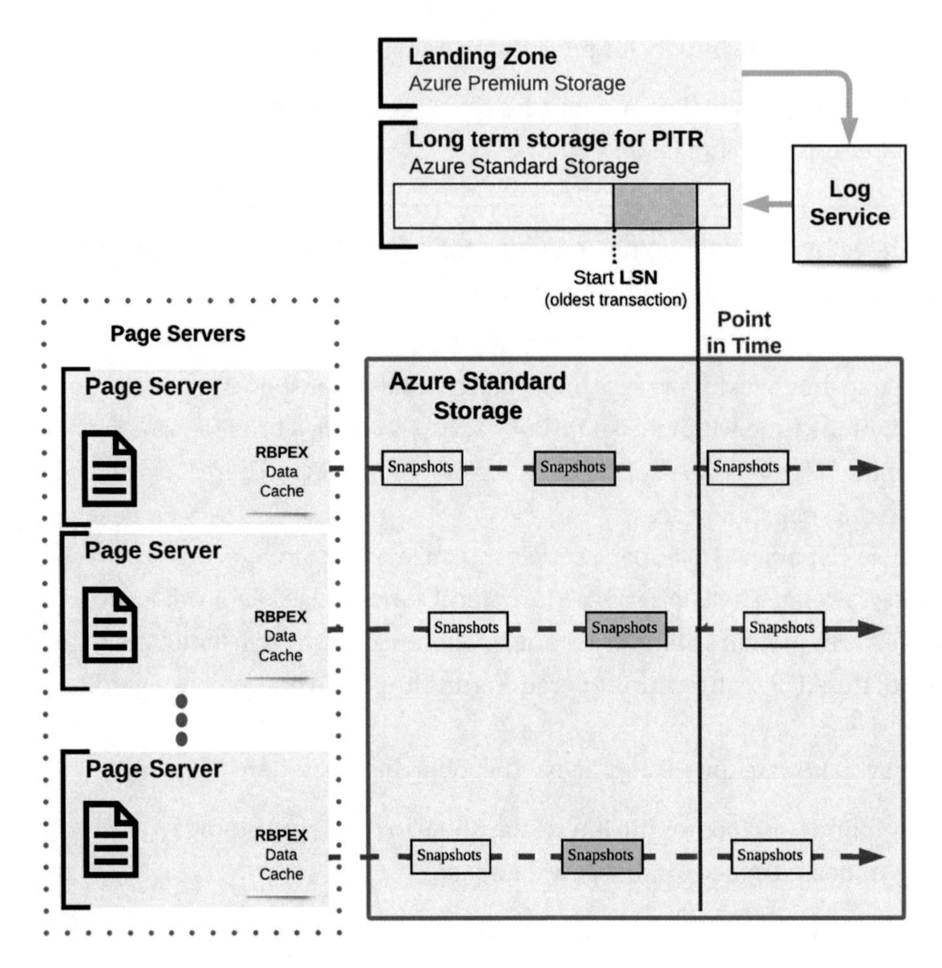

Figure 2-17. *How point-in-time restore works in Hyperscale*

How Do the Tiers Work Together?

All connection requests and read-write transactions come through the compute nodes. If the application requests a read-only workload and `ApplicationIntent=ReadOnly` is specified within the connection string, then the server will enforce the intent at connection time and redirect traffic toward the secondary compute node—provided there is at least one available.

Bear in mind that secondary replicas do not communicate directly with the primary compute node. They consume logs from the log service in the same way as page servers do. The secondary replica only gets those logs applied corresponding to the pages that already exist in the secondary replica's buffer pool, or its local RBPEX cache. Other transaction logs may be applied only in the case where there is a read request that needs pages that require those specific logs.

Figure 2-18 shows compute nodes routing read-write and read-only connections.

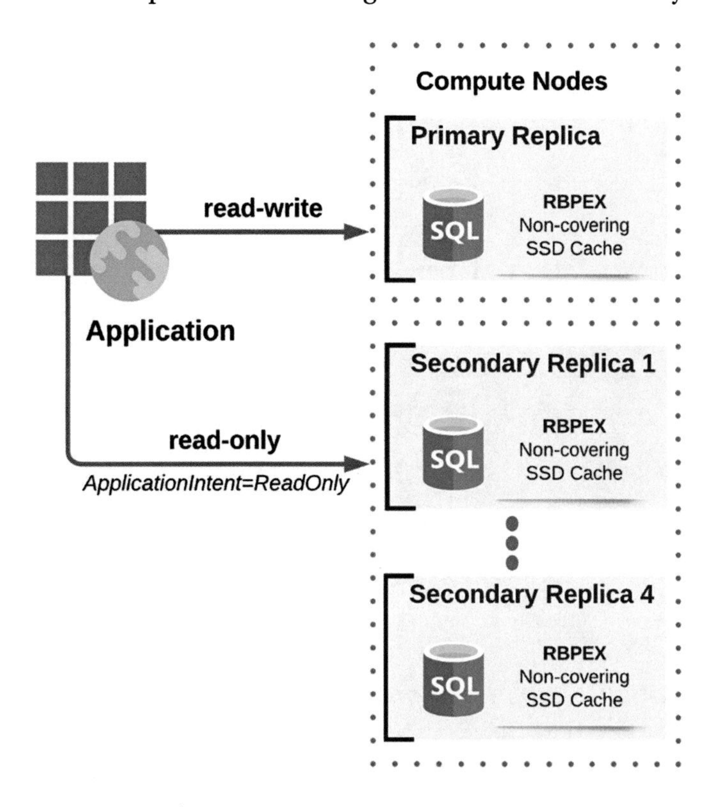

Figure 2-18. *Compute node connection routing*

All other read transactions that do not request a read-only workload will go through to the primary compute node. As previously mentioned, compute replicas do not store the full database in their local cache. Compute replicas use their buffer pool and their RBPEX cache (noncovering cache) to store data pages, though only partially. If pages are not available within the compute nodes' buffer pool (in memory) or their RBPEX cache, the next step is to check the page servers. Those reads will be counted as remote reads and will be somewhat slower than reading those pages from compute nodes directly. Figure 2-19 shows how the primary compute node and page servers interact.

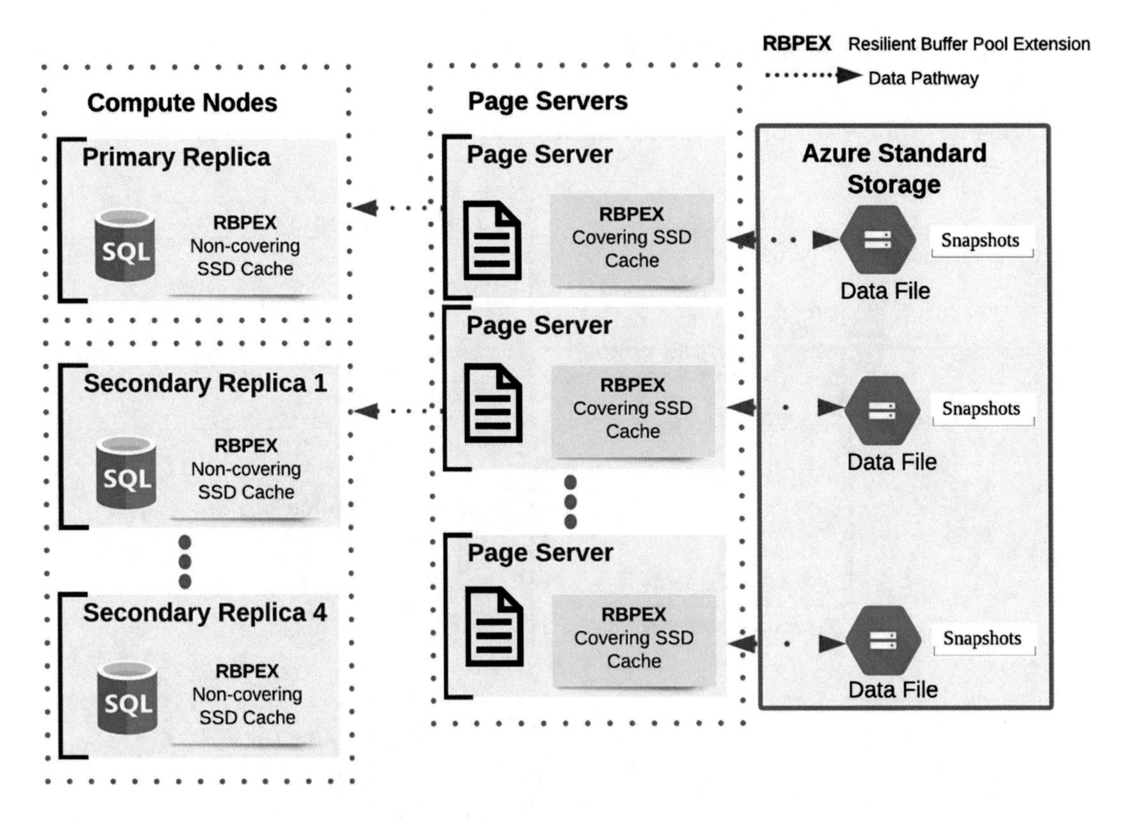

Figure 2-19. *Relationship between primary compute node and page server*

The compute nodes use their RBPEX caches to store information about requested pages. If the page requested does not exist on the primary compute node's buffer pool or local RBPEX cache, then the primary node needs to know which page server it should send `GetPage@LSN` request to.

After the LSN request is sent to the right page server, for a page server to be able to return the page, the log sequence number should be up-to-date.

Only once the page server is up-to-date with the LSN will it return the page to the primary replica. Otherwise, it will send a `GetLogBlock` request to the Log service, then get and apply that log, and finally return the page to the primary replica. This waiting time will be counted toward the `PAGEIOLATCH` wait type. Figure 2-20 shows Log service's Hyperscale tier interaction with compute nodes and page servers.

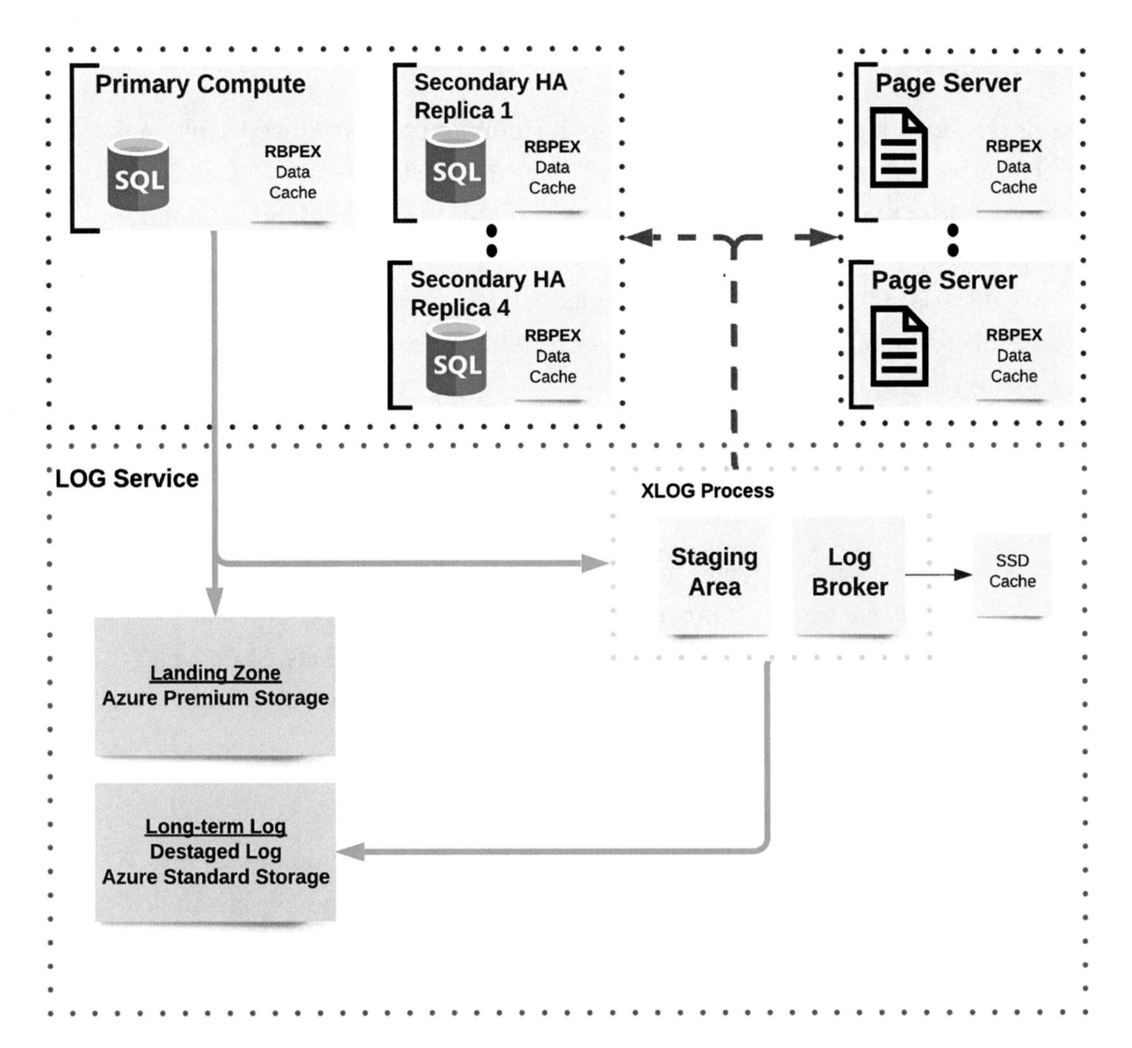

Figure 2-20. *Log service's Hyperscale tier interaction*

In terms of the write requests, all requests from the primary compute node will go directly to the log service, which is a separate tier in the Hyperscale architecture. From the primary compute node, the log is written to a landing zone that sits on Azure Premium Storage.

For the purposes of point-in-time restore processes, the Log service tier has its long-term storage, which sits on Azure Standard Storage. This way, the Log service can serve the log from Azure Premium Storage with faster I/O.

The XLOG process is a part of the Log service tier. The process uses its memory and SSD cache to hold and process the incoming log blocks. The Log Broker then serves the secondary replicas and page servers with log changes.

Summary

In this chapter, we gave you an overview of the foundations and concepts that make up the Hyperscale architecture. We looked at the Resilient Buffer Pool Extension and how it provides the ability to leverage fast SSD storage to enable a larger and more resilient cache.

We discussed the remote durable storage service, as well as the new architecture's shared-disk design, shared version store, and ability to recover more quickly after transaction failures.

We explained the reasons behind decoupling the compute nodes from the storage layer and the database log service into the four tiers creating the Hyperscale multitiered architecture. We discussed the four tiers and what they enable.

- *Compute nodes*: Hyperscale provides one primary compute node for read-write workload processing, including up to four HA and read scale-out secondary replicas and up to 30 named replicas. Moreover, geo replicas can be added for disaster recovery scenarios.
- *Log service*: The XLOG process and Log Broker ensure writes are available to secondaries quickly.
- *Page servers*: Page servers are the second caching layer in the Hyperscale architecture and they present part of the storage layer that contains transient and cached data used by the compute nodes and log service.
- *Storage tier*: This provides the actual persistent storage layer to the page servers and enables an improved backup and restore functionality.

We described how the Azure SQL Hyperscale service tier improves performance, scalability, and availability at a lower cost.

In the next chapter, we start the process of planning to deploy a typical production Hyperscale environment. We will look at some of the features of Hyperscale in more detail.

PART II

Planning and Deployment

CHAPTER 3

Planning an Azure SQL DB Hyperscale Environment

Now that we've completed a short tour of the SQL on Azure landscape and taken a high-level look at how the Hyperscale architecture differs, it's time to look at planning a production Azure SQL DB Hyperscale environment.

At the end of Chapter 1, we deployed a simple Azure SQL DB Hyperscale instance that was ready to be used. However, in a real-life production deployment of Azure SQL Hyperscale, we're likely to have additional nonfunctional requirements that we'll need to satisfy. In this chapter, we'll take a look at some of the key considerations you'll need to make to satisfy these additional requirements, including high availability, disaster recovery, connectivity, authentication, security, and monitoring.

Considerations When Planning for Hyperscale

The considerations you'll need to make when planning your Hyperscale deployment can be grouped into the following categories, which we'll go over in some detail:

- *Reliability*: How will you meet the database availability requirements, including enabling high availability? How will you meet the backup and disaster recovery requirements of the Hyperscale database?

Z. Barać and D. Scott-Raynsford, *Azure SQL Hyperscale Revealed*,
https://doi.org/10.1007/978-1-4842-9225-9_3

- *Security (including network connectivity)*: How will the clients connect to the Hyperscale database? How will you keep the database and the data it contains secure and accessed in a secure fashion? How will you monitor the security of your database and identify and address any security risks?
- *Operational excellence*: How will you monitor the performance and other signals from the Hyperscale database?

Understanding and planning for these requirements before you deploy your Azure SQL Hyperscale database will result in better outcomes. In most cases, your Hyperscale database can be adjusted after deployment, but not meeting critical requirements such as security or cost can have a large impact on your business.

Tip You may have noticed that the previous categories align to some of the Azure Well-Architected Framework (WAF). This is intentional as when you're designing a solution in Azure, you should be reviewing and consulting the Well-Architected Framework. For more information on the WAF, see `https://aka.ms/waf`.

The Azure SQL Database Logical Server

Before we review these considerations in detail, it is important to understand the concept of a logical server. This is because some considerations will apply only to the logical server or only to the Azure SQL Database.

When we deployed our first Hyperscale database in Chapter 1, you may have noticed that a SQL Server resource appeared in the resource group alongside our database, as shown in Figure 3-1.

Name ↑↓	Type ↑↓
hyperscaledb (sqlhyperscalerevealed01/hyperscaledb)	SQL database
sqlhyperscalerevealed01	SQL server

Figure 3-1. The Azure SQL logical server resource

The SQL Server resource is known as a *logical server*. We will refer to this resource simply as the *logical server* throughout this book. A single logical server can host multiple Azure SQL Database instances (up to 5,000 at the time of writing). It can host multiple Hyperscale databases or single databases or a mixture of both.

A logical server can also host multiple elastic pools, with the maximum being determined by the number of eDTUs or vCores in each elastic pool. But because this book focuses on Hyperscale (which can't be contained in an elastic pool), we won't go into more detail on this.

Tip The limits of Azure SQL Database are constantly increasing, so by the time you read this, the number of databases that can be hosted by a logical server may have increased. For the current limits of a logical server, please review `https://learn.microsoft.com/azure/azure-sql/database/resource-limits-logical-server`.

Figure 3-2 shows a conceptual diagram of the relationship between a logical server and the Azure SQL Databases or Azure SQL elastic pools that use it.

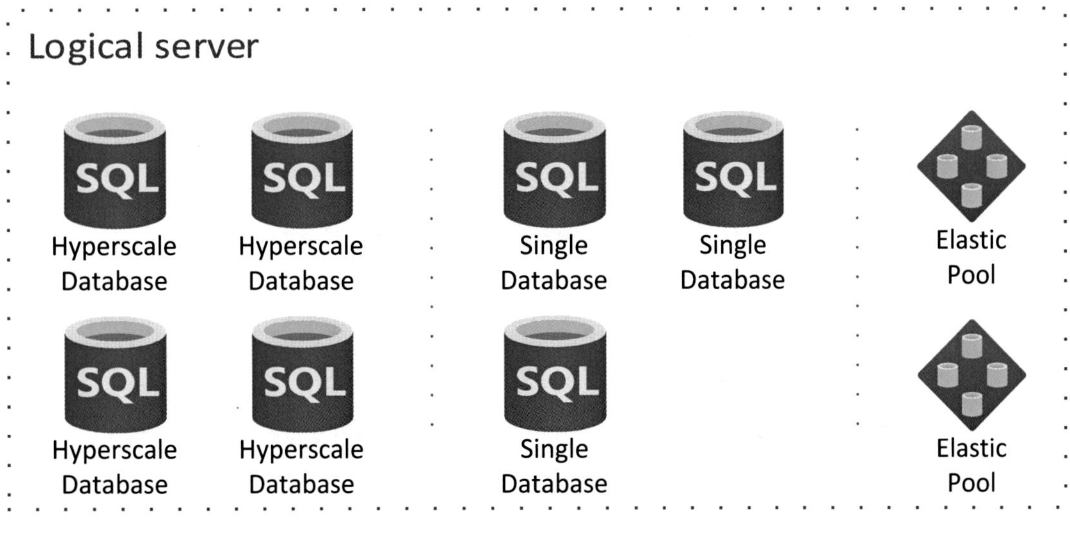

Figure 3-2. *Conceptual diagram of a logical server*

If you're familiar with running Microsoft SQL Server in an on-premises environment or in a virtual machine in a public cloud, then you might be tempted to conflate a Microsoft SQL Server with the Azure SQL logical server. However, they are different in practice and shouldn't be confused.

- *Microsoft SQL Server*: Contains the compute and storage resources for the hosted databases.
- *Azure SQL logical server*: Provides only the network connectivity, identity, security, and backup policies for the databases. The compute and storage resources for the databases hosted by the server are part of each Azure SQL Database (or Azure SQL elastic pool).

Note There is no guarantee that the compute and storage of each database in a logical server is physically located close together, although they will be within the same Azure region.

All Azure SQL Database instances within a logical server share the following:

- *Network connectivity and endpoints*: The public and private endpoints on which databases that are part of this logical server will be accessible. We can also configure outbound connectivity restrictions and whether clients will have connections proxied via an Azure SQL Database gateway or redirected to the database node.
- *Authentication settings*: To enable applications or users to authenticate using Azure Active Directory identities.
- *Azure MI*: The account in Azure Active Directory that the logical server and databases can use to access Azure resources.
- *Transparent data encryption key*: To configure if the logical server and databases should use service-managed or customer-managed keys for transparent data encryption.

Important Because many of the concepts in this section apply to the logical server, they will also apply to Hyperscale databases, single databases, and elastic pools. Where a concept or feature is specific to Hyperscale databases and doesn't apply to single databases or elastic pools, it will be noted.

Now that we have defined what a logical server represents, we can go into more detail about the key decisions we have to make when planning to deploy a Hyperscale database.

Considerations for Reliability

The reliability of the solution is the ability to recover from failures and continue to function. This also includes unexpected failures and planned outages, such as updates.

Defining the reliability of a solution requires considering two key requirements.

- *Availability*: The desired availability of the solution. This is usually expressed as a percentage of time a system needs to be available. For example, 99.99 percent availability is about 52 minutes of downtime per year or 8 seconds per week. The *higher* this percentage is, the more available the system is.
- *Disaster recovery*: The ability to recover from a system outage and the ability to recover damaged or lost data; usually defined by specifying a recovery time objective (RTO) and recovery point objective (RPO). These are usually specified in minutes, hours, or sometimes days. The *lower* the RTO, the faster a solution can be recovered in a disaster (e.g., a regional outage). The *lower* the RPO, the less data that may be lost in a disaster/data loss scenario.

Generally, increasing the availability of a solution or lowering the RTO/RPO requires increasing the cost and/or complexity of the solution. It also requires designing for these requirements as the architecture to achieve 99 percent availability is likely to be different from one that will achieve 99.999 percent availability. It is often more difficult (but not impossible) to change a design after it has been deployed to meet improved resilience requirements, but it is better to plan ahead by understanding these requirements and designing your Azure SQL Hyperscale database deployment accordingly.

In this section, we'll cover in detail the Azure SQL Hyperscale database features that help us meet availability and disaster recovery requirements.

- *High-availability replicas*: These replicate your Hyperscale database within the same Azure region but use the same page servers as the primary replica. This can improve the availability of your database.
- *Named replicas*: These replicate data to another Hyperscale database in the same logical server within the same region. Similar to high-availability replicas, the named replica shares the page servers as the primary replica. The named replica will appear as a different database with a different name. This can affect the availability and RTO of your solution.

Note Although named replicas are included under reliability, they are not typically used to increase resilience. Rather, they're traditionally used to enable read scale-out scenarios and to improve performance efficiency.

- *Geo replica*: This replicates data to another Hyperscale database in a different logical server within the same or a different region. This is the only replication model that can protect against regional Azure outages. This can affect the availability and RTO of your solution.
- *Backup*: This protects against data loss. This affects the RPO of your solution, but if you're not using replication, it also affects the availability and RTO.

We will now look at each of these topics in detail.

High-Availability Replicas

When you deploy a Hyperscale database, you can choose the number of high-availability secondary replicas you want to deploy, from zero to four. These high-availability secondary replicas are deployed within the same Azure region as the primary and share the page servers. This means that no data copy is required to add new replicas. The secondary replicas help prevent unplanned disruption to your workload.

Tip Even if you do not deploy any high-availability secondary replicas with your Hyperscale database, your workload is still protected against disruption due to planned maintenance events. This is because the service will automatically create a new replica before initiating a failure. For more information, see `https://learn.microsoft.com/azure/azure-sql/database/service-tier-hyperscale?view=azuresql#database-high-availability-in-hyperscale`.

High-availability secondary replicas can be used as read replicas. To connect to a high-availability secondary replica, you would need to set the `ApplicationIntent` to `ReadOnly` in the connection string. If you have more than one secondary replica, you can't choose which replica to connect to.

You can change the number of high-availability replicas after creating your Hyperscale database. You can also choose whether these replicas are distributed across availability zones within a region, for regions that support availability zones.

Tip This section assumes you are familiar with the concept of availability zones within Azure. If you're not familiar or would like to brush up, please review `https://learn.microsoft.com/azure/availability-zones/az-overview`.

The following are the two options that control how your secondary replicas are deployed within the region:

- *Locally redundant*: The high-availability secondary replicas will be deployed in the same region, but not necessarily in a different zone to the primary. *This is the default high-availability mode.* However, if your primary database and all replicas are deployed to the same availability zone and that zone has an outage, your workload will be disrupted.
- *Zone redundant*: The high-availability secondary replicas will automatically be created in different zones than your primary replica. This provides your database with resilience to an unplanned outage of an entire availably zone.

Important At the time of writing, zone-redundant high availability is still in preview. Zone-redundant high availability can be deployed only with Gen5 hardware in a region that supports availability zones. Zone-redundant replicas can also not be enabled for named replicas.

Figure 3-3 shows the potential impact of using locally redundant high-availability secondary replicas when an outage of an availability zone occurs. This shows that the workload will be disrupted and would need to be failed over to a different region, if you had enabled geo-replication.

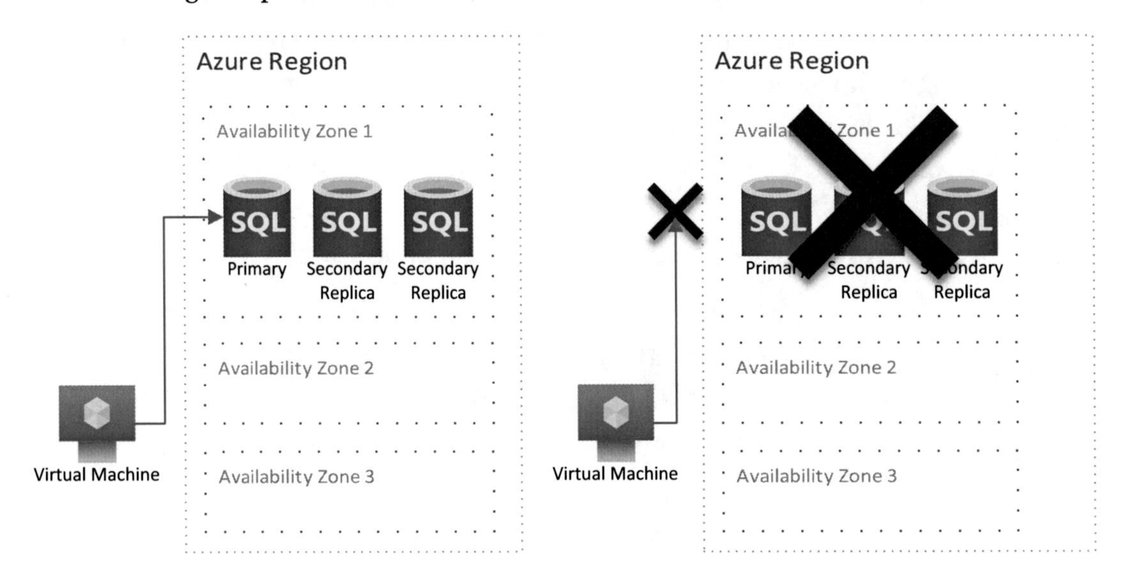

Figure 3-3. *Two high-availability locally redundant secondary replicas in a zone outage*

Figure 3-4 shows what occurs when the availability zone containing the primary replica is disrupted. The application simply connects to one of the secondary replicas in a different zone and continues operating without disruption.

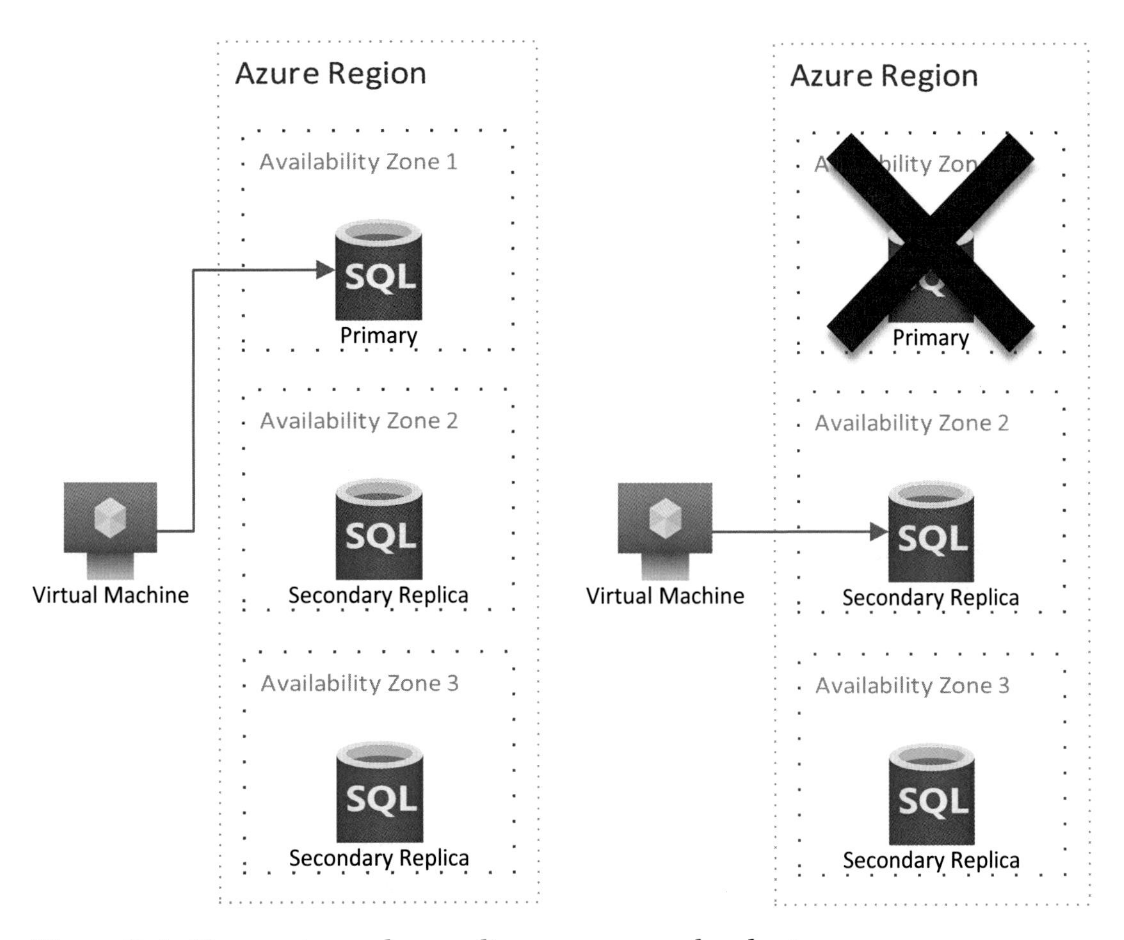

Figure 3-4. *Zone outage when replicas are zone redundant*

When choosing the level of high availability you want to implement, it is a decision based on resource cost versus resilience as choosing the number of replicas does not increase the complexity of the solution.

Figure 3-5 shows creating a new Hyperscale database with two high-availability secondary replicas in a zone-redundant fashion.

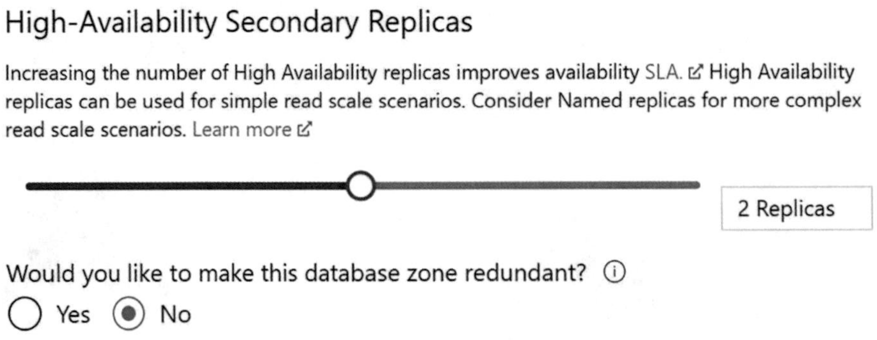

High-Availability Secondary Replicas

Increasing the number of High Availability replicas improves availability SLA. High Availability replicas can be used for simple read scale scenarios. Consider Named replicas for more complex read scale scenarios. Learn more

2 Replicas

Would you like to make this database zone redundant?

Yes No

Figure 3-5. *Creating a hyperscale database with two zone-redundant replicas*

Important High-availability secondary replicas do not protect against a regional outage. To protect against that, you need to implement geo-replication, which is covered next.

Named Replicas

Named replicas are like high-availability (HA) secondary replicas. Like HA replicas, they share the underlying page servers of the primary database. But they do differ in the following ways:

- They appear as a separate read-only database in the Azure Portal.
- They can be connected to the same logical server as the primary or to a different logical server in the same region. This enables different authentication methods to be used for the named replica compared to the primary Hyperscale database.
- The resources on the named replica can be scaled separately from the primary.
- Scaling the resources on the replica does not impact connections to the primary database, and vice versa.
- Workloads running on a replica will be unaffected by long-running queries on other replicas.

Important Although named replicas can increase availability if you aren't using high-availability replicas, it is not the primary use case. They are typically used to enable read scale-out scenarios and to improve hybrid transactional and analytical processing (HTAP) workloads.

Figure 3-6 shows the conceptual architecture for named replicas. The two named replicas are replicas of the Hyperscale database, but the first is attached to logical server 1, and the second is attached to logical server 2.

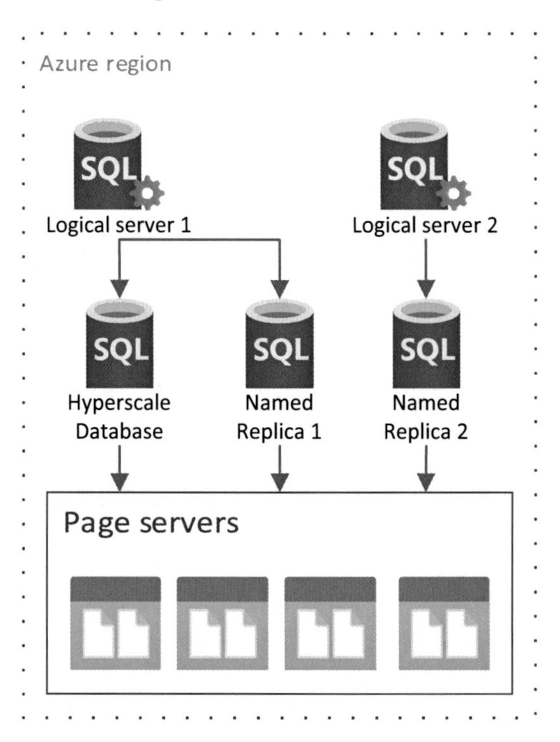

Figure 3-6. *Named replica conceptual architecture*

Figure 3-7 shows the Replicas blade on an Hyperscale database in the Azure Portal. A named replica exists called `hyperscaledb_replica`. Clicking the "Create replica" button allows us to create new named replicas (or geo replicas).

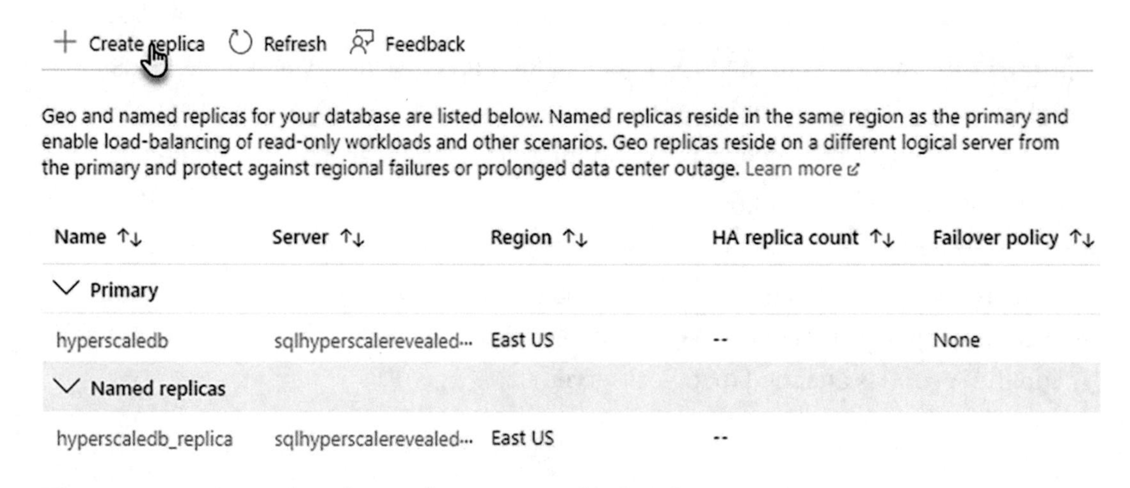

Figure 3-7. *Named replicas of a Hyperscale database*

Figure 3-8 shows the named replica creation step from an existing Hyperscale database.

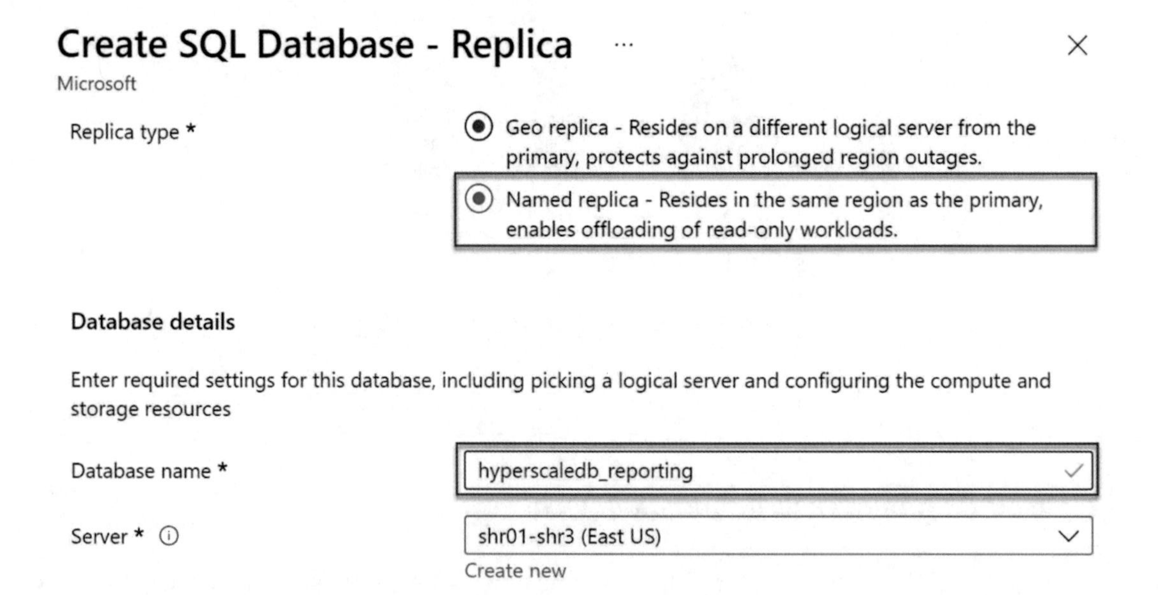

Figure 3-8. *Creating a Hyperscale named replica*

Geo Replica

The geo replica feature of Azure SQL Database Hyperscale provides a more traditional replication implementation with the data being copied from the primary to a separate set of page servers for the replica.

Geo replicas have the following properties:

- They must be created on a different logical server from the primary Hyperscale database.
- Page data is asynchronously copied from the primary Hyperscale database to the replicas.
- The database name of the replicas is always the same as the primary Hyperscale database.
- They maintain a transactionally consistent copy of the primary Hyperscale database.

Figure 3-9 shows the conceptual architecture when using geo-replication with a Hyperscale database.

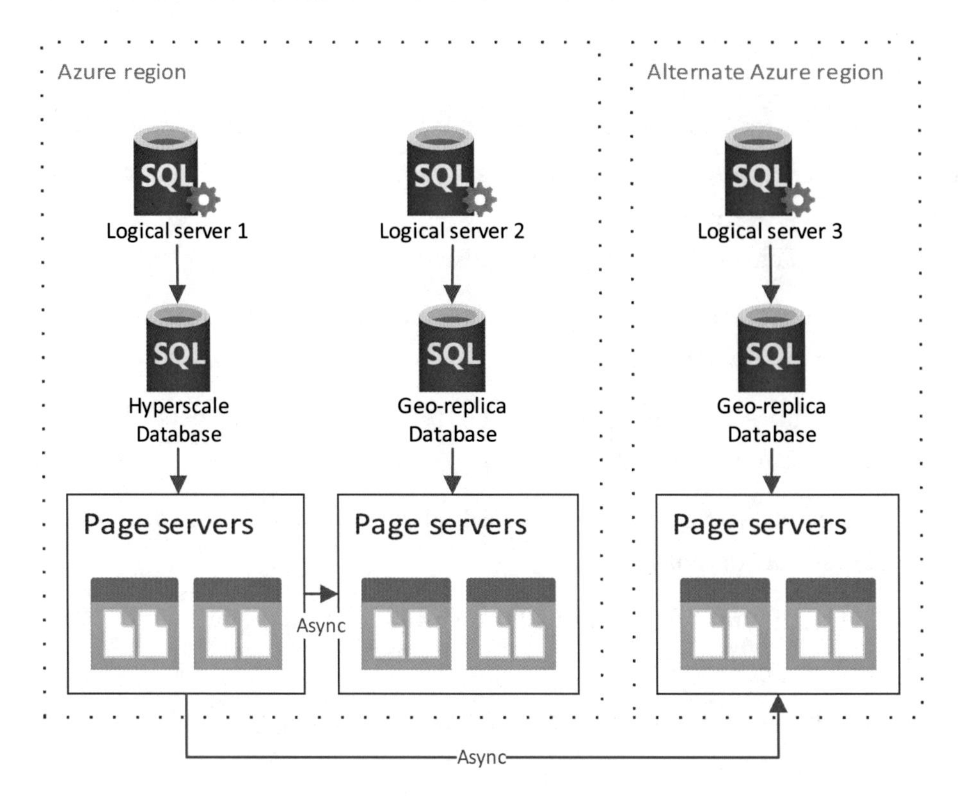

Figure 3-9. *Geo-replication options for a Hyperscale database*

Geo replicas are typically used to mitigate against the outage of an entire Azure region and provide all the necessary infrastructure to perform geo-failovers.

Backup

Performing database backups is the simplest and most common way to prevent data loss. It should not be a surprise that performing automated database backups is a basic feature of Azure SQL Databases. However, because of the unique Hyperscale storage architecture, Hyperscale database backups have some special properties, listed here:

- *Snapshot-based*: Hyperscale backups are performed by the underlying Azure storage platform by taking a snapshot.
- *Backup performance*: Using snapshots to back up the database results in it being nearly instantaneous and does not impact the compute performance of the primary or secondary replicas, regardless of how much data is stored.

- *Restore performance*: Restoration of a Hyperscale database is performed by reverting to a previous snapshot that is available within the retention period. This results in the restoration process within the same Azure region to complete within minutes, even for large, multiterabyte databases.

  Restoring to a new Hyperscale database is also extremely rapid because of the new storage architecture. So, creating copies of an existing database from a backup for development or testing purposes is very fast and will often be complete in minutes.

- *Backup scheduling*: Because Hyperscale databases are backed up using storage snapshots, there are no traditional full, differential, and transaction log backups.

  When you enable Hyperscale database backups, you can choose the redundancy of the storage used to store your backups. There are three options available that will affect the resilience of backups.

- *Locally redundant backup storage*: Three copies of your backup are stored in the same region as your Hyperscale database. They may all be stored in the same zone, which means that in the case of a zone failure, your backup will be unavailable. This is the cheapest option as it uses Azure storage LRS.

- *Zone-redundant backup storage*: Three copies of your backup are stored in the same region as your Hyperscale database, but they are spread across all zones within the region. This makes your backup resilient to a zone failure in the region. However, a regional outage will make your backup unavailable. This is more expensive because it uses Azure storage ZRS.

- *Geo-redundant backup storage*: Like locally redundant backup storage your backup, three copies are stored within the same region as your Hyperscale database. However, the backup is also asynchronously replicated to the paired Azure region. It will make your backup resilient to local, zonal, and regional outages. This uses Azure storage RA-GRS.

- *Geo-zone-redundant backup storage*: This provides both geo-redundant and zone-redundant storage for your backups and enables the highest possible backup resilience. This is the most expensive because it uses Azure storage RA-GZRS.

Tip To enable the geo-restore feature of Azure SQL Databases, you must be using geo-redundant or geo-zone-redundant backup storage for your Hyperscale database. Geo-restore allows you to restore your Hyperscale database to the paired Azure region in the case of a regional outage.

The remaining decision you need to make when considering backups on your Hyperscale database is to define the short-term retention policy (STR). This can be between 1 and 35 days, but it defaults to 7.

Important Long-term retention (LTR) policies are not currently supported for Hyperscale databases.

Figure 3-10 shows how to configure the backup retention policies for a Hyperscale database in the Azure Portal using the Backups blade of the logical server.

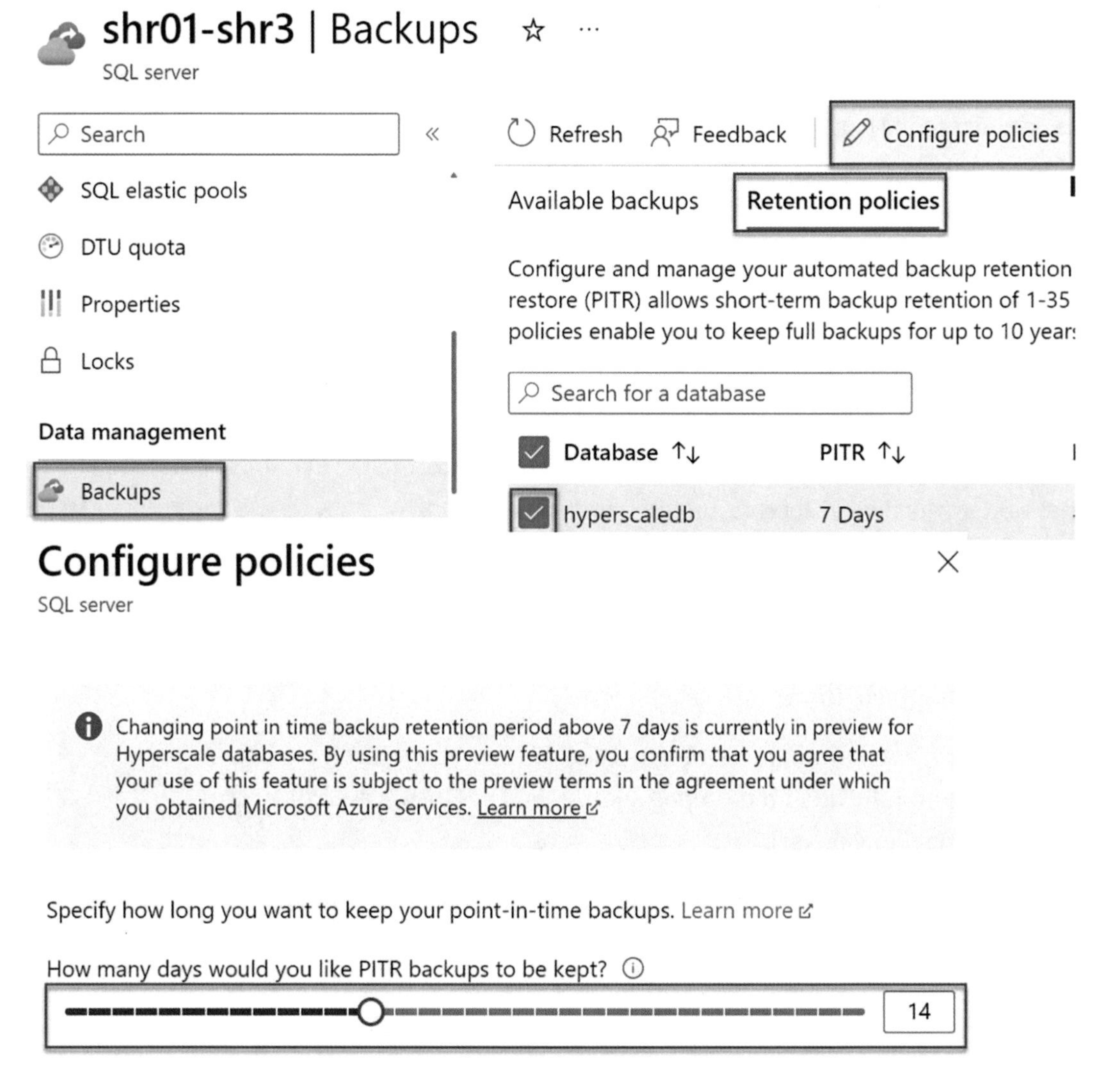

Figure 3-10. *Configuring backup retention policies*

When planning how we should configure backup for our Hyperscale database, it comes down to balancing cost with our resilience requirements as the new Hyperscale storage architecture eliminates the issue of performance impact when performing a backup.

Considerations for Network Connectivity

In most cases, when we deploy a Hyperscale database, there will be clients that need to connect to it. Usually, the clients will be any applications that use the database. However, there may also be other Azure services that also need to access the database. For example, Azure Data Factory might be performing extract, transform, load (ETL) processing, PowerBI could be providing data visualization, and Azure Purview could be providing data cataloging services.

Occasionally we also have services external to Azure and software-as-a-service (SaaS) applications that need direct access to the database. It is also common for services and applications on-premises or in other clouds to need access to the database.

There are two network connectivity options for Azure SQL Databases that are defined by the logical server that the database is part of.

- *Public endpoint*: The logical server is exposed on a public IP address within the Azure public IP address space. It is usually with an IP-based firewall or limits on the Azure virtual networks that traffic can originate from.
- *Private endpoint*: The logical server is assigned a private IP address and fully qualified domain name (FQDN) inside an Azure virtual network (VNet).

A database can be accessible on one or both endpoint types, depending on how you've configured the logical server. A logical server can also be connected to multiple VNets at the same time.

Figure 3-11 shows the conceptual architecture of a logical server both with a public endpoint on the Internet and a private endpoint in an Azure virtual network.

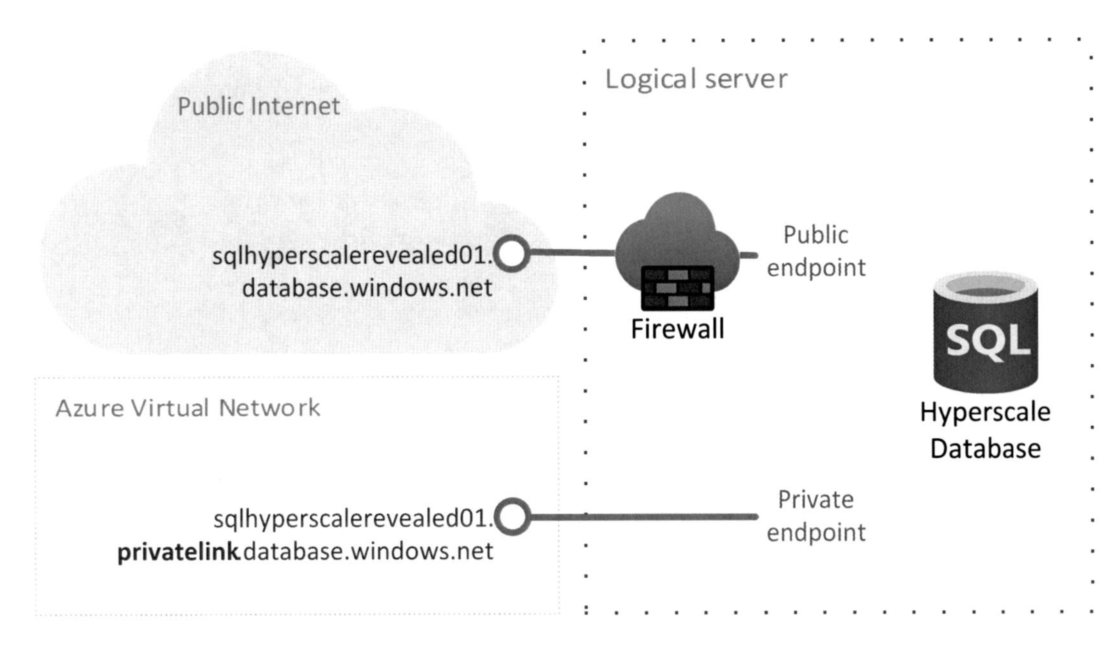

***Figure 3-11.** Network connectivity options for a logical server*

With each client that will need to use your database, you will need to determine whether public or private connectivity can be used and is required.

Important Some clients, such as external SaaS applications, may not be able to use a private endpoint. So, if you have a requirement for a SaaS application that can't use a private endpoint to access the database, then you'll have to factor this into your planning.

Public Endpoint

Every Azure SQL logical server has a public endpoint with a DNS name on the Internet. However, this can be disabled completely or filtered based on the originating traffic. For logical servers hosted in the Azure Public cloud, the fully qualified domain name (FQDN) will be as follows:

yourservername`.database.windows.net`

For logical servers in Azure Government or Azure Sovereign clouds, the FQDN will differ. You can check the FQDN of your logical server in the Azure Portal on the Overview blade of the logical server, as depicted in Figure 3-12.

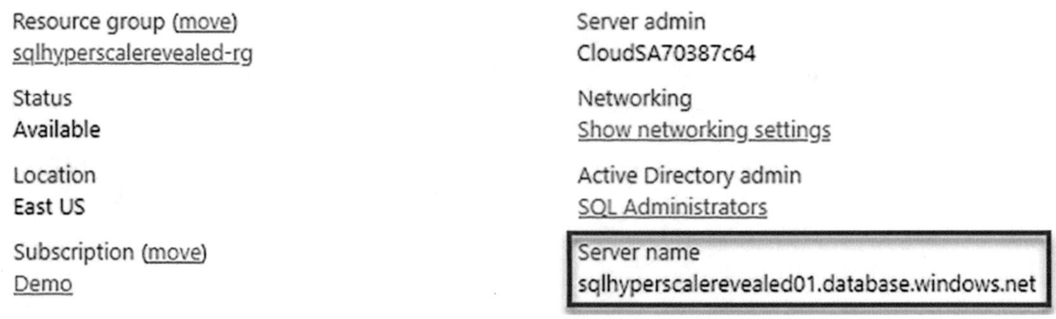

Resource group (move) sqlhyperscalerevealed-rg	Server admin CloudSA70387c64
Status Available	Networking Show networking settings
Location East US	Active Directory admin SQL Administrators
Subscription (move) Demo	Server name sqlhyperscalerevealed01.database.windows.net

***Figure 3-12.** Locating the FQDN of the logical server in the Azure Portal*

You can also use Azure PowerShell or Azure CLI to return the FQDN of the logical server. For example, to return the FQDN of the logical server using PowerShell, use the following:

```
(Get-AzSqlServer -ServerName yourserver).FullyQualifiedDomainName
```

This is the output when the command is run using Azure Cloud Shell with PowerShell:

```
PS /home/     > (Get-AzSqlServer -ServerName sqlhyperscalerevealed01).FullyQualifiedDomainName
sqlhyperscalerevealed01.database.windows.net
```

If you're not familiar with Azure Cloud Shell, review `https://learn.microsoft.com/azure/cloud-shell/overview`.

To get the FQDN using the Azure CLI, use the following:

```
az sql server show --name yourserver --resource-group yourresourcegroup
--query fullyQualifiedDomainName
```

This is the output when the command is run using Azure Cloud Shell with Bash:

```
     @Azure:~$ az sql server show --name sqlhyperscalerevealed01 --resource-group
 sqlhyperscalerevealed-rg --query fullyQualifiedDomainName
"sqlhyperscalerevealed01.database.windows.net"
```

Public Endpoint Access Rules

When we enable the public endpoint on the logical server, we must enable network access rules on it. If we do not enable any network access rules on the public endpoint when it is enabled, then the logical server will block all traffic by default. This is a desirable behavior that improves our default security posture.

Figure 3-13 shows the conceptual architecture of a logical server with the public endpoint disabled or enabled with no access rules defined.

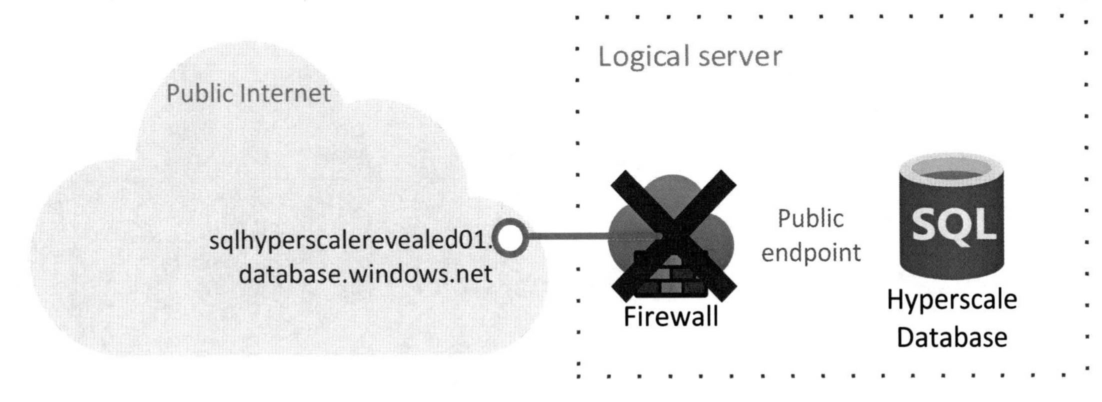

Figure 3-13. *Network connectivity of a logical server with no access controls defined*

There are several different types of network access rules that can be applied in combination to the public endpoint.

- *Virtual networks*: VNets allow you to block traffic that does not originate from one or more subnets from one or more Azure VNet. Each Azure VNet must have the `Microsoft.Sql` service connection enabled on it. Figure 3-14 shows the conceptual architecture when a VNet access control is applied.

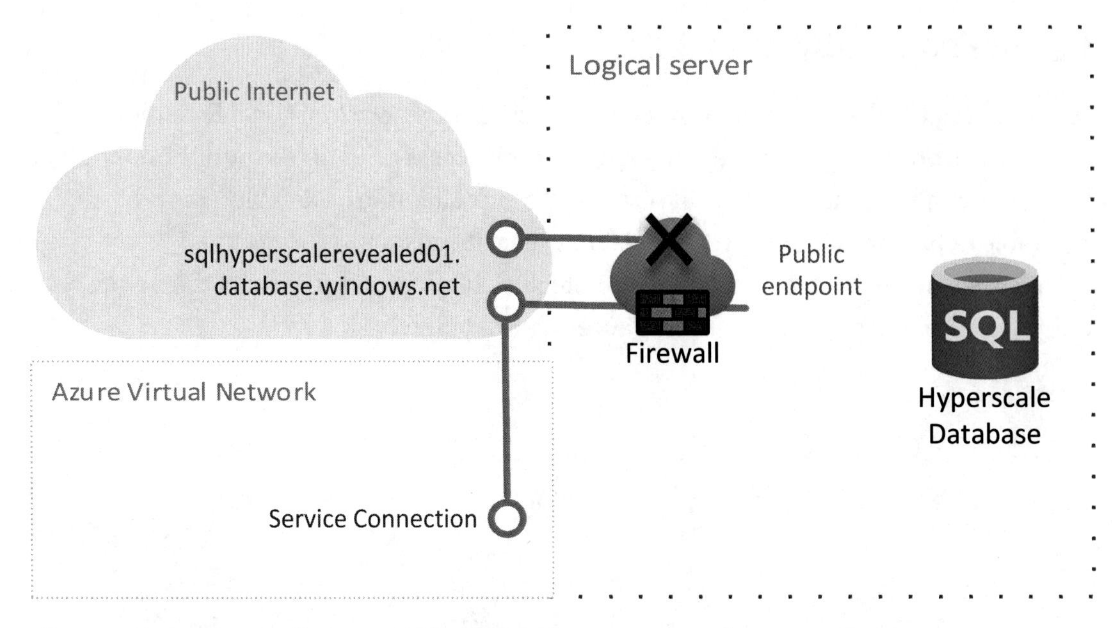

Figure 3-14. *Logical server with a VNet access control*

- *Firewall rules*: This allows you to define IP address ranges that are granted access.

Tip If you're trying to access the logical server from your machine using a tool like SQL Server Management Studio (SSMS), then you'll need to add your client IP address to this list. Otherwise, your client will be denied access. Be sure to remove your client IP address once you've completed your work.

- *Allow Azure services and resources to access this server*: This setting allows any service or resource running in Azure to access this service. This includes services run by other Azure customers. In general, this is not a good idea as it grants too much access to the logical server.

Tip It is strongly recommended to always enable network access controls on the public endpoint to restrict as much access to the database as possible. It is a good idea to perform a security risk assessment to ensure that the controls are as restrictive as possible while still meeting business requirements.

Figure 3-15 shows the Networking blade in the Azure Portal that is seen when configuring public access controls on a logical server.

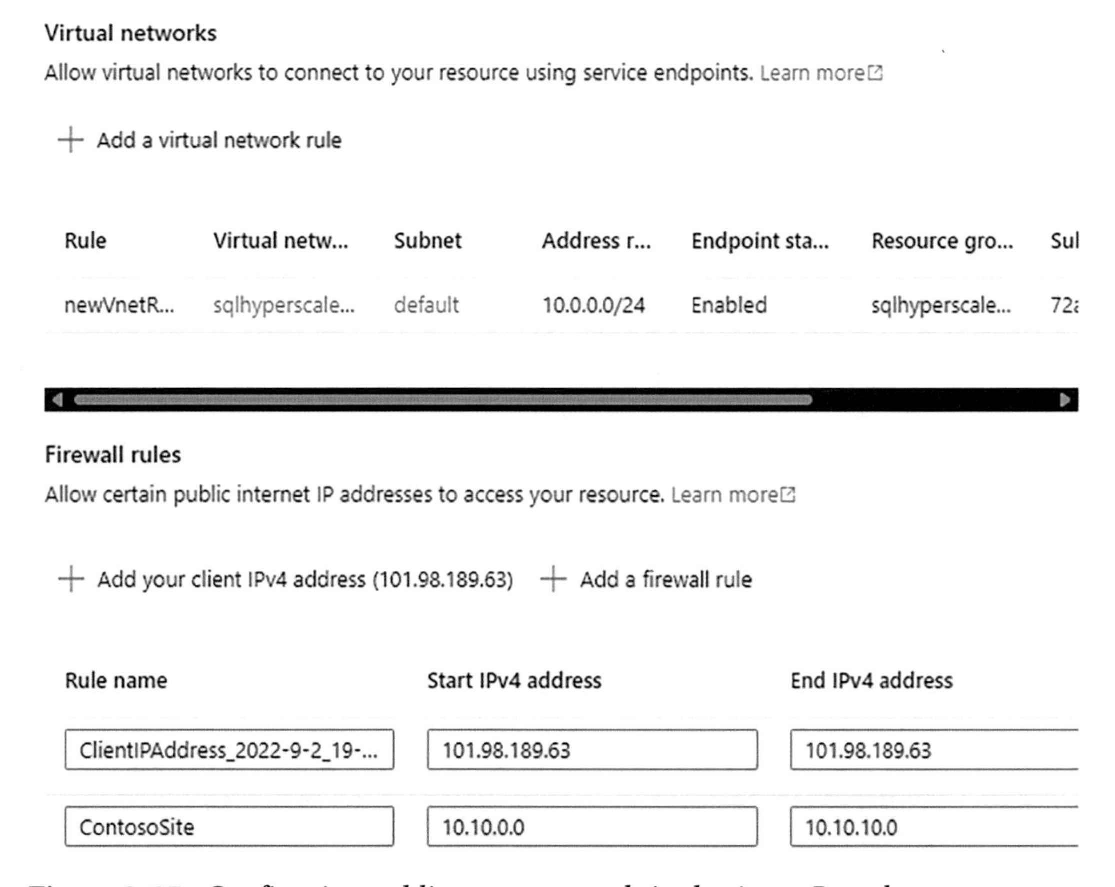

Figure 3-15. *Configuring public access controls in the Azure Portal*

We will cover different methods of configuring these settings in the following chapter.

Public Endpoint Protection

Even when an Azure SQL Database instance is exposed on a public endpoint, it still has multiple layers of protection on it.

- *Azure default infrastructure-level DDoS protection*: This automatically protects the Azure infrastructure against different distributed denial-of-service (DDoS) attacks. This is always on.

Tip Azure provides a more extensive set of DDoS features as part of the Azure DDoS Standard product. However, this needs to be enabled on a VNet, so it cannot be enabled on the public endpoint.

- *Firewall rules and VNet service endpoints*: This allows you to filter traffic to the server based on the originating IP address range or from specific Azure VNets.
- *Authentication*: Your connection needs to be authenticated with the logical server using a SQL username and password or Azure Active Directory username and password. This is discussed in more detail in the "Authentication Settings" section.
- *Connection encryption*: The connection is encrypted using TLS 1.2. Older versions of TLS can be allowed, but this is not recommended.

Traffic traveling to this endpoint will travel over the public Internet, unless it originates from within Azure. If your security or compliance requirements require that your database traffic does not traverse the public Internet, then you should not use a public endpoint.

Important If the traffic to your public endpoint originates from within the Azure network, then it will not leave Azure, but rather it will travel over private Azure private backbone network infrastructure.

Private Endpoints

Private endpoints allow your logical server (and therefore Hyperscale database) to have a private IP address within an Azure VNet. This allows you to ensure that your database traffic never leaves your virtual network environment and doesn't traverse a public network. This is often required for sensitive workloads or workloads that must meet strict compliance requirements.

A single logical server can have private endpoints in multiple VNets.

Enabling a private endpoint for a logical server requires the use of a number of additional Azure resources.

- *Network interface*: This is a virtual network interface that is attached to the logical server and connected to a specific subnet in the Azure VNet. It will be dynamically assigned a private IP address within the virtual network subnet address space.
- *Private DNS zone*: This is a DNS zone that is assigned to the Azure VNet to allow a private FQDN name to be used to access the logical server within the VNet. The zone will usually be automatically configured as `privatelink.database.windows.net` for logical servers in the Azure public cloud.

  If there are multiple logical servers with private endpoints in a single VNet, only one private DNS zone with the name `privatelink.database.windows.net` will be created.

  The private DNS zone will contain a single DNS A record for each logical server connected to the VNet and will have the private IP address assigned to the network interface.
- *Private endpoint*: This is the resource that governs the other resources and ties them all together. You can use this resource to configure the DNS and Application Security group settings, as well as visualize the private endpoint and its dependencies. Figure 3-16 shows the private endpoint's Insights blade to show the dependency view.

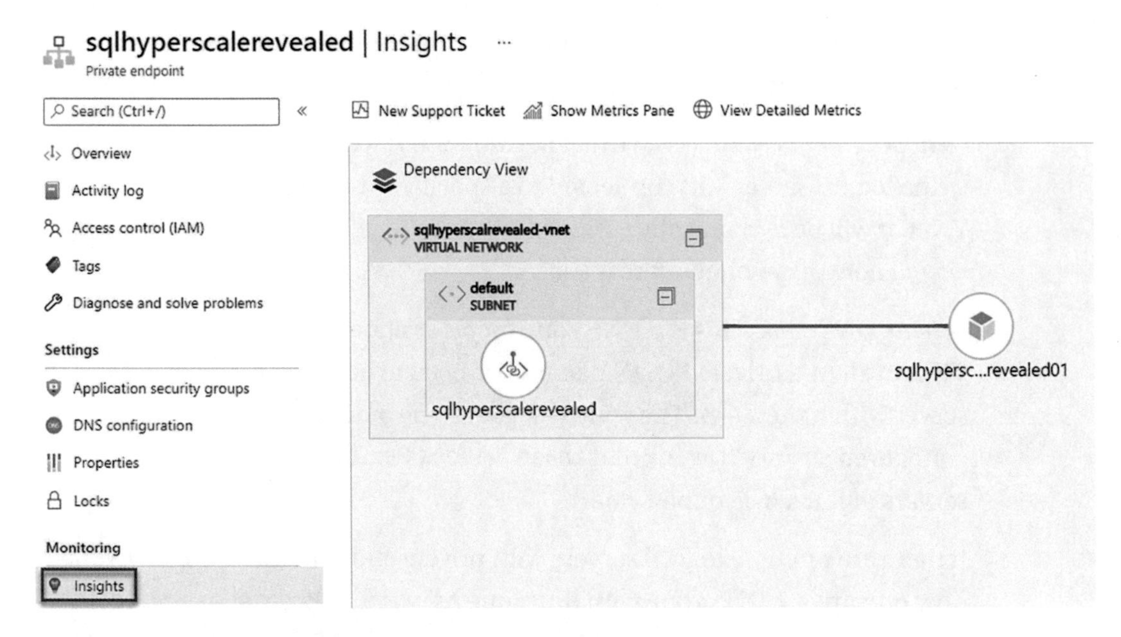

Figure 3-16. *Visualizing a private endpoint using insights*

- *Application security group (optional)*: These can be used to group virtual machines and define network security policies based on those groups. This allows you to limit access to the logical server private endpoint.

These resources are automatically provisioned for you when you configure this in the Azure Portal, but if you're deploying your logical server and the connectivity to it via infrastructure as code (IaC), then you need to ensure your code does it. We will provide more detail on deploying a Hyperscale database using IaC in Chapters 9 to 11.

Tip The private endpoint technology that enables the private endpoint in the VNet is known as Azure Private Link. You can read more about private links at `https://learn.microsoft.com/azure/private-link/private-link-overview`.

Outbound Networking

By default, Azure SQL Database instances in your logical server have no restrictions on their outbound network connectivity. However, in some scenarios, you may want to restrict the outbound connectivity of your Hyperscale database to only be allowed to connect to specific fully domain names. You can configure a list of allowed domain names that the databases in the logical server can connect to. Figure 3-17 shows configuring outbound networking restrictions in the Networking blade of the SQL server resource in the Azure Portal.

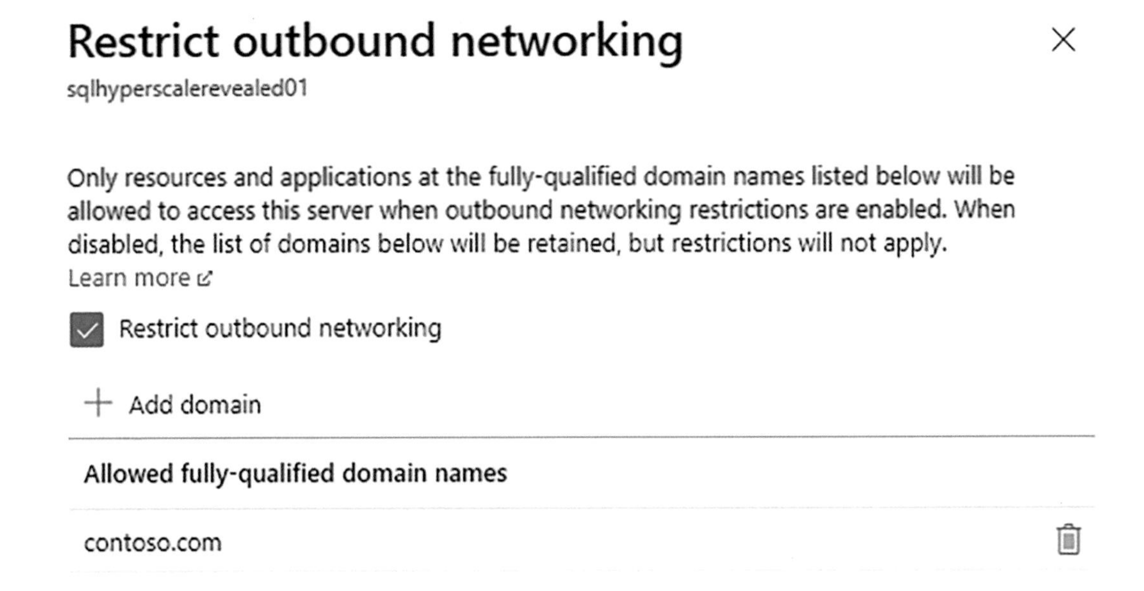

***Figure 3-17.** Configuring outbound networking restrictions*

Connection Policy

When a client connects to the Azure SQL Database, there are actually two different methods that can be used to form the connection.

- *Redirect*: The client connects to the Azure SQL Database gateway. The gateway then redirects the connection to the database cluster hosting the actual Azure SQL Hyperscale database. The destination IP address of the connection is changed from that of the Azure SQL Database gateway to the IP address of the cluster. This is the default connection policy for connections originating inside Azure.

- *Proxy*: The client connects to the Azure SQL Database gateway, and the connection is proxied through to the database cluster hosting the Azure SQL Hyperscale database. This is the default connection policy for connections originating outside Azure. This connection method results in slightly higher latency and a minor reduction in throughput.

Important Currently, when a logical server is connected to a virtual network via a private endpoint, it also uses proxy mode for traffic entering from the virtual network. This may impose a very slight performance penalty. This is likely going to be noticed only for very *chatty* workloads. This requirement is likely to change in future.

Figure 3-18 shows a logical architecture of the Azure SQL Database gateway and the default connectivity methods for connections from inside and outside of Azure.

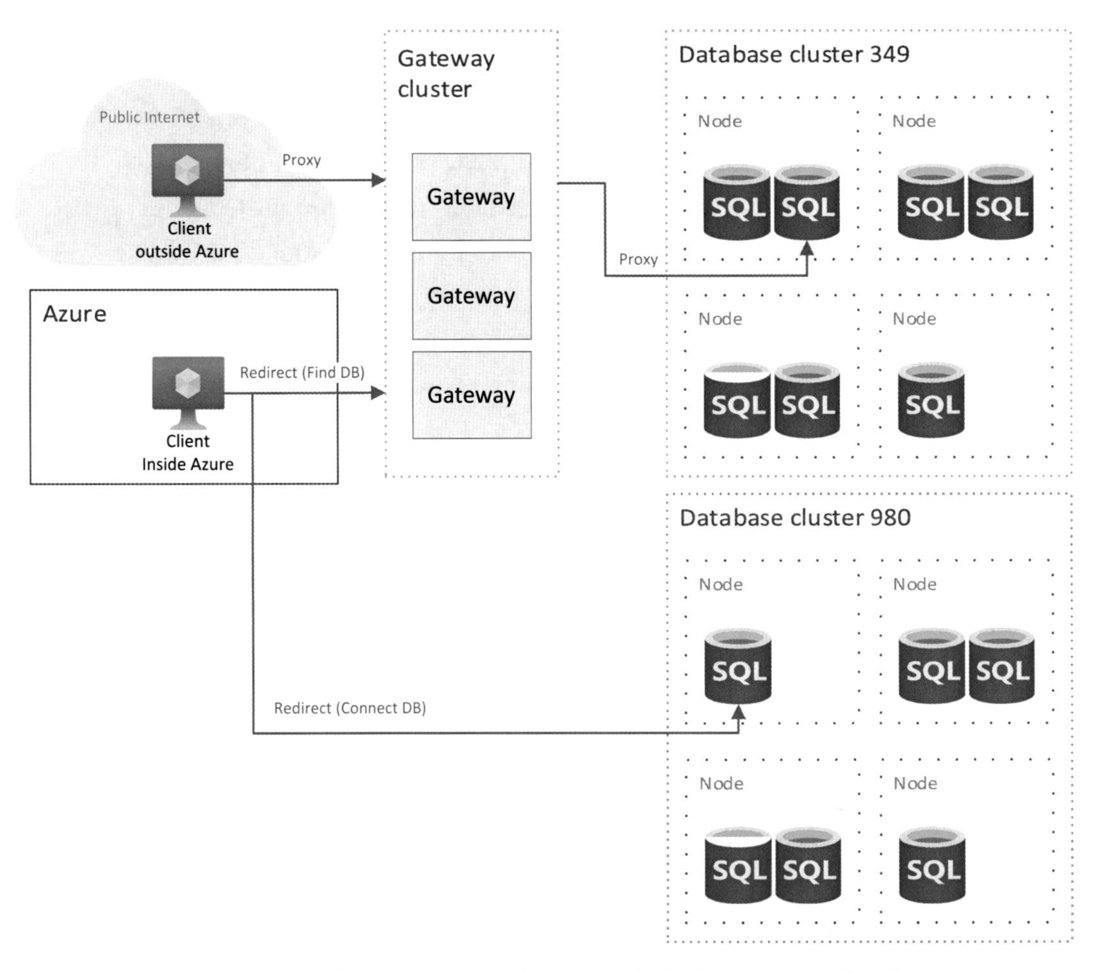

***Figure 3-18.** Conceptual architecture showing default proxy and redirect connection methods*

In general, the redirect policy is preferred because of its lower latency and increased throughput, but it means that your connection will need to be allowed to connect to almost any IP address in Azure. This may not be possible from an on-premises environment or another cloud environment where the network administrator wants to limit the outbound connectivity to a limited known list of IP addresses. When a proxy connectivity policy is used, the list of possible IP addresses that a client will need to connect to is limited to only a few within each region.

Tip The Azure SQL Database gateway is not something you deploy or manage. It is a highly available, resilient, and scalable service within each region. Each regional gateway has a limited list of IP addresses that are assigned, making it easier to whitelist these on a firewall when using Proxy mode. You will not have any visibility or control over it; it is simply there to manage your connectivity. However, it is important to understand that it exists and how it affects latency. You can find a list of the Azure SQL Database gateway IP addresses at `https://learn.microsoft.com/azure/azure-sql/database/connectivity-architecture`.

It is possible to override the default connection policy behavior and force all connections to use one method, either redirect or proxy regardless of whether the connection originates from within Azure. However, if you force connections to always use proxy mode, then you should expect a performance impact from the higher latency and lower throughput.

Connection Encryption

Azure SQL Database logical servers support in-transit encryption using Transport Layer Security (TLS). The minimum allowed version at the time of writing this is 1.2, but 1.0 and 1.1 can be enabled, but this is not recommended as they are considered less secure. Figure 3-19 shows the configuration of encryption in transit settings on a logical server.

Encryption in transit
This server supports encrypted connections using Transport Layer Connections (TLS). Any login attempts from clients using a TLS version less than the Minimum TLS Version shall be rejected. For information on TLS version and certificates, refer to connecting with TLS/SSL. Learn more
Minimum TLS version
TLS 1.2

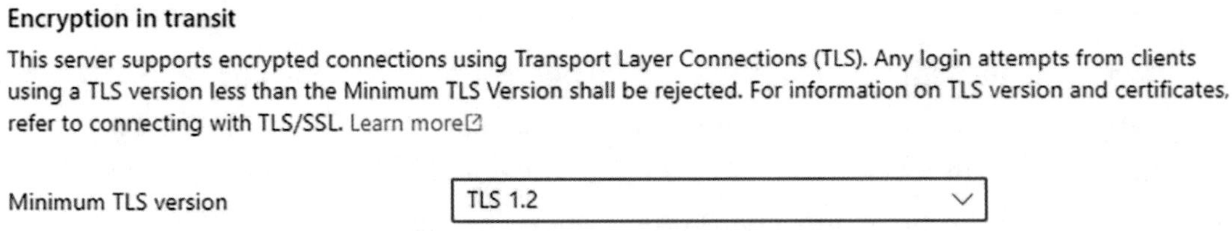

Figure 3-19. *Configuring minimum TLS version in the Azure Portal*

If a client tries to connect using a TLS version that is below what is allowed by your logical server, they will receive an error.

```
Error 47072: Login failed with invalid TLS version
```

Clients that do not support or require an encrypted connection can still connect, but in this case, it is recommended to not set a minimum TLS version.

Understanding Network Connectivity Requirements

We need to understand the requirements and limitations of the clients that connect to the Hyperscale database. Depending on the nature of the data we are storing in the database and any compliance standards we need to meet, it is common to limit the network connectivity options to use only private endpoints.

For most production workloads, it is recommended to use private endpoints whenever possible. However, this does increase the complexity and cost of the solution architecture. For example, implementing the additional networking infrastructure and bastion services to allow access to the database will add to both the Azure resource cost as well as the added person time in deploying and managing it.

Requiring the use of private endpoints to access a database also makes the solution slightly more rigid with deployment and changes requiring additional planning and orchestration. For example, any change to routing within the virtual network infrastructure may impact the accessibility of the database from other virtual networks, so additional care and testing are required.

However, for nonproduction databases or databases that aren't storing sensitive data, it may not be worth the added complexity and cost to require a private endpoint. A public endpoint using a service connection within the Azure Virtual Network might be considered good enough.

You may also choose to allow different applications or services to use different network connectivity methods to access your database. For example:

- Your application tier may connect via a public endpoint that requires the use of a service connection in an Azure VNet subnet.
- Database administrators will use an Azure Bastion in an Azure VNet to connect to a private endpoint in an Azure VNet.

As always, balancing the functional and nonfunctional requirements with the cost and complexity of the solution is the task of the solution and database architects collaborating with business stakeholders.

Common Considerations for Network Connectivity

The following are some common considerations you should make when planning the networking connectivity for the database:

- What compliance requirements does the workload need to meet?
- Can your applications and services be placed inside an Azure VNet?
- Can your database clients use Azure Active Directory authentication? If yes, do they support multifactor authentication (MFA) if it is being used?
- Do any of your database clients require older TLS protocols such as TLS 1.1?
- Are any of your workloads sensitive to network latency and require the lowest possible network latency and highest throughput possible?
- Do you need to connect to the database via the public Internet from an on-premises environment where the network administrator requires IP whitelisting the addresses?

Being able to articulate requirements within these considerations will help you define the required network connectivity to your Hyperscale database.

Considerations for Security

Security is a key consideration for every deployment of Azure SQL Hyperscale database. The security considerations usually need to be balanced with cost and complexity. Security and network connectivity requirements are also closely tied together, so many options for network connectivity also affect the overall security of the design. However, in this section, we'll focus only on features that impact security but aren't covered by network connectivity.

There are four areas that need to be considered when planning the security of your Azure SQL Hyperscale database.

- *Authentication*: This specifies how clients will authenticate to the logical server. This will primarily be a decision about whether to use Azure Active Directory (Azure AD) to authenticate connections. However, it may also cover the use of multifactor authentication and conditional access when authenticating using supported clients, for example, SQL Server Management Studio.

- *Microsoft Defender for SQL*: Enabling this feature provides advanced threat protection and vulnerability assessments for your logical server and Azure SQL databases.
- *Logical server identity*: This securely provides an Azure AD identity for your logical server to use to connect to Azure and access Azure resources. This eliminates the need to store usernames and passwords in the database for functions that need to access Azure resources.
- *Transparent data encryption*: This controls whether a *service*-managed or customer-managed encryption key is used to encrypt the data in the database, backups, and logs at rest. Transparent data encryption can also be disabled on a per-database basis.
- *Ledger*: For some workloads, ledger functionality is required to provide assurance over the data for audit. The ledger functionality is not discussed in detail as it is beyond the scope of this book.

Now we will cover each of these design considerations in more detail.

Authentication

Authentication is the process of verifying the identity of the user (or service) that the client is using to connect to the logical server. During the login process, the logical server will verify the provided credentials with a directory service.

Tip *Authentication* should not be confused with *authorization*, which is the process that determines whether an authenticated identity is allowed to perform an operation.

Azure SQL Database logical servers currently supports the following:

- *SQL login only*: User account details, including credentials, are stored in the Azure SQL database. An administrator SQL login is specified when the logical server is created and will be granted the `db_owner` role in each user database.

- *Azure AD authentication only*: User account details are stored in the Azure Active Directory that provides the identity source for the subscription containing the logical server. A user or group in Azure AD needs to be selected to be the administrator during database creation or when Azure AD authentication is enabled on the server. This user or all users in the group will be granted the `db_owner` role in each user database.

Tip Using a group account as the administrator enhances manageability so is recommended.

- *Both SQL login and Azure AD authentication*: Both methods of authentication are supported.

Tip If you are using replication with either a geo replica or named replica (covered later in this chapter), then you will need to configure the same SQL authentication methods and have sufficient permissions on both the primary and on the replica logical servers.

Figure 3-20 shows the Azure Active Directory blade for the logical server in the Azure Portal being configured to support only Azure AD authentication. An Azure AD group called SQL Administrators has been defined as the admin.

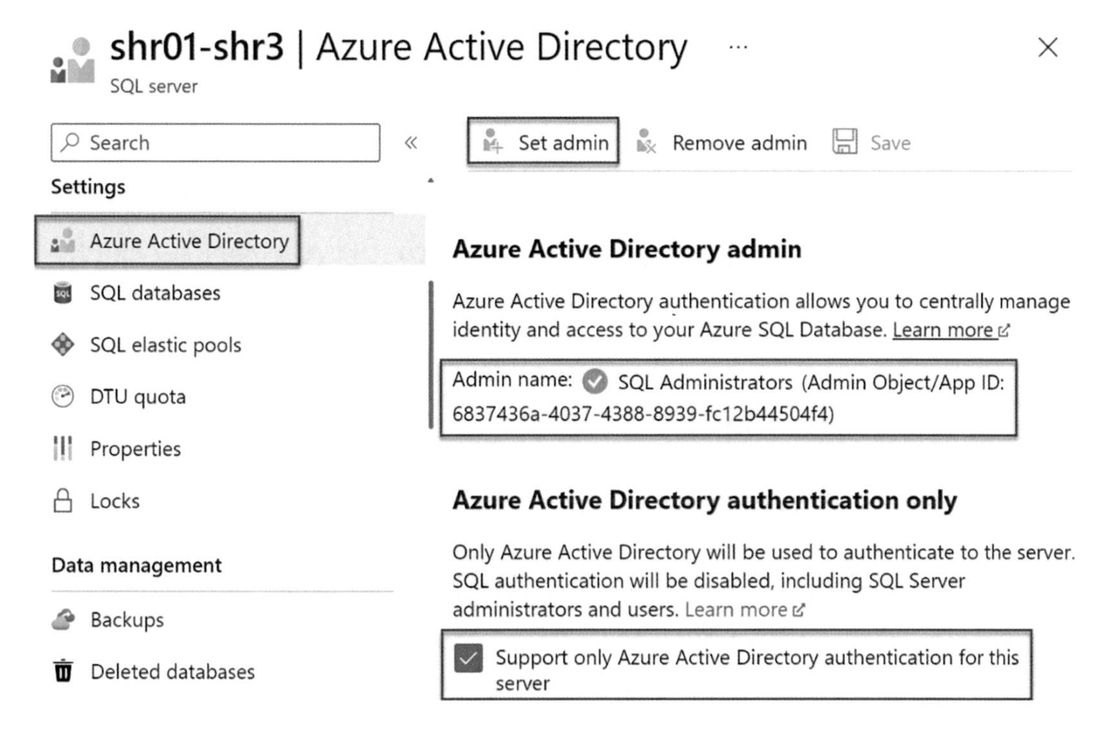

Figure 3-20. *Configuring Azure Active Directory authentication*

There are many benefits to using Azure AD authentication in your logical server.

- Central management of user identities across servers, reducing proliferation of accounts
- Easier password rotation from a single place
- Database permissions management using Azure AD groups
- Increases security by reducing the need to store passwords when integration with Windows authentication is used
- Supports token-based application for applications connecting to the database
- Supports multifactor authentication (MFA) for some tools (such as SSMS) for increased security
- Supports conditional access to enforce additional authentication constraints, for example, allowing logins only from specific locations based on IP address

It is recommended to use Azure AD authentication because of the many security benefits. However, some applications and tools may not support it, especially the MFA part, so you will need to assess your workloads and plan for their requirements.

Tip For more information on using Azure AD authentication with your logical server database, please see `https://learn.microsoft.com/azure/azure-sql/database/authentication-aad-overview`.

Microsoft Defender for SQL

Microsoft Defender for SQL is part of the wider Microsoft Defender for Cloud suite of products. It provides two key security functions.

- *Vulnerability assessments*: These help you discover, track, and remediate potential vulnerabilities in your database. They provide actionable tasks to resolve security issues and enhance your database defense.

  Figure 3-21 shows the results of a Microsoft Defender for SQL vulnerability assessment running against a logical server deployed from the examples in this book.

Vulnerability assessment findings

ID ↑↓	Security Check ↑↓	Applies to ↑↓	Severity ↑↓
VA2063	Server-level firewall rules should not grant excessive access	1 of 1 resources	High
VA2065	Server-level firewall rules should be tracked and maintained at a strict mi...	1 of 1 resources	High
VA1143	'dbo' user should not be used for normal service operation	1 of 1 resources	Medium
VA2130	Track all users with access to the database	1 of 2 resources	Low

Figure 3-21. *A Defender vulnerability assessment*

- *Advanced threat protection*: This detects anomalous activities indicating unusual and potentially harmful attempts to access or exploit your database. It continuously monitors your database for suspicious activities and provides immediate security alerts on potential vulnerabilities, Azure SQL injection attacks, and anomalous database access patterns. Advanced threat protection alerts provide details of suspicious activities and recommend action on how to investigate and mitigate the threat.

Important Microsoft Defender for SQL requires an additional monthly charge. For pricing details, please refer to `https://azure.microsoft.com/pricing/details/defender-for-cloud`.

For more details on Microsoft Defender for SQL, refer to `https://learn.microsoft.com/azure/azure-sql/database/azure-defender-for-sql`.

There are several ways to enable Microsoft Defender for SQL.

- During the deployment of a new Azure SQL Database on the Security page (unless it has already been enabled on the logical server). Figure 3-22 shows the Security page when creating a new Azure SQL Database on a server that does not have Microsoft Defender for SQL enabled on it.

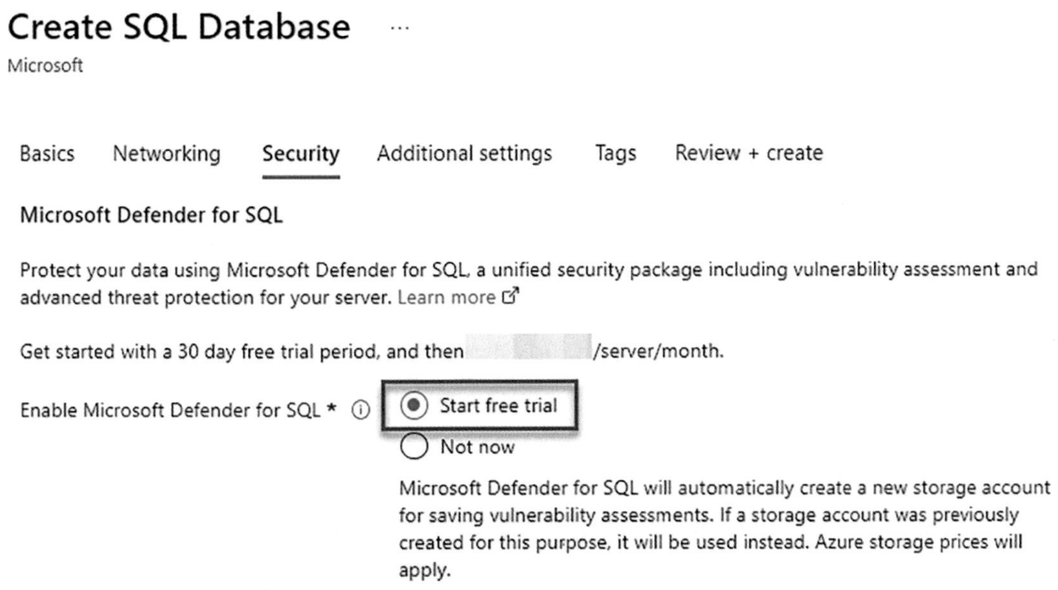

Create SQL Database ...
Microsoft

Basics Networking Security Additional settings Tags Review + create

Microsoft Defender for SQL

Protect your data using Microsoft Defender for SQL, a unified security package including vulnerability assessment and advanced threat protection for your server. Learn more

Get started with a 30 day free trial period, and then /server/month.

Enable Microsoft Defender for SQL * ⓘ
(•) Start free trial
( ) Not now
Microsoft Defender for SQL will automatically create a new storage account for saving vulnerability assessments. If a storage account was previously created for this purpose, it will be used instead. Azure storage prices will apply.

***Figure 3-22.** Enabling Microsoft Defender for SQL at database creation*

- After deployment on the Microsoft Defender for Cloud page.

- Enabling it automatically on any Azure SQL Database in a subscription by enabling it through Microsoft Defender for Cloud in the Azure Portal in the "Workload protections" section.

When Microsoft Defender for SQL is enabled on a logical server, a new storage account is created in the same resource group as the logical server that will be used to store vulnerability assessments. It is possible to configure the logical server to use a different storage account rather than creating one.

Logical Server Identity

It is common for a logical server to need to be able to access resources within Azure. To enable this, we use an Azure Managed Identity instance assigned to the logical server. Managed identities are much more secure than the alternative, which is storing usernames and passwords in your database.

The most common use case of a managed identity in an Azure SQL Database is to access an Azure Key Vault to obtain encryption keys or certificates. However, there many other possible use cases.

There are two types of managed identities that we can use.

- *System managed*: The life cycle of the identity is managed by the logical server. When the server is created or the system-managed identity is enabled, the identity will be automatically created in Azure AD using the name of the SQL Server resource as the name of the identity. The identity will be deleted automatically when the logical server is deleted.

  A logical server can have only a single system managed identity assigned to it.

- *User managed*: The life cycle of the identity is managed by you, or some process your organization has created. It exists separate from the life cycle of the logical server. If the logical server that the user-managed identity (UMI) is assigned is deleted, the UMI remains.

  A single UMI can also be used as the identity for multiple logical servers (or even other Azure resources). This can help to reduce the complexity of a solution.

  A logical server can have *multiple* user-managed identities assigned to it with a primary identity specified to use by default when more than one is available.

Figure 3-23 shows the Identity blade in the Azure Portal for a logical server, with a single user-managed identity assigned. A system managed identity can also be assigned here.

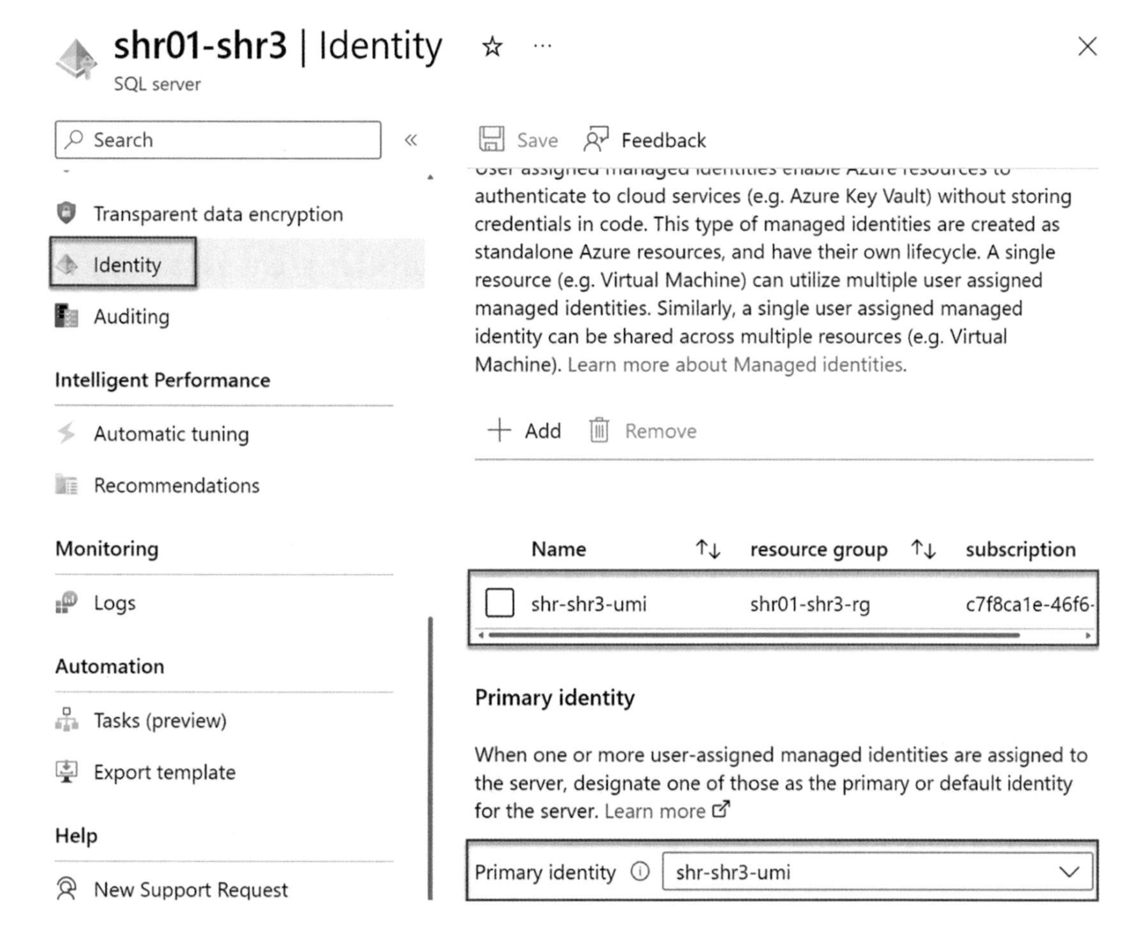

Figure 3-23. *Configuring managed identities for a logical server*

Managed identities in Azure can be granted RBAC roles just like any other identity. This gives the logical server and the databases it contains the same access to Azure as the identity that is being used has. The support for multiple UMIs within a single logical server allows you to reduce the scope of each UMI to only the roles needed.

Transparent Data Encryption

Azure SQL Database provides transparent data encryption over all data at rest, including the database, logs, and backups. By default, a service-managed key is used to encrypt this data. This is usually acceptable for most organizations, but some organizations might require they provide and manage their own encryption key—a customer-managed key. This could be to meet compliance or other business requirements.

Tip There are also several other benefits to using customer-managed keys. For more information, see `https://learn.microsoft.com/en-us/azure/azure-sql/database/transparent-data-encryption-byok-overview`.

Using a customer-managed key does come with additional operational overhead because an Azure Key Vault or Azure Managed Hardware Security Module (HSM) is required to store the key. When a customer-managed key is required, the following additional tasks need to be performed (ideally as part of an IaC deployment):

- An Azure Key Vault or Azure Managed HSM is required. The Azure Key Vault must be configured with soft-delete and purge protection.
- An encryption key will need to be created in the Azure Key Vault or Azure Managed HSM to be used for encryption.
- The identity assigned to the logical server needs to be granted `get`, `wrapKey`, and `unwrapKey` permissions to the Azure Key Vault or Azure Managed HSM and/or the encryption key (depending on the Key Vault security policy in use).

Tip If your Azure Key Vault is configured to use Azure RBAC, then you can simply grant the logical server identity the Key Vault Crypto Service Encryption User role.

- If you are using the firewall with Azure Key Vault, then you must enable the "Allow trusted Microsoft services to bypass the firewall" option.

Once the customer-managed key has been set, the databases will be updated to use it. Figure 3-24 shows a logical server configured to use a customer-managed key in an Azure Key Vault.

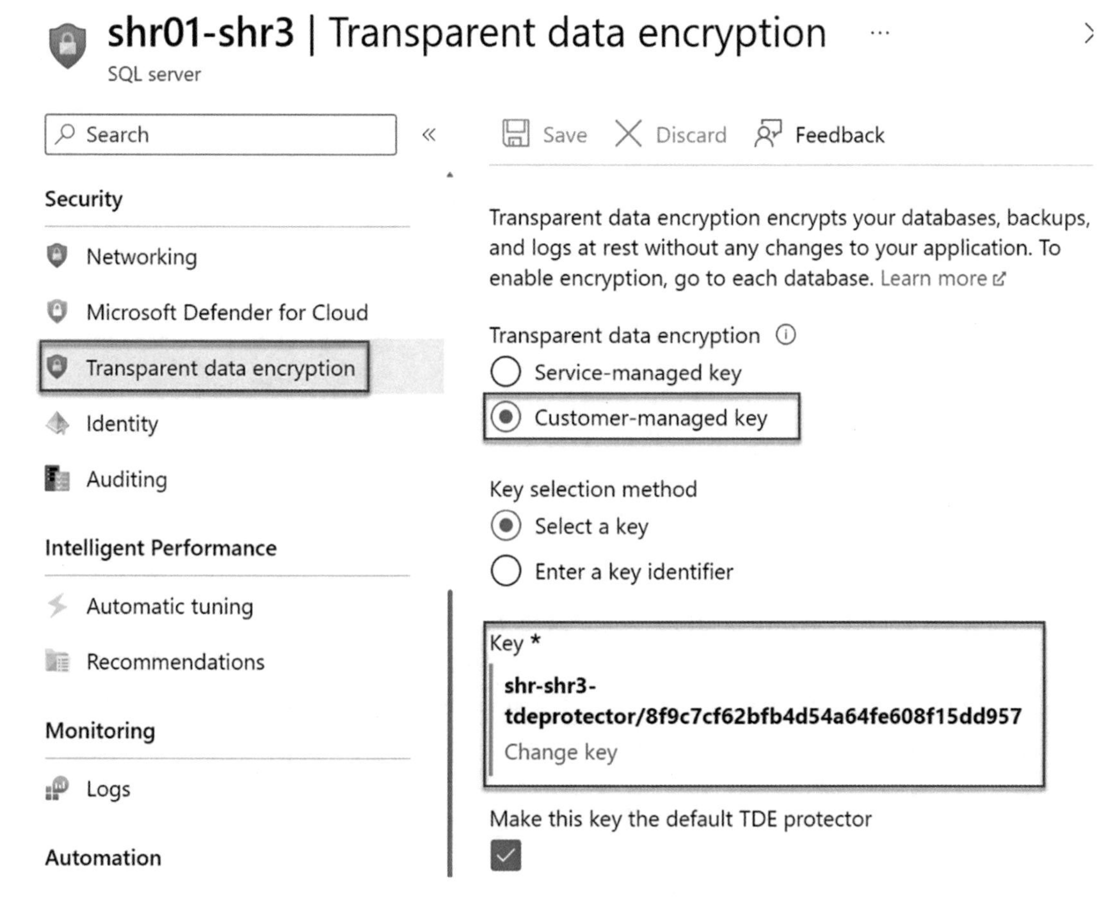

Figure 3-24. *Customer-managed key configuration*

Ledger

The ledger feature of Azure SQL Database helps establish trust around the integrity of data stored in the database by providing tamper-evidence capabilities in your database.

Figure 3-25 shows the Security step in the creation of an Azure SQL Database instance with the ledger configuration section highlighted.

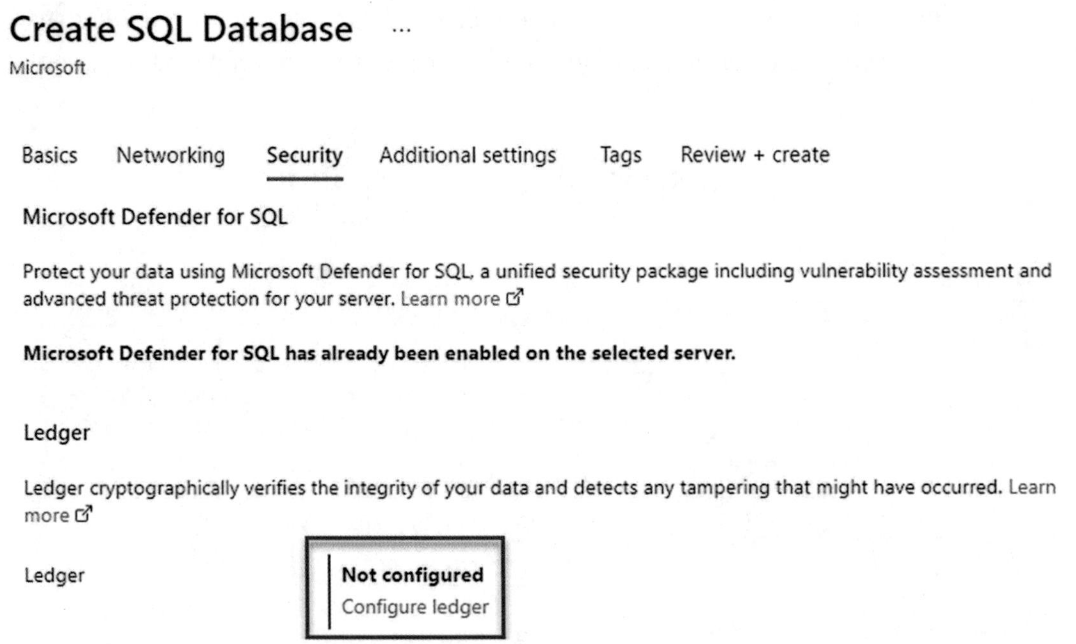

Create SQL Database ...
Microsoft

Basics Networking Security Additional settings Tags Review + create

Microsoft Defender for SQL

Protect your data using Microsoft Defender for SQL, a unified security package including vulnerability assessment and advanced threat protection for your server. Learn more

Microsoft Defender for SQL has already been enabled on the selected server.

Ledger

Ledger cryptographically verifies the integrity of your data and detects any tampering that might have occurred. Learn more

Ledger **Not configured**
Configure ledger

Figure 3-25. *Configuring the ledger when deploying a new database*

Configuring a ledger will require you to specify an Azure Storage account or an Azure Confidential Ledger (in preview at the time of writing) to store the generated digests. Otherwise, you can manually generate the digests and store them in your own location.

Figure 3-26 shows the configuration settings for storing the ledger digests.

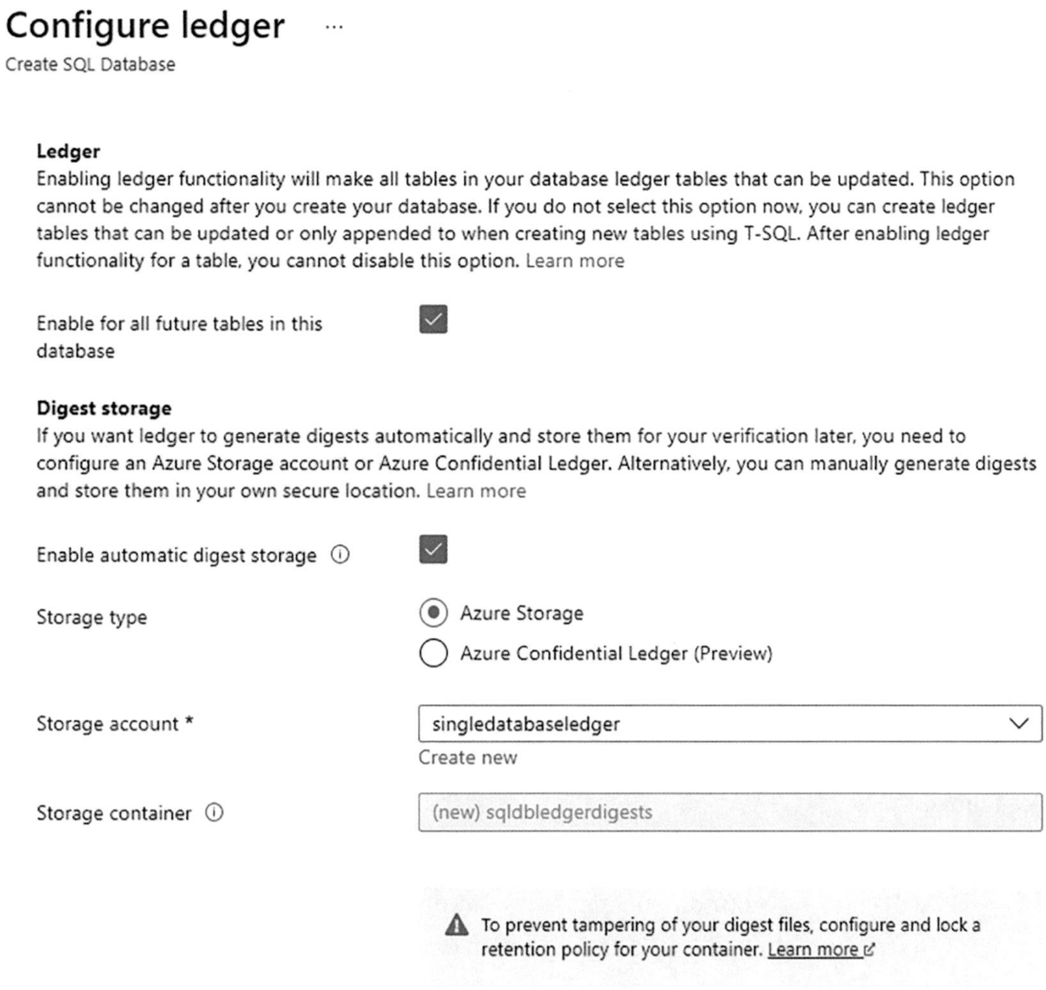

Configure ledger ...

Create SQL Database

Ledger

Enabling ledger functionality will make all tables in your database ledger tables that can be updated. This option cannot be changed after you create your database. If you do not select this option now, you can create ledger tables that can be updated or only appended to when creating new tables using T-SQL. After enabling ledger functionality for a table, you cannot disable this option. Learn more

Enable for all future tables in this database

Digest storage

If you want ledger to generate digests automatically and store them for your verification later, you need to configure an Azure Storage account or Azure Confidential Ledger. Alternatively, you can manually generate digests and store them in your own secure location. Learn more

Enable automatic digest storage

Storage type: Azure Storage / Azure Confidential Ledger (Preview)

Storage account *: singledatabaseledger

Create new

Storage container: (new) sqldbledgerdigests

To prevent tampering of your digest files, configure and lock a retention policy for your container. Learn more

***Figure 3-26.** Configuring a confidential ledger using an Azure Storage account*

Once the database has been deployed, the ledger digest storage can be changed by going to the Ledger blade in the SQL Database instance.

Considerations for Operational Excellence

There are many factors that affect operational excellence, meaning the processes that ensure our database continues to operate correctly. However, the main consideration for operational excellence when planning a Hyperscale deployment is monitoring.

Note There are other elements to operational excellence, such as DevOps practices and IaC. However, these are things that don't need to be designed into our Hyperscale architecture (there are some exceptions, but they're beyond the scope of this book).

In this section, we'll look at how we can monitor our Hyperscale database.

Diagnostic Settings

Every Azure SQL Database instance (including the Hyperscale tier) continuously produces diagnostic logs and performance metrics. To retain these logs so that diagnostics, audit, and performance analysis can be performed on the database, they need to be streamed to an external storage service. The recommended service for this is an Azure Log Analytics workspace.

Azure Log Analytics is part of the Azure Monitor suite of tools. It provides a single storage and analytics platform to ingest, analyze, and visualize diagnostics and performance metrics from one or more resources within Azure.

When planning for monitoring of your Hyperscale database, you need to consider the following:

- *Log Analytics workspace region*: Which log analytics workspace will you send your diagnostics to? Sending logs to a workspace in a different region will have an impact on cost.
- *Retention period*: How long do you need to retain the diagnostic and audit logs for the Hyperscale database? By default, data is retained for 31 days in an Azure Log Analytics workspace. But you may need to increase this for compliance or business reasons. Increasing the retention period of diagnostic information will increase the cost.

Tip You can control the retention period for each diagnostic source to help you manage costs in a more granular fashion.

- *Diagnostics and metrics to collection*: Hyperscale databases emit a lot of different diagnostic record types as well as several different metric counters. Collecting all these records will increase the cost of your log analytics workspace. Determining the diagnostic information that is important to your business is important to controlling costs.

Each Azure SQL Database that should send data to a Log Analytics workspace needs to have Diagnostic Settings configured. Figure 3-27 shows the Diagnostic setting for an Azure SQL Hyperscale database in the Azure Portal. It shows all diagnostic categories and metrics being set to a Log Analytics workspace called `sqlhyperscalerevealed-law` in East US (the same region as the Hyperscale database).

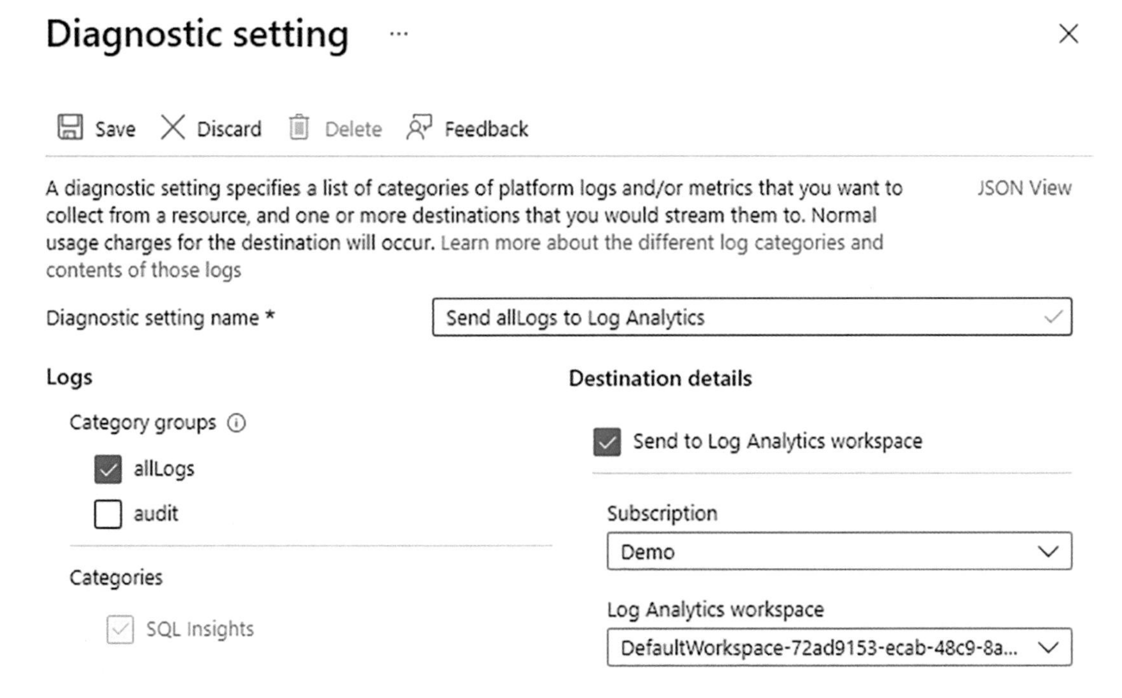

Figure 3-27. *Configuring diagnostic settings for a Hyperscale database*

Azure SQL Auditing

This feature tracks database events and sends them to another Azure service for storage or analysis. These are events that occur within the Hyperscale database engine itself rather than within the Azure control pane. This might be required to meet regulatory compliance and gain insight into anomalies and discrepancies that could indicate suspected security violations.

Depending on the purpose the audit logs are being collected for, you might choose one of the three following destinations:

- *Azure Storage account*: This is a good choice for long-term retention but will require more effort to perform analytics.
- *Azure Log Analytics workspace*: This is a good choice for performing analysis and reporting, but longer-term retention requires using the archive function of Azure Log Analytics workspaces or configuring the desired retention on the workspace.
- *Azure Event Hub*: Use this to send the audit logs to an external system.

You can also choose to enable auditing of Microsoft support operations, which will capture Microsoft support engineers' (DevOps) operations.

Tip For more information on Auditing for Azure SQL Database, see this page: `https://learn.microsoft.com/azure/azure-sql/database/auditing-overview`.

Figure 3-28 shows the audit configuration on a logical server to send database events to a Log Analytics workspace.

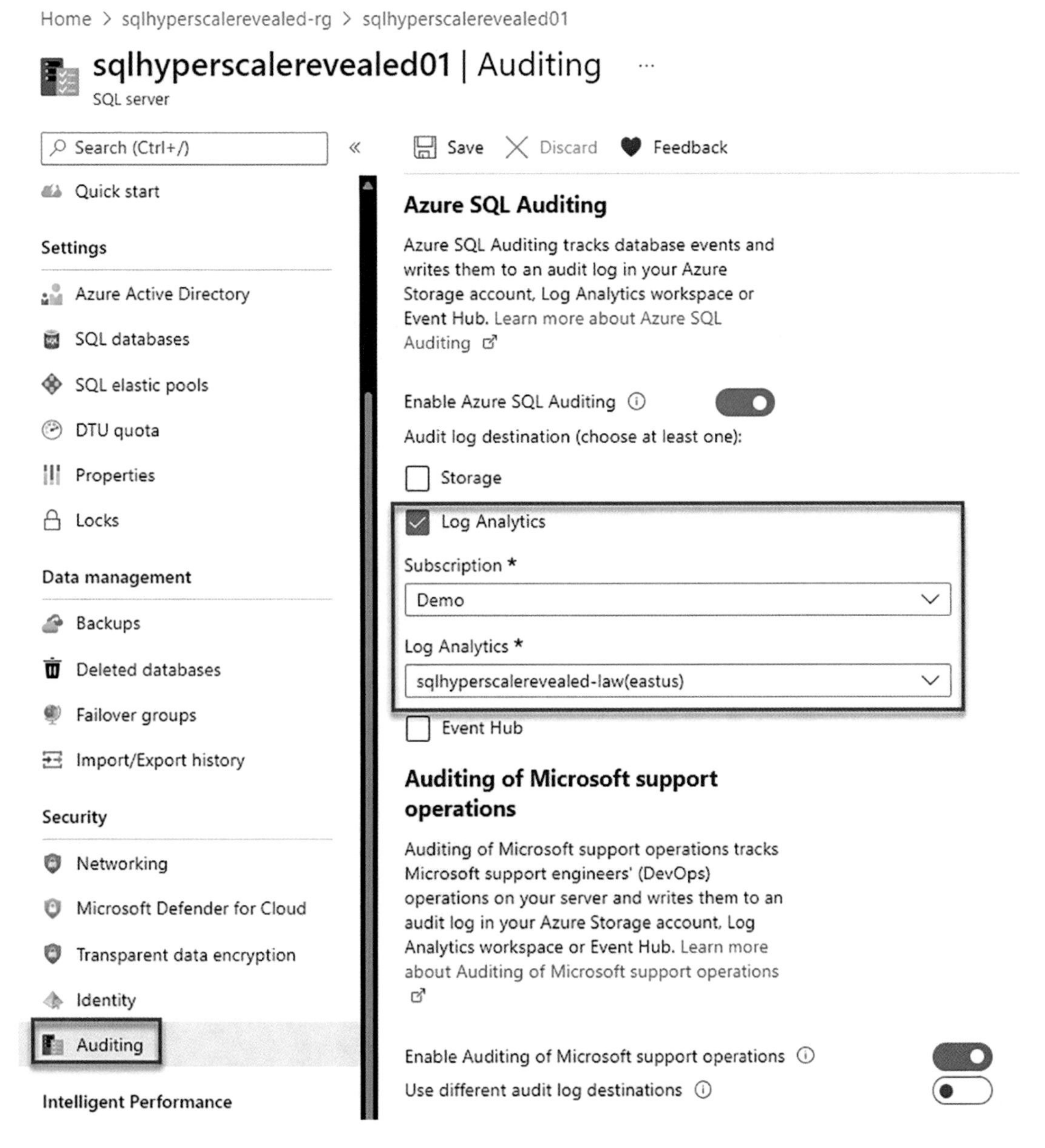

***Figure 3-28.** Sending database events to a Log Analytics workspace*

The main consideration for this is the cost of ingesting and retaining the audit data that is generated by the database. Sending this data out of the region will incur an additional traffic cost.

Summary

As you can see, there are a number of factors we need to consider when planning an Azure SQL Database Hyperscale deployment. Some of these factors are similar to deploying any other Azure SQL Database instance, but others have considerations specific to the Hyperscale architecture. Understanding these differences and being clear on the requirements of the solution are key to ensuring a successful architecture that meets or exceeds business objectives.

In the next chapter, we will design a Hyperscale environment that will satisfy a set of functional and nonfunctional requirements, such as restricting network access, configuring high availability and disaster recovery, and enabling encryption of the database with our own keys. Once we have defined the architecture that will satisfy the requirements, we will deploy a starting environment and then build up our new Hyperscale environment step-by-step over the following five chapters, providing guidance on each feature.

CHAPTER 4

Deploying a Highly Available Hyperscale Database into a Virtual Network

In Chapter 3, we looked at some of the key design decisions that we need to make when deploying Azure SQL DB Hyperscale into our environment. This also gave us the opportunity to identify and examine some of the key features of Hyperscale in more detail. In this chapter and Chapters 5 to 8, we're going to demonstrate these key features of Hyperscale by deploying a more comprehensive example into an Azure environment. This environment will be representative of how a Hyperscale database will usually be deployed in production. You can also perform the steps described in the following chapters in your own Azure subscription to configure the same environment.

Tip In this chapter, we'll be sharing several scripts. These will all be available in the `ch4` folder in the source files accompanying this book and also in this GitHub repository: `https://github.com/Apress/Azure-SQLHyperscale-Revealed`.

We will start by defining the nonfunctional requirements that our example environment should meet. Based on those requirements, we will show an example architecture that will satisfy them.

Z. Barać and D. Scott-Raynsford, *Azure SQL Hyperscale Revealed*,
https://doi.org/10.1007/978-1-4842-9225-9_4

Most environments are likely to contain more than just the Hyperscale database. There are usually Azure resources that need to be available to support the Hyperscale database in an environment, such as Azure VNets, Key Vaults, and Log Analytics workspaces. These services are common to most environments in Azure so are likely already in place. But, for the examples in these chapters, we will deploy them by using a script and call it our starting environment.

We will then go through the steps to deploy and configure the Hyperscale database using the Azure Portal. Once the Hyperscale database is deployed, the next four chapters will demonstrate how to update the configuration to make use of additional features and how to manage a Hyperscale database once it is put into a virtual network.

An Example Hyperscale Production Environment

A typical production Hyperscale environment in Azure will usually contain many other Azure services, aside from the database and applications that use it. These services help the solution meet your nonfunctional requirements.

In the example in this chapter, we're going to be deploying a Hyperscale database that meets the following function and nonfunctional requirements:

- *Networking*: The logical server and the Hyperscale database must be accessible only from a private endpoint connected to a virtual network called `sqlhyperscalerevealed01-vnet`. All traffic to and from the logical server must travel over the private endpoint. The public endpoint on the logical server must be disabled.

Important Almost every subnet in a virtual network will contain Azure network security groups to control the flow of traffic between the Internet, the virtual networks, and the resources connected to it. This is beyond the scope of this book, but it is best practice to restrict traffic within your virtual networks using network security groups. For more information on network security groups, see `https://aka.ms/networksecuritygroups`.

- *Authentication*: Only Azure Active Directory authentication should be allowed to the logical server and the Hyperscale database. Only members of the SQL Administrators security group in Azure Active Directory should have the `db_owner` role on all databases in the logical server.
- *High availability*: The database must be highly available, even if the Azure availability zone where the database is located is unavailable.
- *Disaster recovery*: The database must also be able to recover to a replica in the failover region if the primary region becomes unavailable. The regional replica will be available only in a virtual network called `sqlhyperscalerevealed02-vnet`. The regional replica will not be using availability zones and will back up to locally redundant storage.
- *Backup storage redundancy*: Backups of the Hyperscale database will be available even if the availability zone or the region containing the backup is unavailable.
- *Identity*: A user-assigned managed identity will be created and assigned to the logical servers in both the primary region and the failover region. This is to simplify configuration as both logical servers will share the same identity and Azure Key Vault.
- *Encryption*: A customer-managed key will be used for transparent data encryption (TDE) and stored in an Azure Key Vault.
- *Security*: Automated vulnerability assessments should be created and stored in an Azure Storage account by Microsoft Defender for SQL.
- *Diagnostic and audit logs*: Hyperscale database diagnostic logs and logical server audit logs will be stored in an Azure Log Analytics workspace in the same region as the database.

Tip A service level objective (SLO), recovery time objective (RTO), and recovery point objective (RPO) would normally be defined as part of the nonfunctional requirements.

Figure 4-1 shows a diagram of the architecture we will be deploying throughout this chapter. This architecture satisfies the previous requirements.

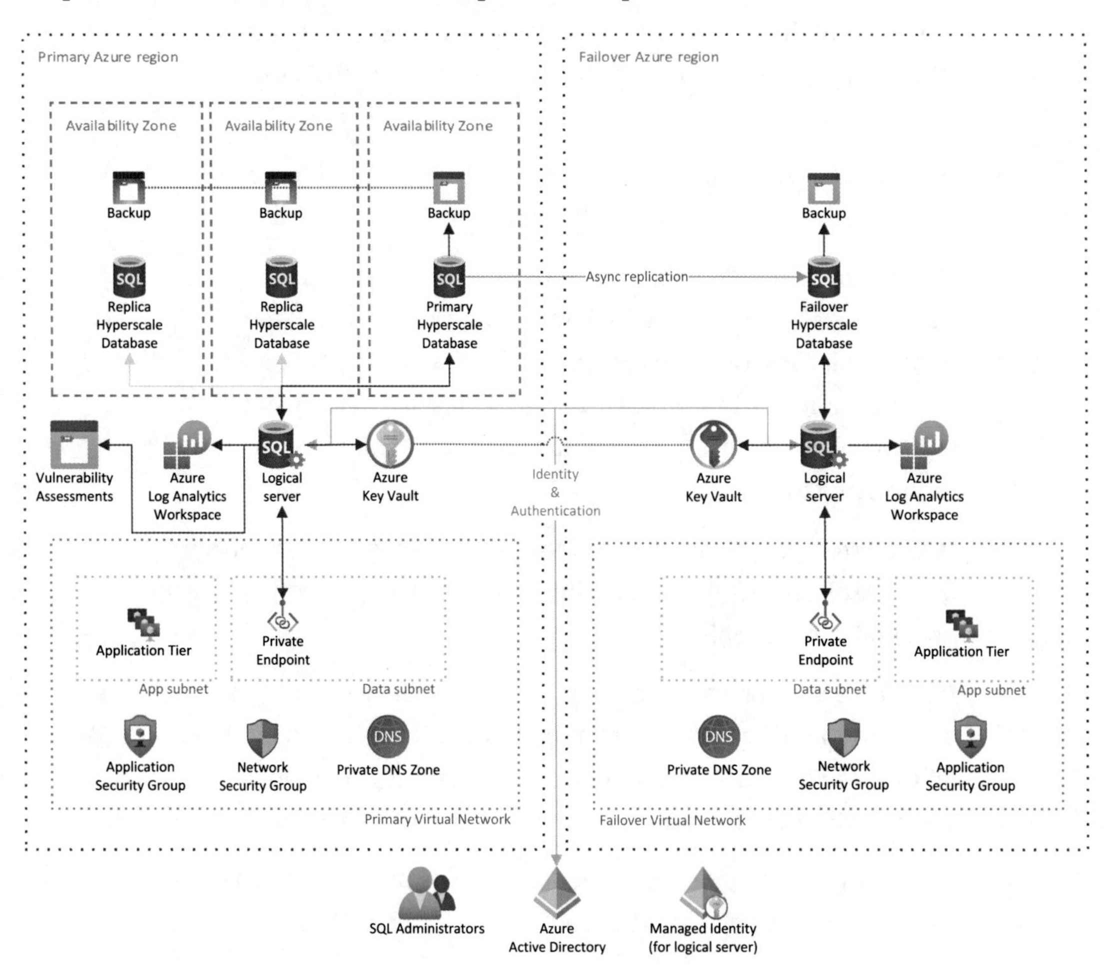

***Figure 4-1.** Example Hyperscale environment*

Important In our example environment failover region, we have only a single replica. In a real-world scenario, if the replicas in the primary region are acting as read replicas, then it is strongly recommended to have at least the same number of read replicas in your failover region. This is because in the case of a complete regional failover, we would need enough request capacity on the read replicas to service the application. Scaling out the read replicas to increase read request capacity in a regional outage is risky as many other Azure customers may be trying to do the same thing.

The Starting Environment

Some of the Azure resources in the example architecture are just supporting services that aren't part of our Azure SQL DB Hyperscale deployment itself. We will refer to these supporting services as the *starting environment.*

If you want to follow along with the examples in this book, then you will need to deploy the starting environment. If you do decide to follow along with the examples in this chapter, then you will need the following:

- A `contributor` (or `owner`) role on the Azure subscription you will be using to perform these tasks.
- Approval from the subscription owner (or whomever is covering the cost or credit) to deploy these resources for test purposes.

Important The Azure resources that will be deployed to your subscription will cost money to run (unless you've got credit on your subscription). If this is for the purposes of learning, then you should take care to dispose of your environment after completing any exercises to minimize cost. Instructions for cleaning up the environment are provided at the end of this chapter, as well as a simple script you can use.

You do *not* need to follow along with these examples to gain value from them, but doing so will provide you with more practical experience with Azure and Azure SQL Hyperscale.

These services are still needed for our Hyperscale database, but we'll deploy them via a PowerShell script (using Azure Cloud Shell) to simplify the process. The following are the supporting services that we will deploy via a script:

- *Resource groups*: The Azure resource groups that will contain the supporting services. The resource group in the primary region will be called `sqlhr01-rg` and the failover region `sqlhr02-rg`. These aren't shown on the architecture diagram for the sake of simplicity.
- *Azure virtual networks*: The primary region `sqlhr01-vnet` and the failover region `sqlhr02-vnet`. The subnets `App` and `Data` will also be created. The virtual networks are required to host the application tier as well as the private endpoints for the logical servers.
- *Azure Key Vault*: This will be used to store the customer managed key for TDE on the Hyperscale database. The Azure Key Vault is called `sqlhr`. Azure Key Vaults are automatically replicated to the paired region.

Important It is recommended that you deploy a second Key Vault into the failover region when using customer-managed TDE, even though Azure Key Vault is geo-replicated.

- *Azure Log Analytics Workspaces*: The Azure Log Analytics workspace in the primary and secondary regions where diagnostic logs will be sent. The workspaces will be called `sqlhr01-law` for the primary region and `sqlhr02-law` in the secondary region.

We will also use the Azure East US as the primary region and West US 3 as the failover region. However, if you're following along with the examples, you can choose whichever regional pair you choose, from the list on this page: `https://learn.microsoft.com/azure/reliability/cross-region-replication-azure#azure-cross-region-replication-pairings-for-all-geographies`.

The virtual machine scale set in the application tier won't be deployed as part of the example. It is just shown for reference and to demonstrate how we might architect a multitier application using Azure SQL DB Hyperscale.

Figure 4-2 shows a diagram of the starting environment.

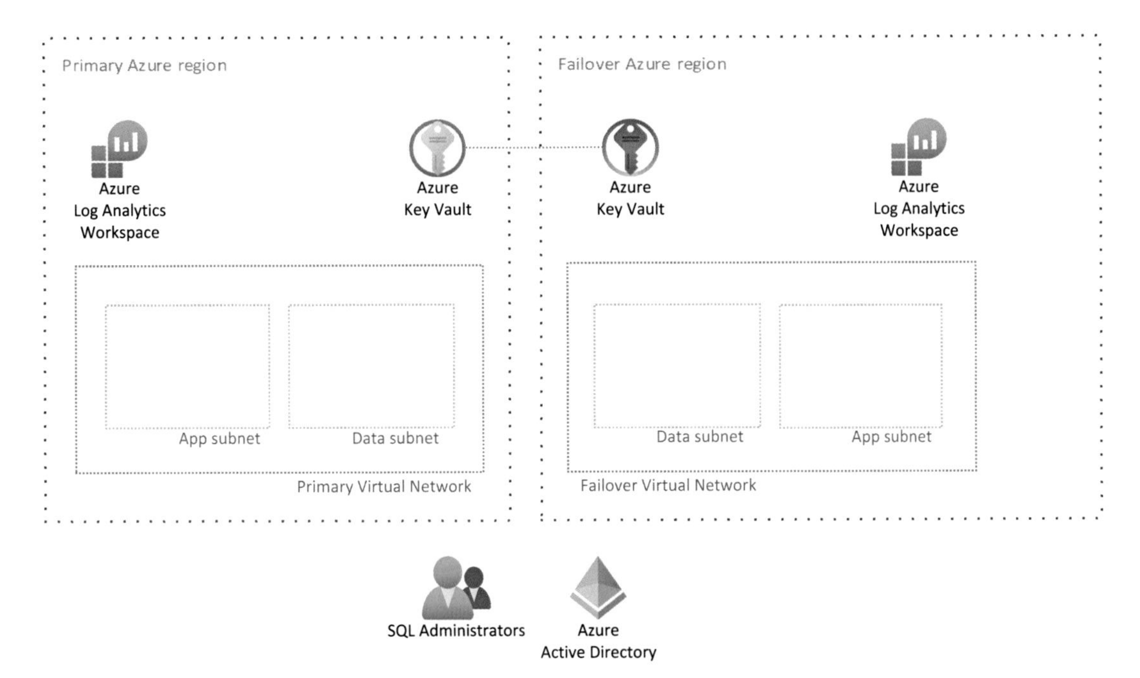

Figure 4-2. *The starting environment*

Of course, in a real-world Azure environment, we would see many other Azure resources deployed, performing all manner of functions. There may be multiple applications and infrastructure-as-a-service (IaaS), platform-as-a-service (PaaS), and software-as-a-service (SaaS) resources all connected to our Hyperscale database. But for simplicity we are deploying only the basics needed for our demonstration.

The Starting Environment Deployment Script

We are going to run a script from the Azure Cloud Shell to deploy the starting environment. The script can be found in the `ch4` folder in the files provided for this book and is called `New-SQLHyperscaleRevealedStartingEnvironment.ps1`.

Tip The scripts and other files accompanying this book can be found in the GitHub repository: `https://github.com/Apress/Azure-SQL-Hyperscale-Revealed`.

The script accepts a few parameters that allow you to customize the environment for your needs.

- `PrimaryRegion`: This is the name of the Azure region to use as the primary region. If you choose to use a different Azure region, you should ensure it is availability zone enabled. If you don't specify this parameter, it will default to East US.

  For a list of availability zone-enabled regions, see this page: `https://learn.microsoft.com/azure/availability-zones/az-overview#azure-regions-with-availability-zones`.

- `FailoverRegion`: This is the name of the Azure region to use as the failover region. If you don't specify this parameter, it will default to West US 3.

- `ResourceNameSuffix`: This is the suffix to append into the resource group and resource names. If you want to use your own four-character code for the suffix instead of a randomly generated one, then omit this parameter and specify the `UseRandomResourceNameSuffix` switch instead.

Important Some Azure resource names must be globally unique. For example, Azure Storage Accounts and Azure Key Vaults both require globally unique names. Therefore, we need to be careful that any names for these resources are unique.

To ensure the resource names are all unique, the script will suffix a four-character string into each resource name to reduce the chance that we will have a conflict. However, even with the suffix, name conflicts are still not impossible, so be on the lookout for error messages that indicate a name conflict has occurred.

The `New-SQLHyperscaleRevealedStartingEnvironment.ps1` script deploys the environment using Azure Bicep, so it can be run multiple times as long as you have specified `ResourceNameSuffix` to the same value each time. If you have allowed the script to randomly generate a resource name suffix by setting `UseRandomResourceNameSuffix`, then each time you run the script, it will create a new environment.

Deploying the Starting Environment Using the Azure Cloud Shell

Now that we know what the starting environment script will do, it is time to run it to create the basic environment we'll use throughout this chapter.

We are going to use the Azure Cloud Shell to run the script. This is because the Azure Cloud Shell has all the necessary tools pre-installed and has the Azure CLI and the Azure PowerShell modules already connected to your Azure account.

Tip If you're not familiar with Azure Cloud Shell or would like to know more about it, see `https://aka.ms/CloudShell`.

We are going to use PowerShell in Azure Cloud Shell for this process, because the `New-SQLHyperscaleRevealedStartingEnvironment.ps1` is written in PowerShell.

To create the starting environment, follow these steps:

1. Open your browser and navigate to `https://portal.azure.com`.
2. Open the Azure Cloud Shell by clicking the Cloud Shell button.

Tip If you would like a full-screen Azure Cloud Shell experience, you can navigate to `https://shell.azure.com`.

---

3. Ensure PowerShell is selected in the shell selector at the top of the Cloud Shell.

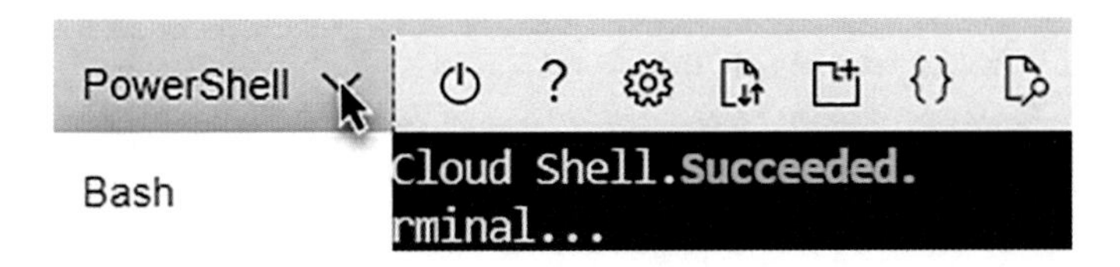

4. Clone the GitHub repository that contains the source code for this book by running the following command:

```
git clone https://github.com/Apress/Azure-SQL-Hyperscale-
Revealed.git
```

5. Before running the script, it is important to ensure the correct Azure subscription is selected by running the following command:

```
Select-AzSubscription -Subscription '<subscription name>'
```

6. Run the script by entering the following:

```
./Azure-SQL-Hyperscale-Revealed/ch4/New-SQLHyperscaleReveale
dStartingEnvironment.ps1 -ResourceNameSuffix '<your resource
name suffix>'
```

7. It will take a few minutes to run but will eventually produce output like Figure 4-3, showing partial output from creating the starting environment, including a list of resources that were created.

```
Name                                 Type                      Value
==================================== ========================= ==========
primaryResourceGroupName             String                    "sqlhr01-shr8-rg"
failoverResourceGroupName            String                    "sqlhr02-shr8-rg"
primaryVirtualNetworkName            String                    "sqlhr01-shr8-vnet"
failoverVirtualNetworkName           String                    "sqlhr02-shr8-vnet"
keyVaultName                         String                    "sqlhr-shr8-kv"
primaryLogAnalyticsWorkspaceName     String                    "sqlhr01-shr8-law"
failoverLogAnalyticsWorkspaceName    String                    "sqlhr02-shr8-law"
```

Figure 4-3. *Partial output from running the script*

8. Navigate back to the Azure Portal and select Resource Groups from the Azure home screen.

9. In the "Filter for any" field box, enter **shr**. This will show the resource groups that were created for the starting environment, as shown in Figure 4-4.

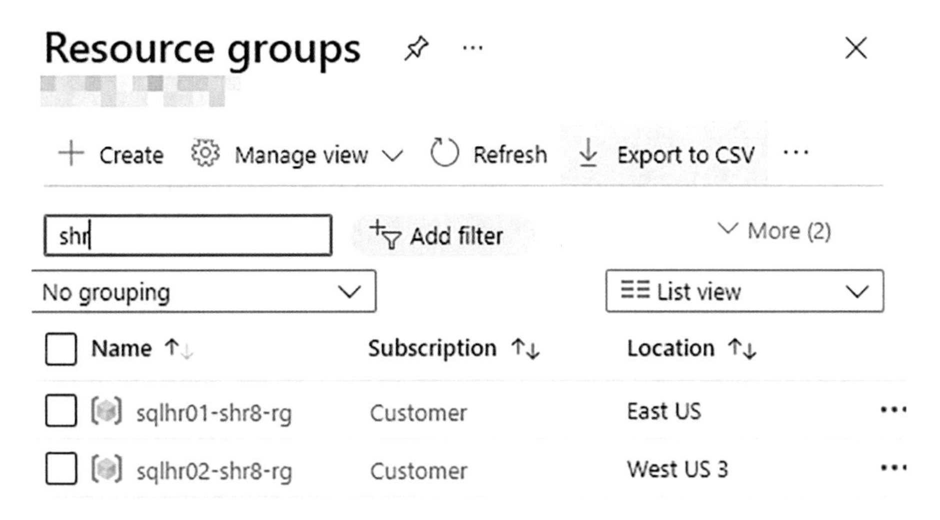

Figure 4-4. Starting environment resource groups

10. You can click into one of the resource groups to see the resources that were created.

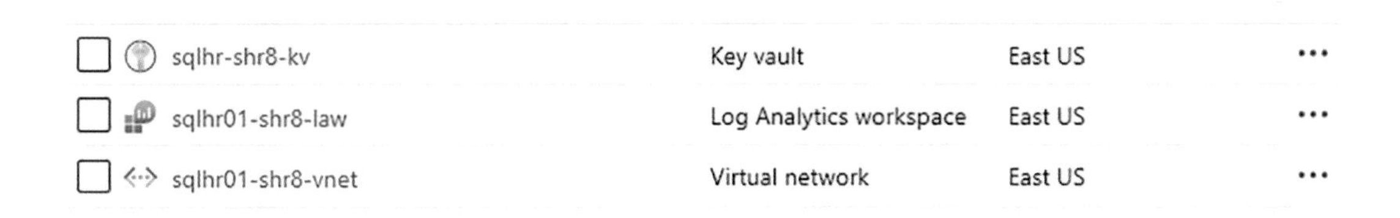

Your starting environment is now ready for you to begin deploying the Hyperscale resources into it. If at any time you want to get rid of this environment, see the instructions at the end of this chapter, or simply delete the two resource groups. You can always re-create the starting environment again by running the script.

Creating a SQL Administrators Group

If you are planning to enable Azure Active Directory as an identity source for your Hyperscale database, then it recommended that you assign the SQL administrator to a security group rather than to an individual user. For the examples in this chapter, we are going to create a SQL Administrators group in Azure Active Directory.

Tip It is not possible to assign more than one user to be the SQL administrator of a logical server; therefore, if you want to enable more than one, you must create a security group with members consisting of the users you want to be admins and assign this group instead.

If you're familiar with the process of creating security groups in Azure Active Directory, then you should find this very straightforward.

Important Your Azure Active Directory administrator might not permit you to create security groups in your tenant. If that is the case, you'll need to request them to create the group for you.

To create the `SQL Administrators` group, follow these steps:

1. Open your browser and navigate to `https://portal.azure.com`.
2. Enter **Search** into the Azure Search box.
3. Select Groups from the search results.

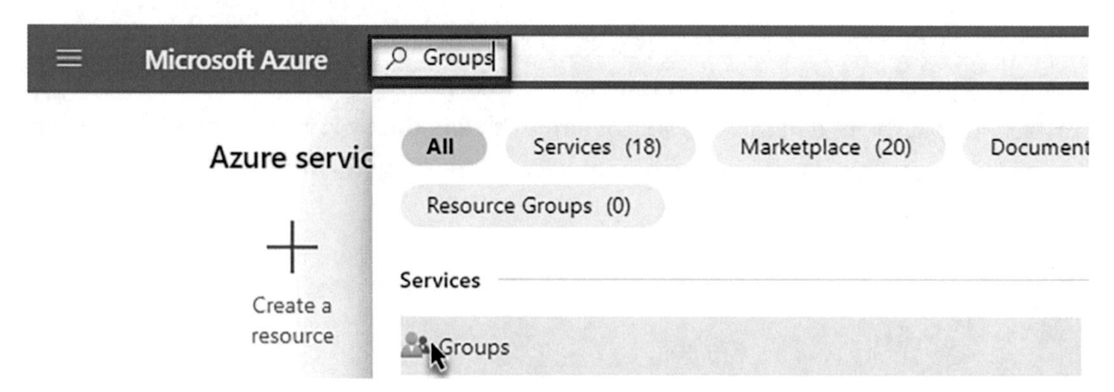

4. Click "New group."

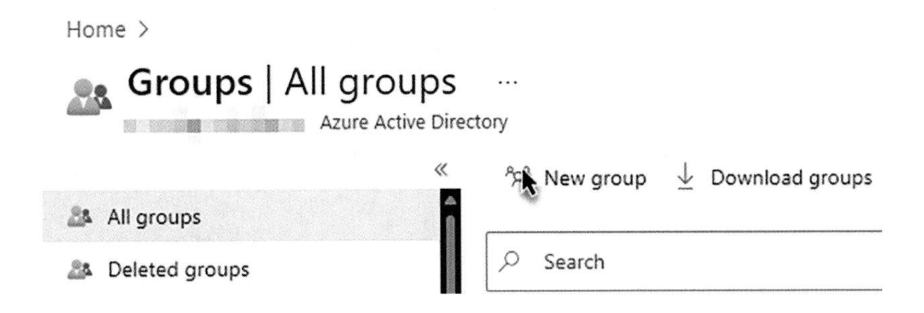

5. Set the group type to Security and enter **SQL Administrators** as the group name. You can use a different group name if you want, but remember the name you select as it will be used when configuring the logical server. You can also set the group description to clarify the purpose of this security group.

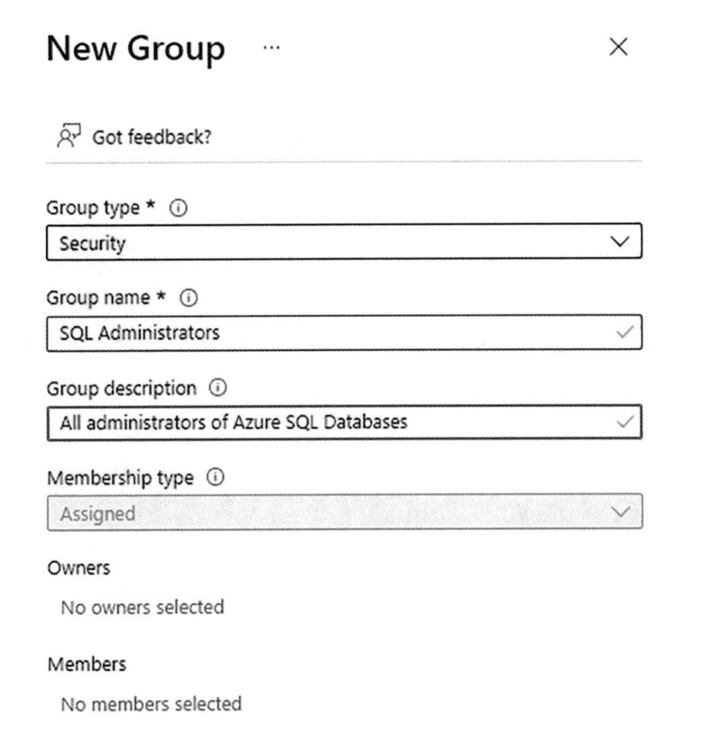

6. Click the link under Owners to configure the user accounts (or other security groups) that own this security group. These accounts can change the membership of the group, so they can assign or remove SQL administrators to all logical servers that are using this group as the list of SQL administrators.

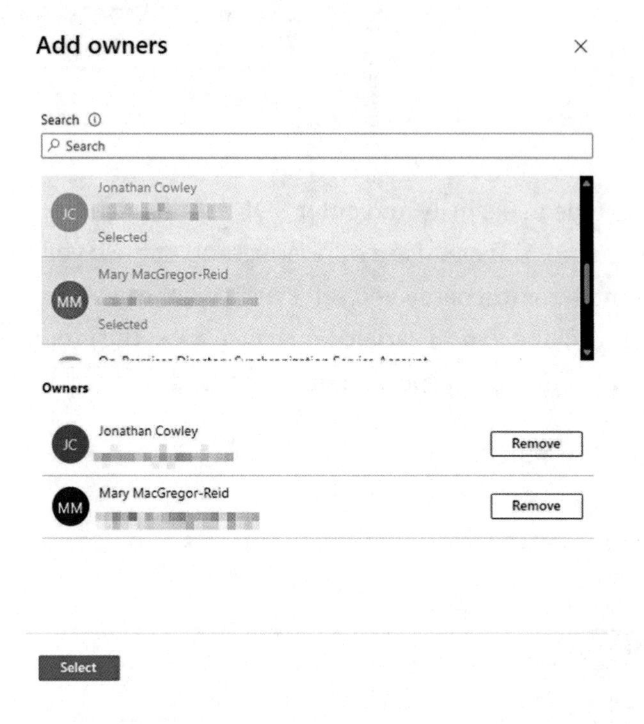

7. Click the link under Members to configure the user accounts (or other security groups) that are members of the SQL Administrators group. These accounts will be granted the `db_owner` role on all databases attached to logical servers where this group is set as the SQL Admin.

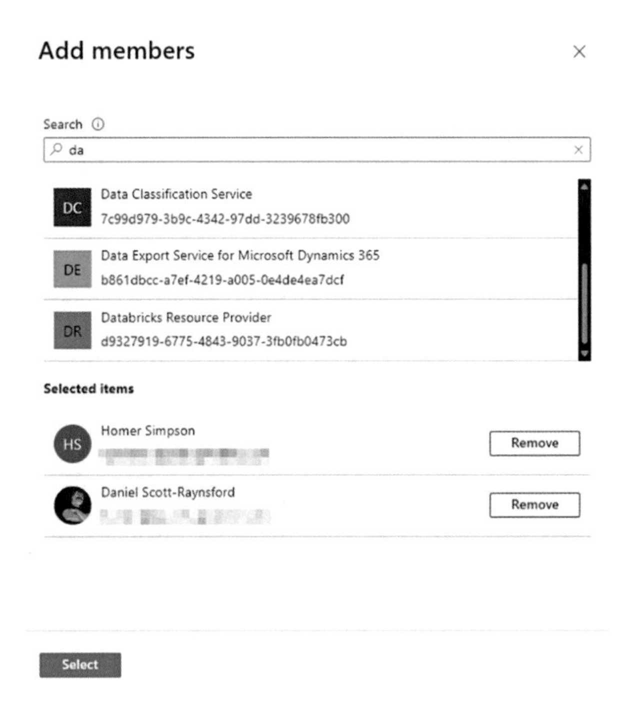

8. Click the Create button to create the security group.

9. After a few seconds, the "Successfully create group" message will appear.

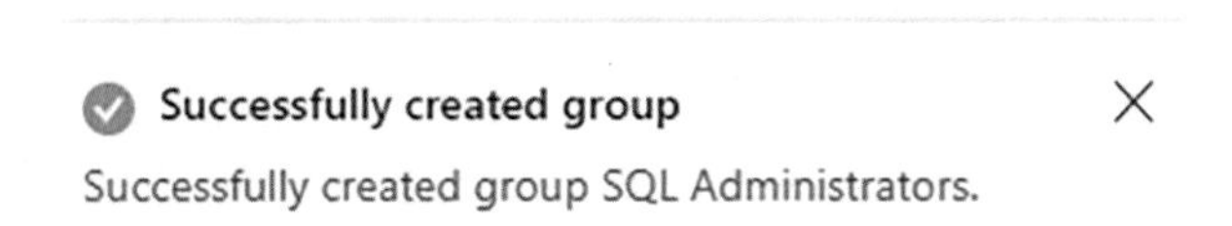

You can now assign this security group as the SQL administrator for your logical servers.

Deploying a Highly Available Hyperscale Database into a Virtual Network

Now that a starting environment has been deployed into Azure, we can begin deploying our Hyperscale database and logical server and connecting it to the virtual network deployed into the primary region.

We will be using a service called Azure Private Link. This will add our logical server to a subnet in the virtual network using a virtual network adapter and make it available as a private endpoint. The virtual network adapter will be assigned a private IP address within the subnet, and a DNS A record will also be created in an Azure DNS zone. The Azure DNS zone will be added to the virtual network as a DNS resolver.

Tip The Azure Private Link service allows many different resources to be attached to a virtual network. For more information on Private Link, see this page: `https://aka.ms/privatelink`.

Although all the previous tasks sound complicated, they are straightforward to perform when using the Azure Portal. The deployment process for an Azure SQL Database and the private endpoint takes care of most of the work for you. However, if you're deploying the logical server and Hyperscale database using code, then you will need to deploy the additional private endpoint components individually as part of your code, which we'll show in Chapters 9, 10 and 11.

The integration of a logical server into a virtual network using a private endpoint can be done either at the time of deploying the logical server or after deployment has been completed. The Networking blade of the SQL Server resource (the logical server) can be used to add an existing logical server into one or more virtual networks but will instead use the Azure Private Link user interface.

Tip It is possible to have a single logical server attached to multiple Azure virtual networks. This can be useful if your Hyperscale database needs to be accessed by multiple applications that are deployed into separate disjointed virtual networks.

We are also going to enable high-availability replicas across availability zones in the primary region. This will ensure at least two read replicas of our primary Hyperscale database are available within the primary region, each in a different availability zone. This will protect our workload against failure of one or two availability zones in the region. Just like the virtual network connectivity with private endpoint, you can configure the high-availability replicas during deployment or after they are complete. For now, we will deploy the high-availability replicas during deployment.

Note The following steps are like those we saw in Chapter 1 when deploying our first Hyperscale database. Therefore, to increase brevity, we are going to omit some of the screenshots that you'll already have seen if running through the steps in that chapter. It is also quite common for the user interface for this process to change slightly as new features are added. So, what you see might slightly differ but should still be similar enough for you to follow along.

Finally, during the deployment, we will also configure the logical server to use only Azure Active Directory (Azure AD) authentication and set the admin as the `SQL Administrators` group.

Deploying the logical server and Hyperscale database and connect them to the virtual network using the Azure Portal will require completing a wizard. This wizard does have several pages. So, we've chosen to break each element of the configuration into its own section.

Basic Configuration of the Database

The first step of the deployment is to configure the logical server and the database compute.

1. Open your browser and navigate to `https://portal.azure.com`.
2. Sign in using credentials that give you access to the Azure subscription you'll be deploying your Hyperscale database into. In this case, we are using the same subscription that we deployed the starting environment into.
3. Click the "Create a resource" button in the home dashboard.
4. Click Create under SQL Database on the "Create a resource" blade.
5. The Create SQL Database blade will open showing the Basic configuration options of our database.
6. Make sure the subscription is set to the name of the subscription you have permission to deploy into and contains the starting environment we deployed earlier.

7. Set the resource group to the primary resource group that was created as part of the starting environment. This will be the `sqlhr01-<suffix>-rg` that is in the Azure East US region.

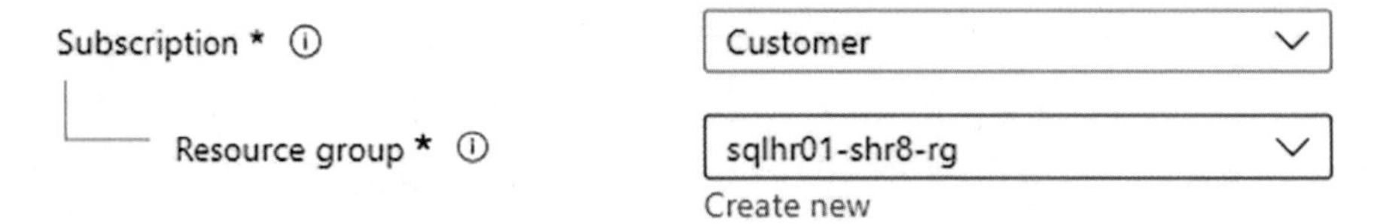
Project details
Select the subscription to manage deployed resources and costs. Use resource groups like folders to organize and manage all your resources.
Subscription *
Customer
Resource group *
sqlhr01-shr8-rg
Create new

8. In the "Database name" box, enter the name of the Hyperscale database that will be created. We're using `hyperscaledb` for this example.

9. We need to create a new logical server by clicking the "Create new" button.

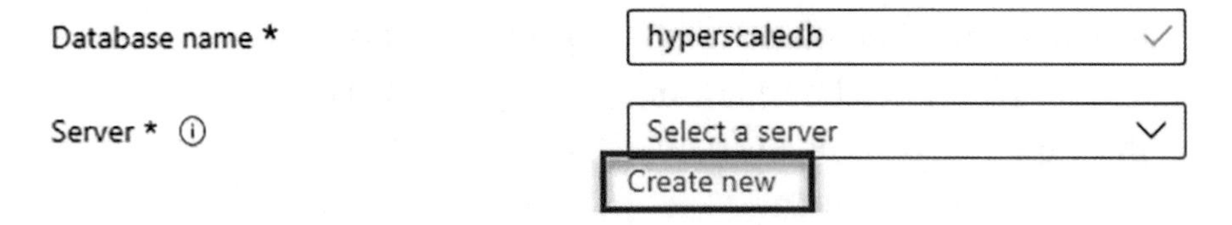
Database details
Enter required settings for this database, including picking a logical server and configuring the compute and storage resources
Database name *
hyperscaledb
Server *
Select a server
Create new

10. Every logical server must have a unique name, not just within our environment, but globally. For this example, we're using `sqlhr01-shr8`, but you should use a name that is likely to be globally unique.

11. Set the location to the Azure region to which you want to deploy the SQL Database Server. This is the primary region where the logical server will be deployed. All databases contained by this logical server will also be in this region. We will set this to East US.

Important Azure SQL Database Hyperscale is enabled in most Azure regions. However, if it is not enabled in the primary region you chose for your starting environment, select an alternate region, or see this page for more information: `https://learn.microsoft.com/azure/reliability/cross-region-replication-azure#azure-cross-region-replication-pairings-for-all-geographies`.

12. Set the authentication method to "Use only Azure Active Directory (Azure AD) authentication."

13. Click the "Set admin" button to select the Azure AD user or group that will be granted the admin role on this SQL Database server. Select the SQL Administrators group that was created earlier.

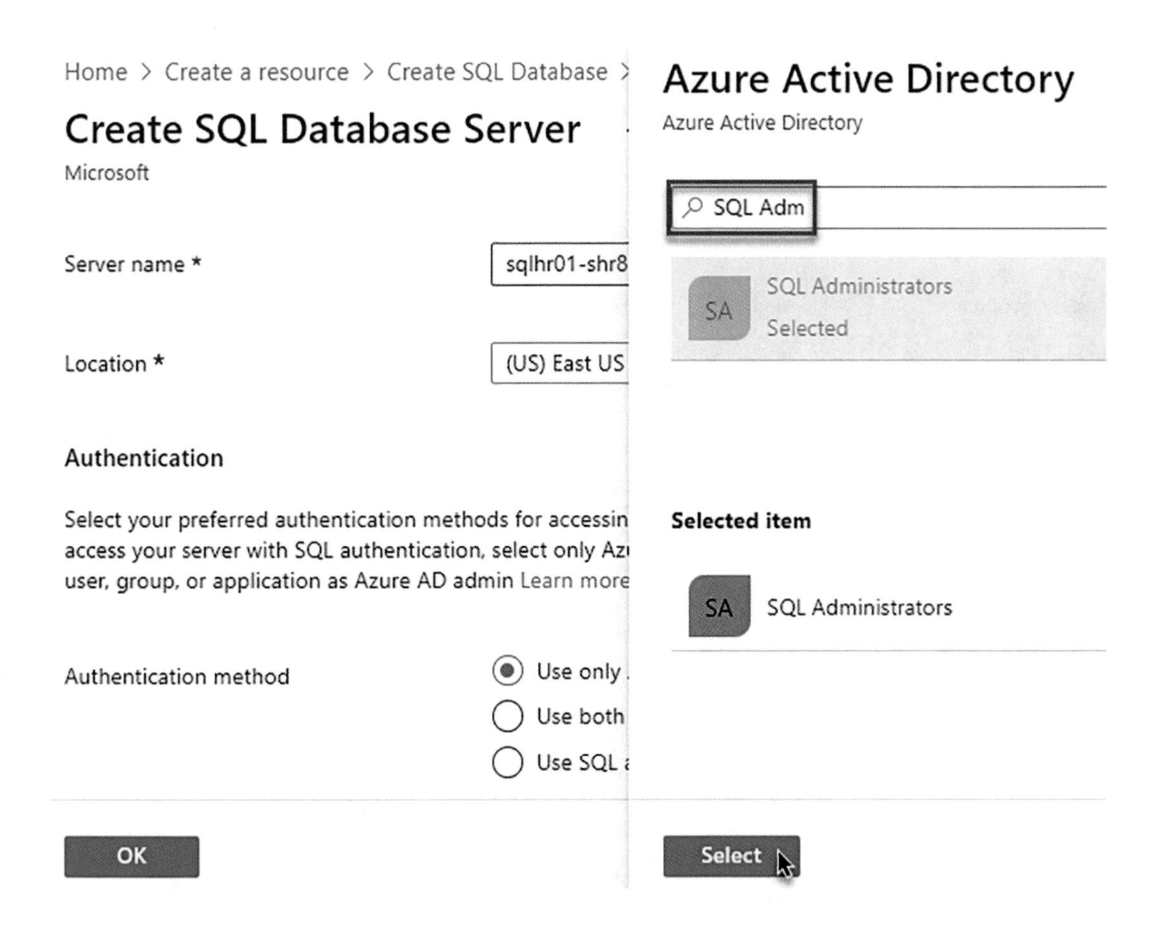

14. Click OK on the Create SQL Database Server form.

15. Ensure that No is specified for "want to use SQL elastic pool" because Hyperscale databases can't be put inside an elastic pool.

Tip You can ignore the workload environment setting because we will always have to override this when specifying a Hyperscale database.

---

16. Click "Configure database" in the "Compute + storage" section to configure the Hyperscale database.

17. For the Service tier, we need to select "Hyperscale (On-demand scalable storage)."

18. Ensure that the "Hardware configuration" is set to Standard-series (Gen5) (or newer if available) or DC-Series (if confidential compute is required) as Hyperscale is supported only on these hardware configurations.

Available hardware configurations

Based on your workload requirements, select from available hardware configurations listed below.

Configuration	Description	Max vCores	Max memory	Max storage
Standard-series (Gen5)	Balanced memory and...	80	415.23 GB	--
Premium-series	Balanced memory and...	128	625 GB	--
Premium-series - me...	Memory optimized	80	830.47 GB	--
DC-series	Enables confidential c...	8	36 GB	--

19. Set the vCores according to your needs. In our case, we'll use two to keep costs down, but we can always scale up later. Each secondary replica will also have this number of vCores assigned to it.

20. Select the number of high-availability secondary replicas to the number of secondary replicas you want in the region. We'll set this to 2 as well. This will mean we have one primary replica and two secondary replicas.

21. Specify Yes for the "Would you like to make this database zone redundant" option.

Note You may have noticed a warning that making the database replicas zone redundant may change the database backup storage settings. This is because when you enable this setting, only zone-redundant backup storage is supported.

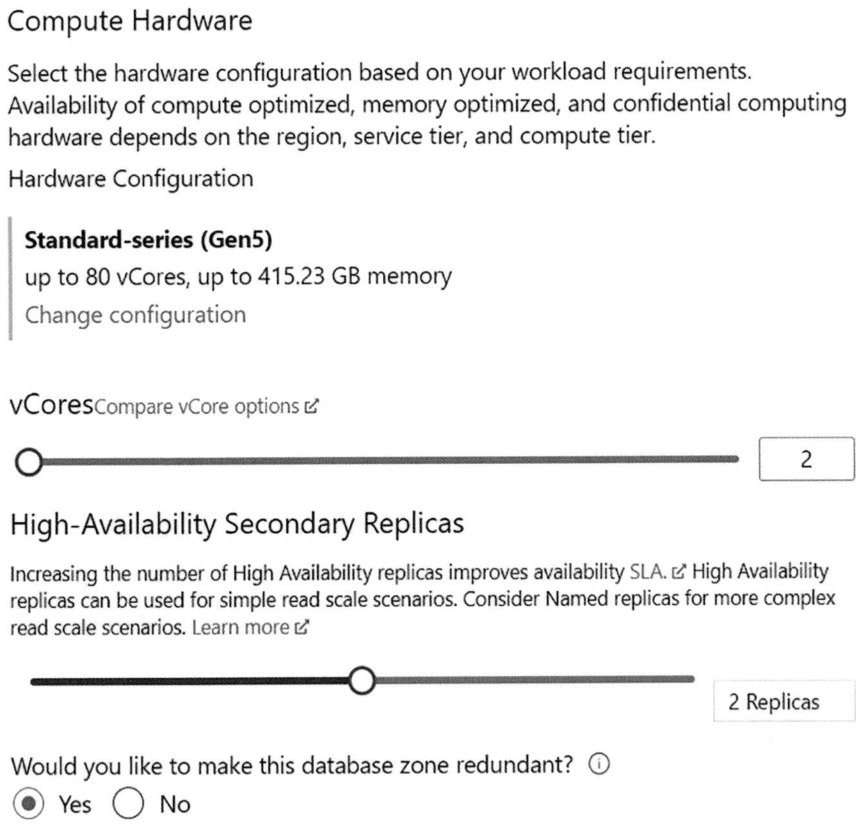

Compute Hardware

Select the hardware configuration based on your workload requirements. Availability of compute optimized, memory optimized, and confidential computing hardware depends on the region, service tier, and compute tier.

Hardware Configuration

Standard-series (Gen5)
up to 80 vCores, up to 415.23 GB memory
Change configuration

vCores Compare vCore options

2

High-Availability Secondary Replicas

Increasing the number of High Availability replicas improves availability SLA. High Availability replicas can be used for simple read scale scenarios. Consider Named replicas for more complex read scale scenarios. Learn more

2 Replicas

Would you like to make this database zone redundant?

Yes No

22. Click Apply to set the database configuration.

23. You will return to the Create SQL Database wizard. You may have noticed that the backup storage redundancy has only two enabled options: geo-zone-redundant backup storage and zone-redundant backup storage. We will choose the geo-zone-redundant backup storage so that our backup is replicated to another region in the case of a regional outage.

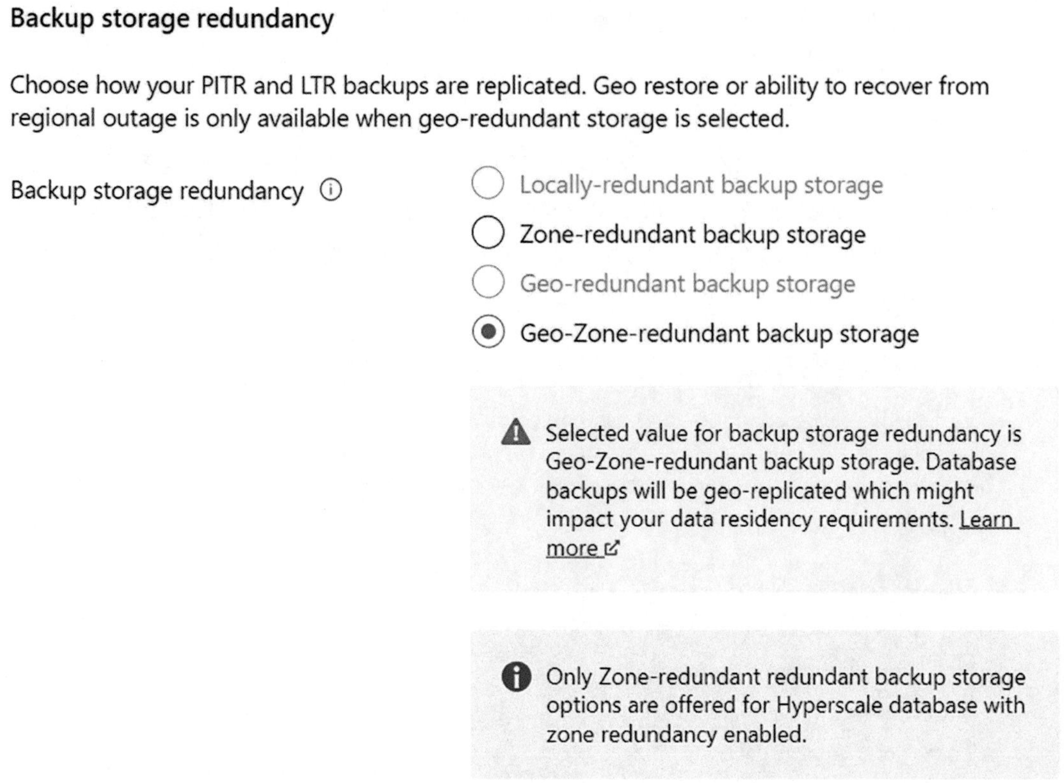

Backup storage redundancy

Choose how your PITR and LTR backups are replicated. Geo restore or ability to recover from regional outage is only available when geo-redundant storage is selected.

Backup storage redundancy ⓘ

- ○ Locally-redundant backup storage
- ○ Zone-redundant backup storage
- ○ Geo-redundant backup storage
- ◉ Geo-Zone-redundant backup storage

⚠ Selected value for backup storage redundancy is Geo-Zone-redundant backup storage. Database backups will be geo-replicated which might impact your data residency requirements. Learn more

ⓘ Only Zone-redundant redundant backup storage options are offered for Hyperscale database with zone redundancy enabled.

24. Click Next to configure the networking.

Now that we've configured the logical server and the Hyperscale database, we can move on to the networking page.

Configuring Network Connectivity

We are going to configure this logical server to connect to a virtual network in the same region, using a private endpoint. We will also disable public network access.

1. As we mentioned earlier, we wanted to enable access to this logical server only from inside a virtual network. So, select "Private endpoint" from the connectivity method options.

2. The private endpoints in virtual networks that this logical server is connected to are now displayed. Because this is a new logical server, it won't currently be connected to any virtual networks. Click the "Add private endpoint" button.

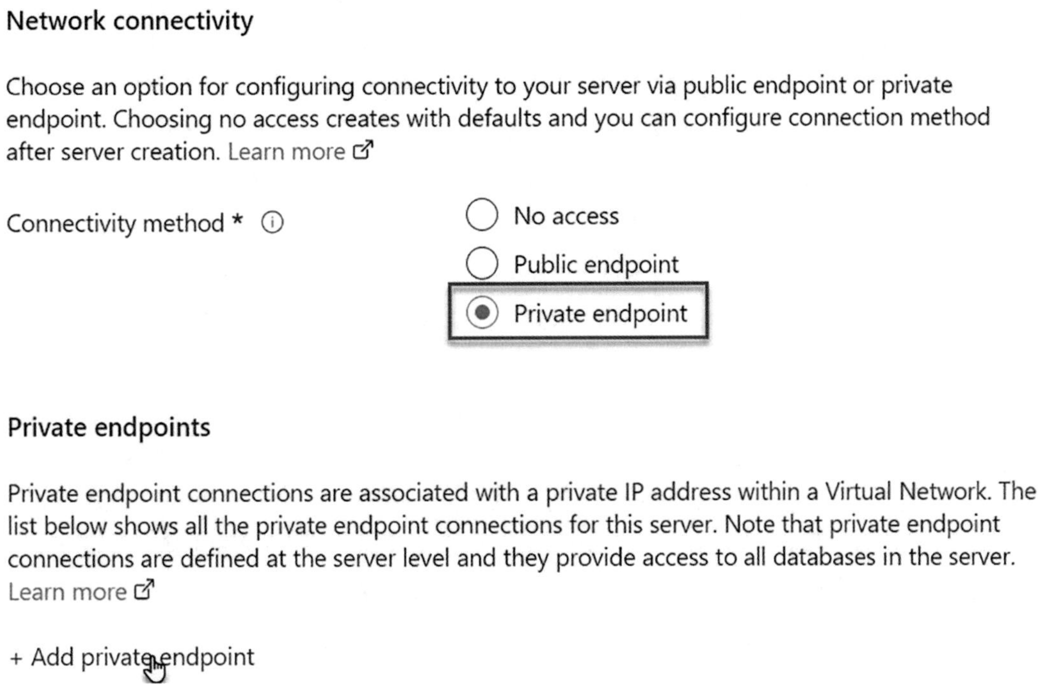

Network connectivity

Choose an option for configuring connectivity to your server via public endpoint or private endpoint. Choosing no access creates with defaults and you can configure connection method after server creation. Learn more

Connectivity method *

- ○ No access
- ○ Public endpoint
- ◉ Private endpoint

Private endpoints

Private endpoint connections are associated with a private IP address within a Virtual Network. The list below shows all the private endpoint connections for this server. Note that private endpoint connections are defined at the server level and they provide access to all databases in the server. Learn more

+ Add private endpoint

3. The private endpoint settings will be displayed, with most of the settings already configured correctly. You will need to set the name to be a unique resource name for the private link itself, for example, sqlhr01-<suffix>-pe. You should also check the following:

    - The resource group is set to your primary resource group, sqlhr01-<suffix>-rg.
    - The location should be East US.
    - The target subresource should be SQLServer.
    - The virtual network is the primary network, sqlhr01-<suffix>-vnet.
    - The subnet is the data_subnet with the primary network.
    - If you want to have the database assigned a private FQDN in the virtual network, then you will need to select "yes" for the "integrate with a private DNS zone" option.

- The private DNS zone should be `(New) privatelink.database.windows.net`. This tells us that a new private DNS zone will be created and attached to the primary virtual network, and an A record for our logical server will be created in the zone. In our case, this will end up being `sqlhr01-shr8.privatelink.database.windows.net`.

Important Although there is an A record that resolves to the private IP address of the logical server, it has an alias that resolves the `sqlhr01-shr8.database.windows.net` public name to the private IP within this virtual network. This means you still connect to the database using `sqlhr01-shr8.database.windows.net` from within the private network. In fact, you must not use `sqlhr01-shr8.privatelink.database.windows.net` to connect to the database, because you will receive a "The target principal name is incorrect" error message. This is because the TLS certificate will not match.

Figure 4-5 shows the create private endpoint page with a completed configuration.

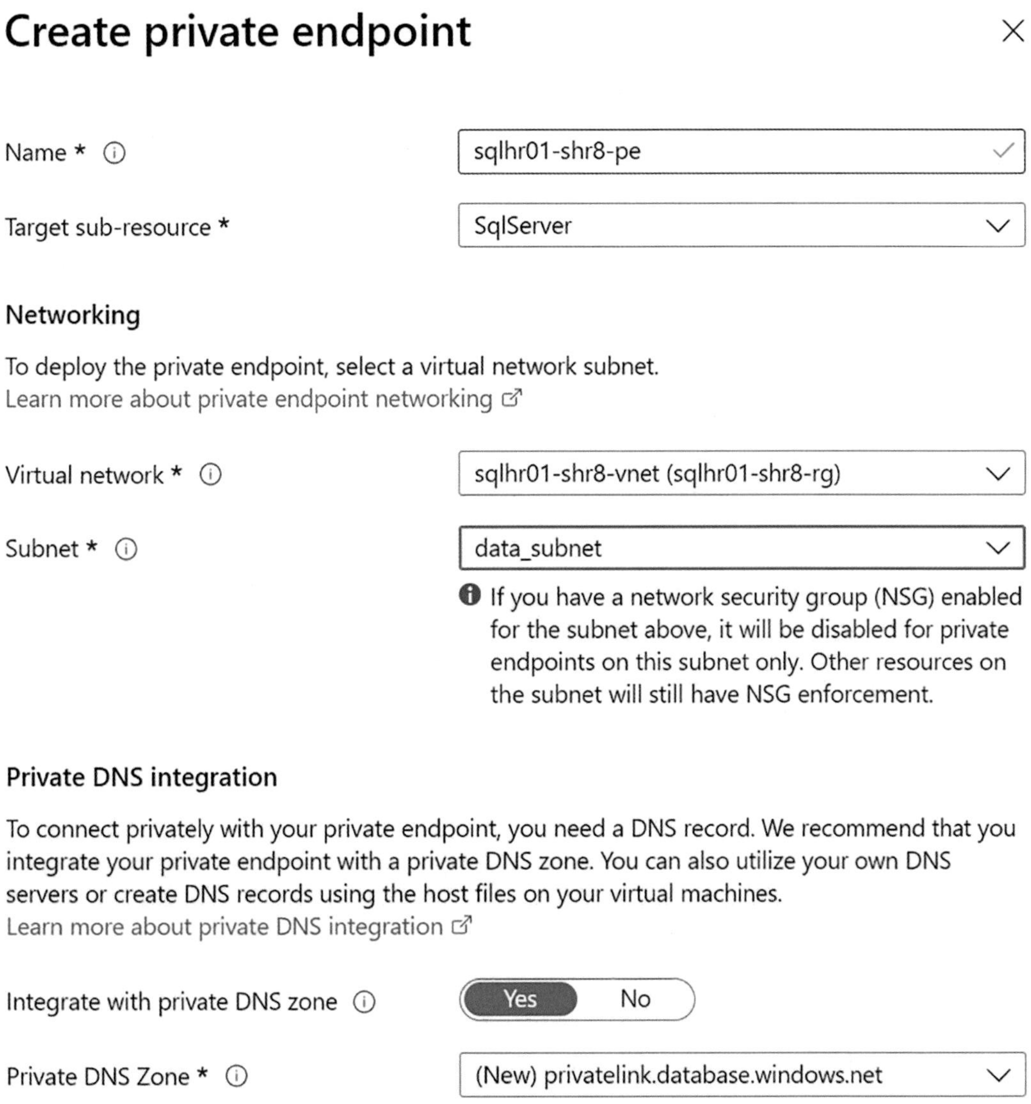

Create private endpoint

Name * sqlhr01-shr8-pe

Target sub-resource * SqlServer

Networking

To deploy the private endpoint, select a virtual network subnet.
Learn more about private endpoint networking

Virtual network * sqlhr01-shr8-vnet (sqlhr01-shr8-rg)

Subnet * data_subnet

If you have a network security group (NSG) enabled for the subnet above, it will be disabled for private endpoints on this subnet only. Other resources on the subnet will still have NSG enforcement.

Private DNS integration

To connect privately with your private endpoint, you need a DNS record. We recommend that you integrate your private endpoint with a private DNS zone. You can also utilize your own DNS servers or create DNS records using the host files on your virtual machines.
Learn more about private DNS integration

Integrate with private DNS zone Yes No

Private DNS Zone * (New) privatelink.database.windows.net

Figure 4-5. *The private endpoint settings*

4. Click OK.
5. Set the connection policy to Default. We will talk about the connection policy in the next chapter.
6. Click the Next button.

Now that we have configured the networking, we can finalize the deployment by setting any security and additional settings.

The Final Configuration Tasks and Deployment

Although for our environment we do want to configure some of the security settings such as enabling Microsoft Defender for SQL, identity, and TDE, we're going to configure these after deployment so that we can focus on each independently.

1. Select "not now" for "enable Microsoft Defender for SQL."
2. We can ignore the Ledger and Identity options for now.
3. Make sure TDE is set to "Service-managed key selected."
4. Click Next to move to the next step.
5. This step allows us to specify a backup file or sample database to use as a template to build this database from. This can be useful for migrating to Hyperscale or for testing and demonstration purposes. We can also configure database collation and define a maintenance window when Azure will be allowed to patch any underlying infrastructure.
6. We'll move onto the next step by clicking Next.
7. If you want, add an `Environment` tag that has a value of `SQL Hyperscale Revealed demo` and is assigned to both the SQL database and SQL database server resources. This will help you identify all resources you've been creating as part of this chapter.

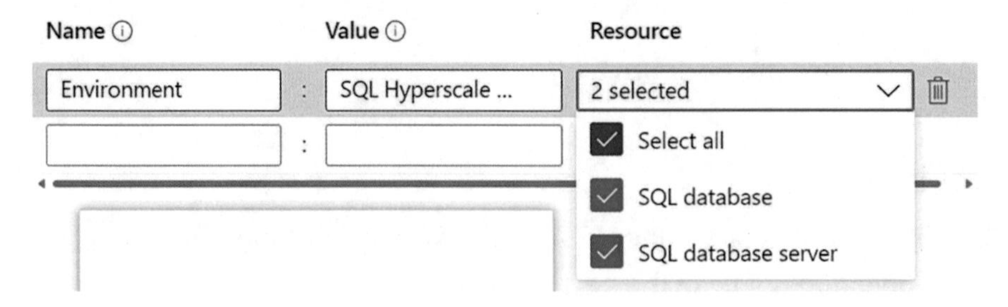

8. Click Next to move to the final step of the deployment.

9. Review the settings for your logical server and database and ensure the costs for running this environment are what you expect. Figure 4-6 shows an example of the summary of costs that are displayed on the settings review screen before committing the deployment.

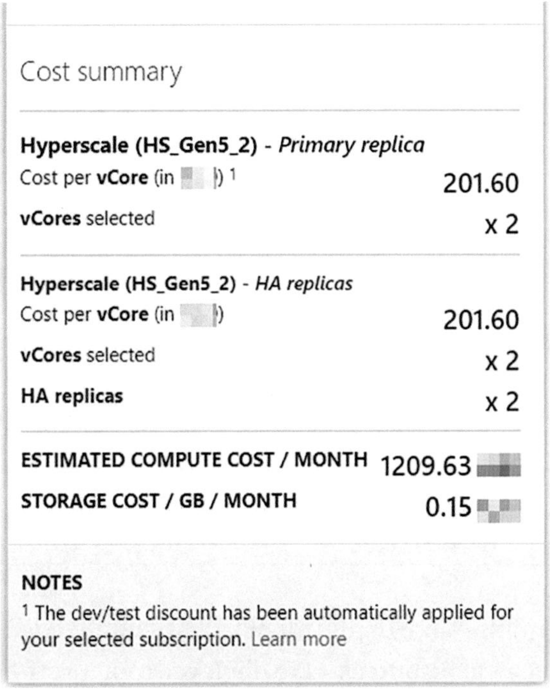

Figure 4-6. *The costs summary for this deployment*

10. Click Create to begin the deployment. If there are no problems in our settings, the deployment will begin (see Figure 4-7).

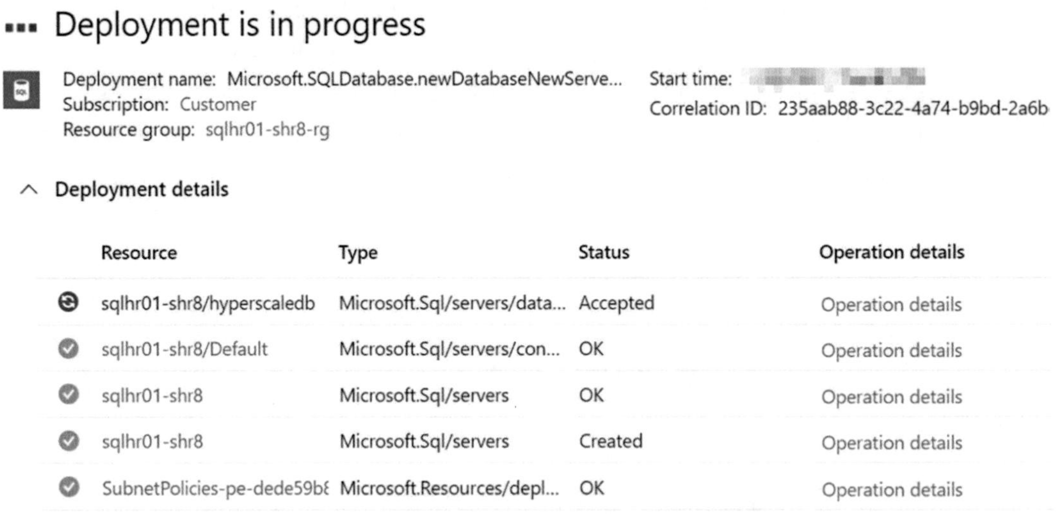

Deployment is in progress

Deployment name: Microsoft.SQLDatabase.newDatabaseNewServe...
Subscription: Customer
Resource group: sqlhr01-shr8-rg

Start time:
Correlation ID: 235aab88-3c22-4a74-b9bd-2a6b

Deployment details

Resource	Type	Status	Operation details
sqlhr01-shr8/hyperscaledb	Microsoft.Sql/servers/data...	Accepted	Operation details
sqlhr01-shr8/Default	Microsoft.Sql/servers/con...	OK	Operation details
sqlhr01-shr8	Microsoft.Sql/servers	OK	Operation details
sqlhr01-shr8	Microsoft.Sql/servers	Created	Operation details
SubnetPolicies-pe-dede59b8	Microsoft.Resources/depl...	OK	Operation details

Figure 4-7. *The deployment progress for the resources*

Tip We could click "download a template for automation," which will produce a declarative IaC template that will allow us to re-create this exact deployment.

---

11. After a few minutes the deployment will be completed. You will notice several resources were deployed, not just the logical server and database but also the resources relating to the private endpoint.

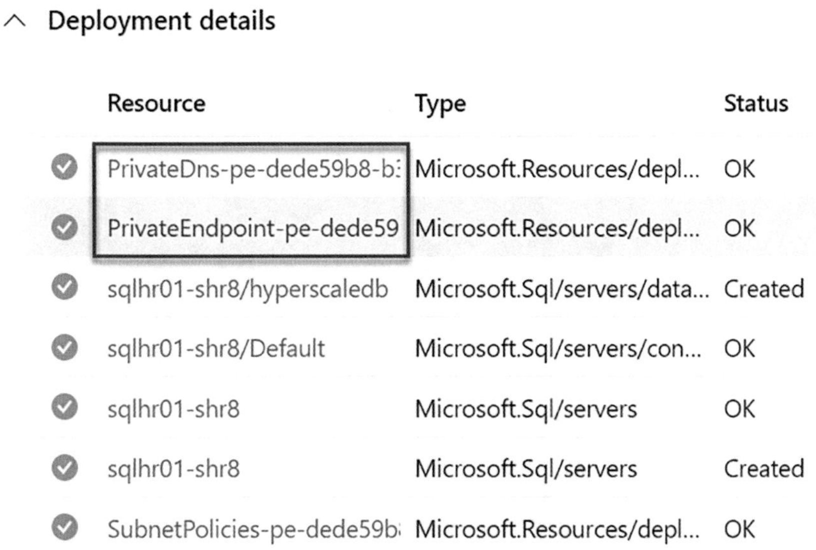

^ Deployment details

Resource	Type	Status
PrivateDns-pe-dede59b8-b:	Microsoft.Resources/depl...	OK
PrivateEndpoint-pe-dede59	Microsoft.Resources/depl...	OK
sqlhr01-shr8/hyperscaledb	Microsoft.Sql/servers/data...	Created
sqlhr01-shr8/Default	Microsoft.Sql/servers/con...	OK
sqlhr01-shr8	Microsoft.Sql/servers	OK
sqlhr01-shr8	Microsoft.Sql/servers	Created
SubnetPolicies-pe-dede59b:	Microsoft.Resources/depl...	OK

This completes the deployment of the new logical server connected to the private network and configured with a two-vCore primary Hyperscale database with two secondary high-availability replicas, spread across three availability zones in the region. Our primary region should now look like Figure 4-8.

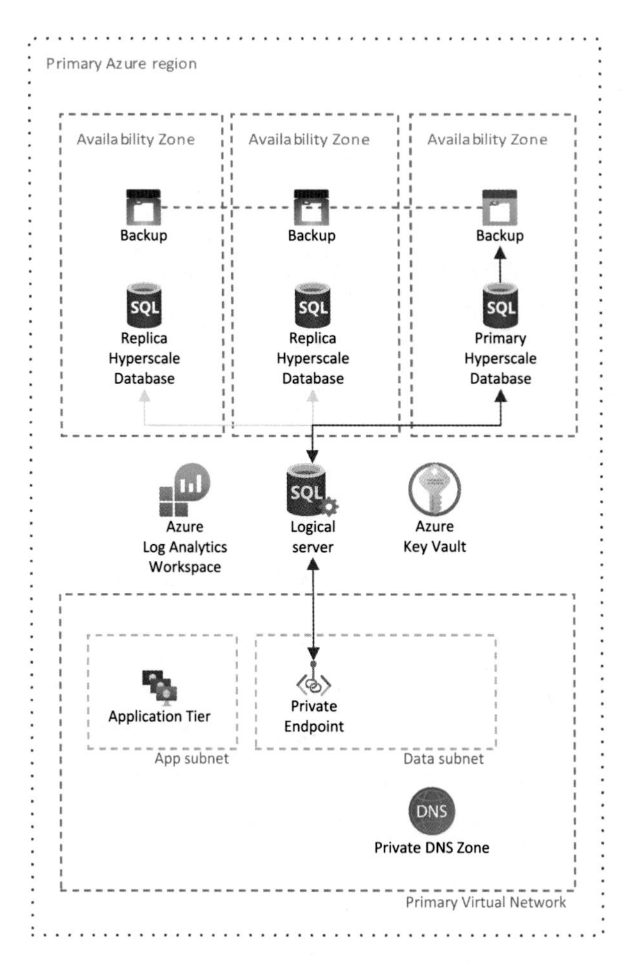

Figure 4-8. *The primary region for our Hyperscale database*

If we navigate to the primary resource group, `sqlhr01-<prefix>-rg`, we should see the new resources have been deployed into it. Figure 4-9 shows the list of resources that should now be displayed in the primary resource group.

Name	Type	Location
hyperscaledb (sqlhr01-shr8/hyperscaledb)	SQL database	East US
privatelink.database.windows.net	Private DNS zone	Global
sqlhr-shr8-kv	Key vault	East US
sqlhr01-shr8	SQL server	East US
sqlhr01-shr8-law	Log Analytics workspace	East US
sqlhr01-shr8-pe	Private endpoint	East US
sqlhr01-shr8-pe.nic.4185c11d-99ff-46e5-ad02-ea57cd1bd···	Network Interface	East US
sqlhr01-shr8-vnet	Virtual network	East US

Figure 4-9. *The resources deployed for the Hyperscale database*

Now that we've completed deployment of the Hyperscale database resources and they've been connected to the virtual network, we can continue to configure the remaining nonfunctional requirements. But first, we should verify that we can connect to the database from within the virtual network.

Deleting the Example Environment

Now that we have deployed a starting environment and used it to deploy some of the Azure SQL DB Hyperscale resources, we might want to delete them so that they're not costing us any money (or credit). This section will show you how to delete the starting environment resource groups.

Note Once Chapter 8 has been completed, we will have deployed a full Hyperscale environment, so if you're following along with these examples, you shouldn't delete the environment until the end of Chapter 8.

It is a good idea to remember to clean up any learning environments you're no longer using as it ensures you're not spending money (or consuming credit) when you're not using it. We can always re-deploy the environment if we need it again. Once we've learned how to do this simply using infrastructure as code the process can be completely automated.

Tip Being able to re-deploy an environment using infrastructure as code has many benefits for both learning and for development, testing, and production. So, developing your ability to leverage this will provide you with a great deal of value. It is strongly recommended that you use infrastructure as code for the deployment of production environments because of the many benefits it provides.

Cleaning up the starting environment, including any resources you deployed following along with the steps in this chapter, can be done by simply deleting the resource groups that were created by the starting environment script. Figure 4-10 shows the resource groups that were created as part of our starting environment and then added to throughout this chapter.

Name ↑↓	Subscription ↑↓	Location ↑↓
sqlhr01-shr8-rg	Customer	East US
sqlhr02-shr8-rg	Customer	West US 3

Figure 4-10. *The starting environment resource groups*

We have provided a handy PowerShell script for you to quickly clean up any starting environments you might have deployed. As before, we are going to use Azure Cloud Shell to run the script. We are going to use PowerShell in the Azure Cloud Shell for this process, because `Remove-SQLHyperscaleRevealedEnvironment.ps1` is written in PowerShell.

Important The `Remove-SQLHyperscaleRevealedEnvironment.ps1` script looks for all the resource groups in the currently selected subscription that have the `environment` tag set to `SQL Hyperscale Revealed Demo` and will offer to delete them for you. If you have created the resource groups manually without this tag, then the script will not locate them and won't be able to delete them.

The `Remove-SQLHyperscaleRevealedEnvironment.ps1` script can be found in the GitHub repository we cloned in the "Deploying the Starting Environment Using Azure Cloud Shell" section earlier in this chapter. The following steps assume that you have cloned the GitHub repository to your Azure Cloud Shell environment.

To delete the starting environment using the script, follow these steps:

1. Open your browser and navigate to `https://portal.azure.com`.
2. Open the Azure Cloud Shell by clicking the Cloud Shell button.
3. Ensure PowerShell is selected in the shell selector at the top of the Cloud Shell.
4. Before running the script, it is important to ensure the correct Azure subscription is selected by running the following command:

   ```
   Select-AzSubscription -Subscription 'subscription name'
   ```

5. Run the script by entering the following:

   ```
   ./Azure-SQL-Hyperscale-Revealed/ch4/Remove-SQLHyperscaleRevealed
   Environment.ps1
   ```

6. Type **yes** and press Enter to confirm deletion of these resource groups. You will still be asked to confirm the deletion of each individual resource group.
7. For each resource group, press Y to confirm deletion.

After a few minutes, the resource groups will be deleted, and you will no longer be charged for any resources you have deployed as part of the examples in this chapter.

Summary

In this chapter, we demonstrated how to do the following:

- Deploy the multi-region starting environment that will be the basis of the example environment over the next four chapters.
- Create an Azure Active Directory security group to assign as the SQL administrators of the logical server and Hyperscale database.

- Deploy a Hyperscale database into a virtual network with a private endpoint and disable the public endpoint.
- Enable Azure AD authentication to the logical server and configure the administrators to be members of the SQL Administrators security group.
- Delete the resource groups and resources created as part of the example environment.

In the next chapter, we will look at some of the options we have for connecting to and managing this Hyperscale database now that it is accessible only from within a virtual network.

CHAPTER 5

Administering a Hyperscale Database in a Virtual Network in the Azure Portal

In Chapter 4, we deployed a logical server with a Hyperscale database and connected it to a virtual network using a private endpoint. The logical server's public endpoint was disabled. This is a common design pattern that is used to allow access to the database only from resources connected to the virtual network infrastructure and to ensure all traffic to and from the logical server travels from the virtual network via the private endpoint.

However, when we have a logical server connected to a virtual network using a private endpoint and have disabled the public endpoint, then we will need to consider how database administrators (DBAs) will connect to the Hyperscale database to perform management tasks.

In this chapter, we will list some of the common ways DBAs can connect to the Hyperscale database in a virtual network. We will also demonstrate a simple and secure approach to enabling management when another, more complex networking infrastructure connects your management machines to the virtual network.

Z. Barać and D. Scott-Raynsford, *Azure SQL Hyperscale Revealed*,
https://doi.org/10.1007/978-1-4842-9225-9_5

Administering a Hyperscale Database in a Virtual Network

Whenever we deploy an Azure SQL logical server and make it accessible only from within a virtual network, we need to plan for how our database administrators (or other authorized users or processes) will connect to it.

Tip In some cases, there will be no reason for database administrators to connect to the database at all. Perhaps your application tier can run any configuration scripts on the database to create tables as part of the deployment. If this is the case, then you may not need to worry about accessing a database that is within a virtual network.

The Hyperscale database that we deployed in Chapter 4 was attached to the `data_subnet` in the `sqlhr01-<prefix>-vnet` virtual network in the primary region. We also disabled public access to the database. This means it is accessible only from services that are within the virtual network or peered networks and there aren't any network security groups that block access to it.

If we try to connect to the database from within the SQL Database query editor from within the Azure Portal, we will receive an error message telling us that to connect we will need to use the private endpoint from within the virtual network. Figure 5-1 shows the error message that is received if you try to connect to the database from outside the virtual network without using the private endpoint.

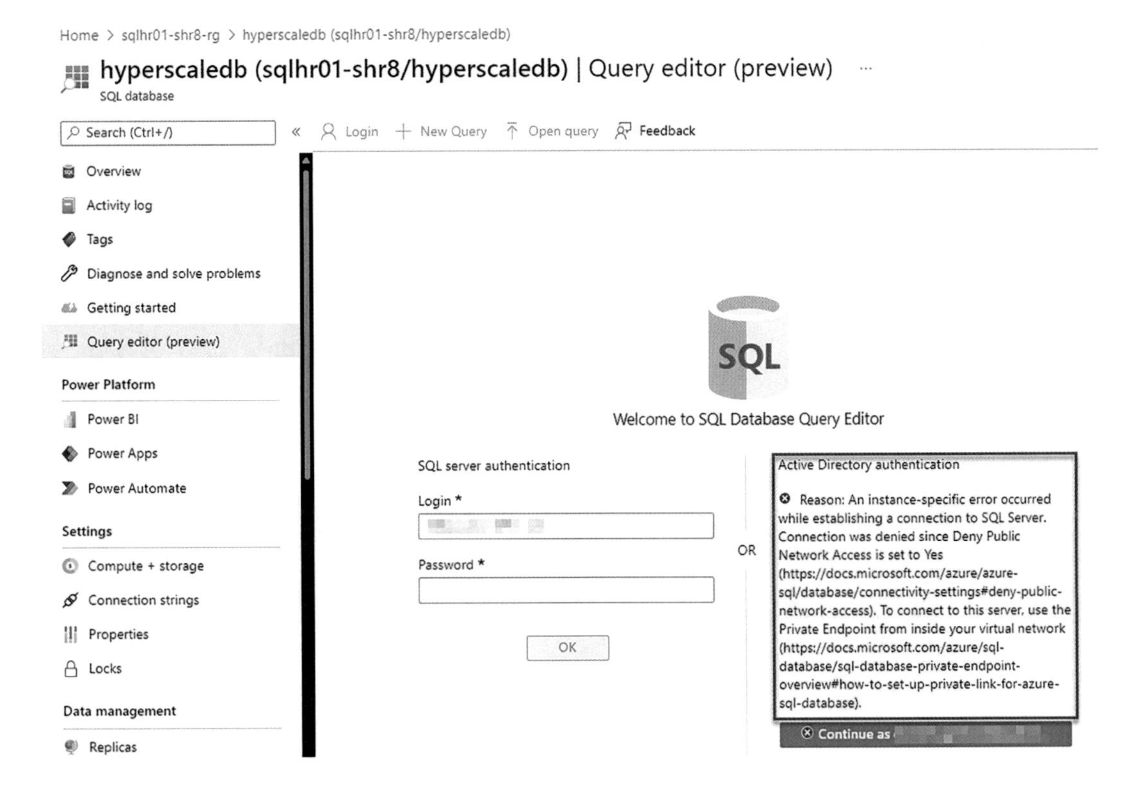

Figure 5-1. *Connection denied to a database with no public endpoint*

This won't be a problem for our application tier because it will be deployed into the `app_subnet` within the same virtual network and can use the private endpoint to connect to the database. But how would we go about administering the Hyperscale database from outside that network?

There are many ways database administrators can still connect to the database, but some of the most common approaches are the following:

- Create a management VM with the appropriate database management tools such as SQL Server Management Studio (SSMS) and Azure Data Studio in the virtual network. Deploy an Azure Bastion service in a new subnet called `AzureBastionSubnet` in the virtual network. Azure Bastion is a fully platform-managed PaaS service that allows RDP/SSH from outside your virtual network via HTTPS from the Azure Portal to the management VM. For more information on the Azure Bastion service, visit `https://aka.ms/bastion`.

Figure 5-2 shows a diagram of a management VM and an Azure Bastion instance in the virtual network for connecting to the database.

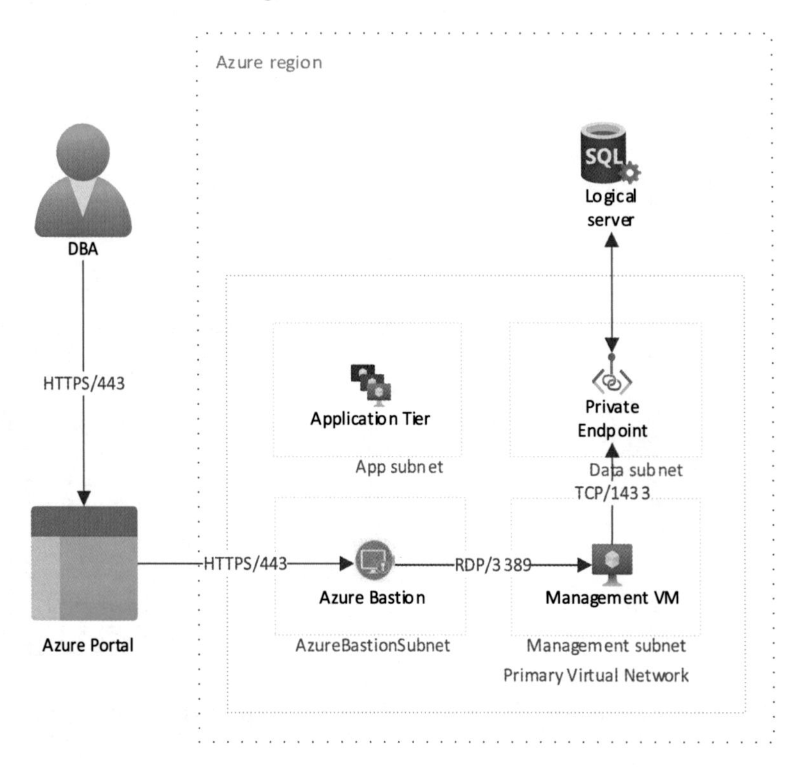

***Figure 5-2.** Azure Bastion and management VM in the VNet*

- Create a management VM in the virtual network but also provide it with a public IP address and allow RDP/3388 and/or SSH/22 traffic to it. The management VM will contain the management tools you need to administer the database such as SQL Server Management Studio and Azure Data Studio.
- Connect your own management machine to the virtual network using a point-to-site VPN Gateway or Azure Virtual WAN service.
- Connect your own management machine to an on-premises network that is connected to the Azure virtual network via an ExpressRoute, site-to-site VPN Gateway, or Azure Virtual WAN service.

The best method for your environment will depend on how your networking is configured and the type of controls your technology and security teams place on it. For the remainder of this chapter, we will be using the management VM with Azure Bastion. This is a good method to use if you can't connect your own management machine directly to the virtual network that contains the private endpoint to your logical server.

Deploying a Management VM and Azure Bastion

Deploying the management VM and Azure Bastion can be done simply in the Azure Portal, but you can use whatever method you choose. The following process assumes you have a fundamental knowledge of how to deploy a virtual machine in Azure and how to configure subnets:

1. In the Azure Portal, navigate to the `sqlhr01-<suffix>-vnet` resource, select the Subnets blade, and add two new subnets.

    - `management-subnet` with a subnet address range of `10.0.3.0/24`
    - `AzureBastionSubnet` with a subnet address range of `10.0.4.0/24`

    Figure 5-3 shows the list of subnets after deploying the new subnet.

Name ↑↓	IPv4 ↑↓	IPv6 ↑↓	Available IPs ↑↓
app_subnet	10.0.1.0/24	-	251
data_subnet	10.0.2.0/24	-	250
management_subnet	10.0.3.0/24	-	251
AzureBastionSubnet	10.0.4.0/24	-	251

Figure 5-3. *Primary region virtual network with new subnets*

2. Create a new Azure virtual machine using an appropriate operating system that will support your choice of tooling. In our case, we are choosing Windows Server 2022 Datacenter: Azure Edition - Gen2. Figure 5-4 shows the virtual machine configuration in the Azure Portal.

Project details

Select the subscription to manage deployed resources and costs. Use resource groups like folders to organize and manage all your resources.

Subscription * ⓘ Demo

Resource group * ⓘ sqlhr01-shr8-rg

Create new

Instance details

Virtual machine name * ⓘ sqlhr01-shr8-mgmt

Region * ⓘ (US) East US

Availability options ⓘ No infrastructure redundancy required

Security type ⓘ Standard

Image * ⓘ Windows Server 2022 Datacenter: Azure Edition - Gen2

See all images | Configure VM generation

Run with Azure Spot discount ⓘ

Size * ⓘ Standard_D2s_v3 - 2 vcpus, 8 GiB memory (NZ$105.96/month)

See all sizes

Figure 5-4. *Creating the management VM*

Tip Always consider the availability of your management infrastructure. In the case of a failure of an availability zone, your database might be highly available, but you might not be able to use your management infrastructure if it was in the failed availability zone.

---

3. The VM should be connected to `management_subnet` and should not have a public IP address attached. It should not allow any public in-bound ports as the traffic will come from the Azure Bastion service, not the Internet. It also doesn't need an Azure Load Balancer. Figure 5-5 shows the network configuration for the management virtual machine.

When creating a virtual machine, a network interface will be created for you.

Virtual network * sqlhr01-shr8-vnet
Create new

Subnet * management_subnet (10.0.3.0/24)
Manage subnet configuration

Public IP None
Create new

NIC network security group None / Basic / Advanced

Public inbound ports * None / Allow selected ports

Select inbound ports Select one or more ports

Figure 5-5. *The management VM networking configuration*

We will leave the remaining VM configuration up to you, but you could consider enabling an auto-shutdown policy to reduce costs when the management VM is not in use. You should also enable automatic guest OS patching if possible.

Next, we need to create an Azure Bastion that can be used to access the management VM. To create this in the Azure Portal, follow these steps:

1. Create a new Azure Bastion service by selecting "Create a resource" in the Azure Portal and locating the Azure Bastion listing.

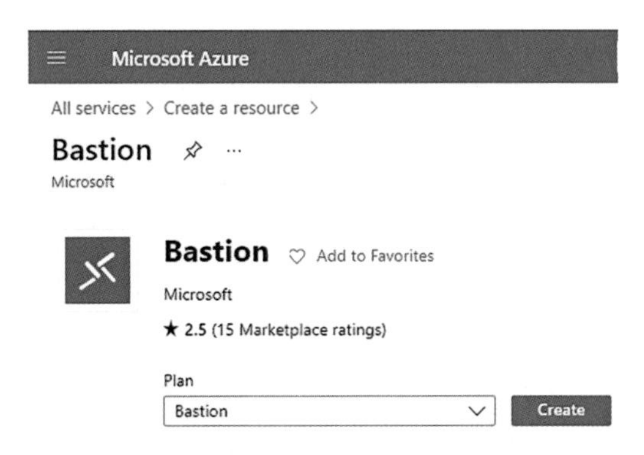

2. Configure the Azure Bastion service in the same region as your primary virtual network. You should select the primary virtual network sqlhr01-<suffix>-vnet subnet AzureBastionSubnet to connect the Azure Bastion to. We selected the basic SKU to reduce costs, but for a production environment you might want to use standard. Figure 5-6 shows the configuration of the Azure Bastion service we are using.

Project details

Subscription * Demo

Resource group * sqlhr01-shr8-rg
Create new

Instance details

Name * sqlhr01-shr-bastion

Region * East US

Tier * Basic

Instance count 2

Configure virtual networks

Virtual network * sqlhr01-shr8-vnet
Create new

Subnet * AzureBastionSubnet (10.0.4.0/24)
Manage subnet configuration

Public IP address

Public IP address * Create new Use existing

Public IP address name * sqlhr01-shr8-bastion-ip

Public IP address SKU Standard

Figure 5-6. *Azure Bastion configuration*

Once the management VM and the Azure Bastion has been deployed, we will have a way to connect to the Hyperscale database safely and securely from outside the virtual network.

Using the Management VM with an Azure Bastion

Now that we have a management VM set up and it is accessible via Azure Bastion, we'll go ahead and log into the management VM.

1. In the Azure Portal, navigate to the management VM resource (see Figure 5-7).

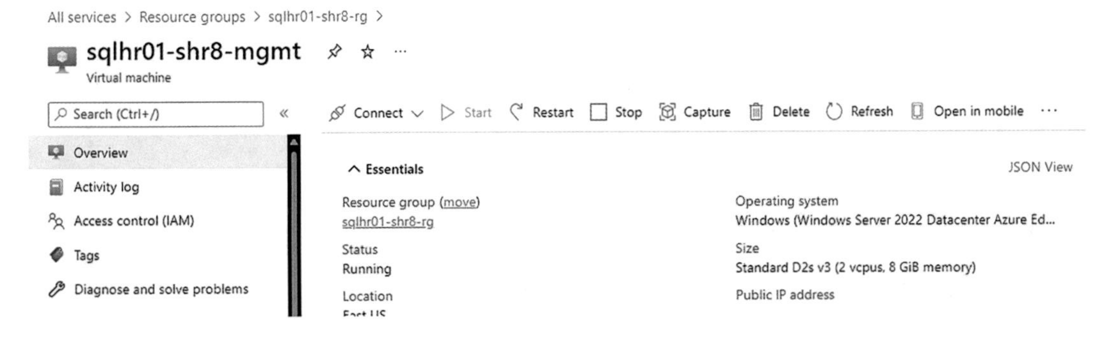

***Figure 5-7.** The management VM resource in the Azure Portal*

2. Select Bastion from the Connect drop-down list (see Figure 5-8).

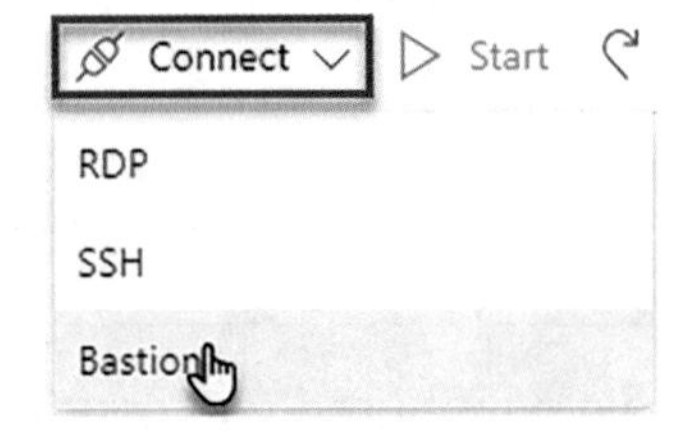

3. Enter the username and password you configured for your management VM and set Authentication Type to Password. A good idea is to add the password into a Key Vault, but this is beyond the scope of this example.

4. Select the "Open in a new browser tab" box.

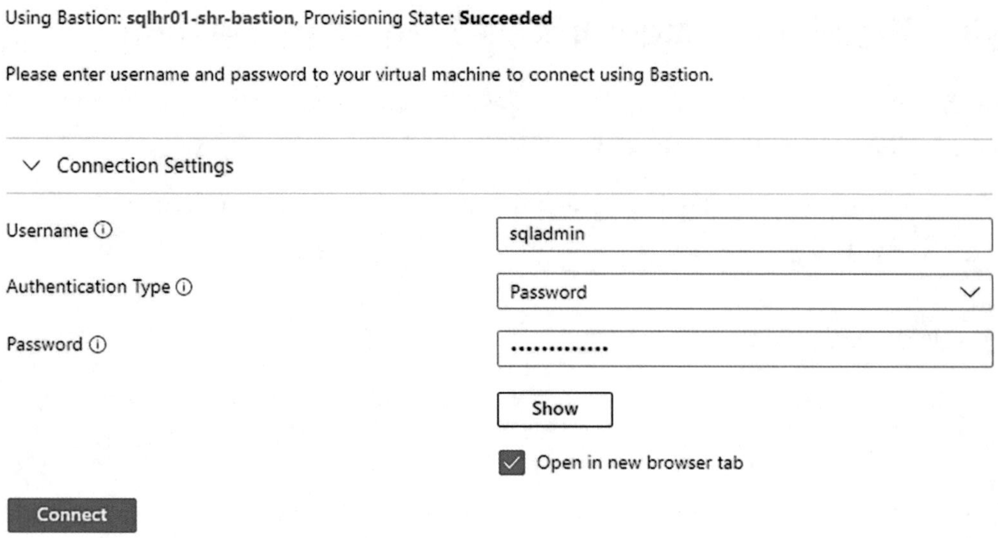

Using Bastion: **sqlhr01-shr-bastion**, Provisioning State: **Succeeded**

Please enter username and password to your virtual machine to connect using Bastion.

Connection Settings

Username: sqladmin

Authentication Type: Password

Password: •••••••••••••

Show

Open in new browser tab

Connect

Figure 5-8. *Connecting to the VM using Azure Bastion*

5. Click Connect. After a few seconds you will be logged into the management VM.

6. Install your SQL management tools onto the management VM. You need to do this only the first time you use the machine. For this example, we're using SQL Server Management Studio.

7. Start SQL Server Management Studio on the management VM.

Tip Depending on how you've configured and connected your Azure VNets, you could use the same management VM and Azure Bastion instance to manage Azure SQL Database instances in other virtual networks. However, it is important to consider what may happen in a regional outage if your management VM and Azure Bastion instance is deployed to the region that is unavailable.

---

8. Connect to the Hyperscale database using the public FQDN of the logical server, set the authentication to Azure Active Directory – Password, and use the username and password of one of the users added to the SQL Administrators group that was created earlier (see Figure 5-9).

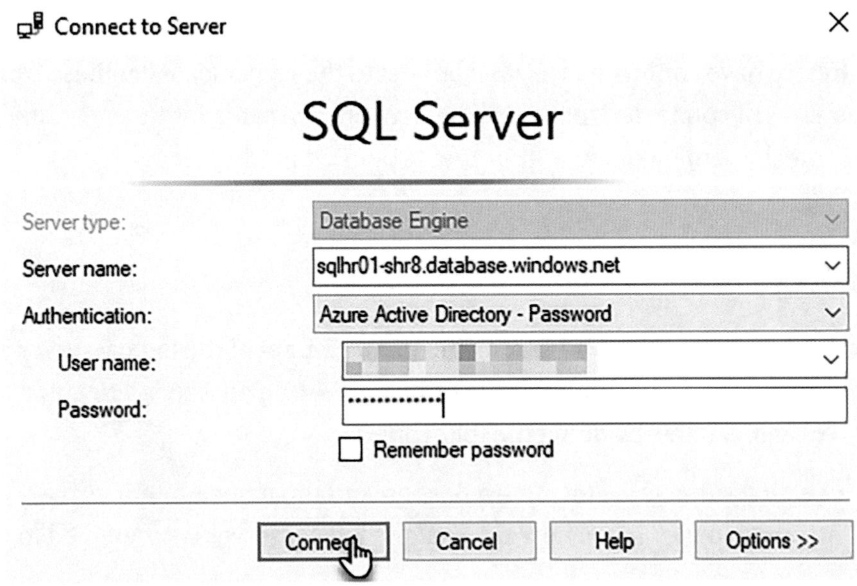

***Figure 5-9.** Connecting to the Hyperscale database using the public FQDN*

9. Once we are connected to the database, we can verify the edition of the database `hyperscaledb` by executing the following query:

```
SELECT DATABASEPROPERTYEX ('hyperscaledb', 'edition')
```

This will report `Hyperscale` if we are using the Hyperscale tier. Figure 5-10 shows the output from the command to query the database edition.

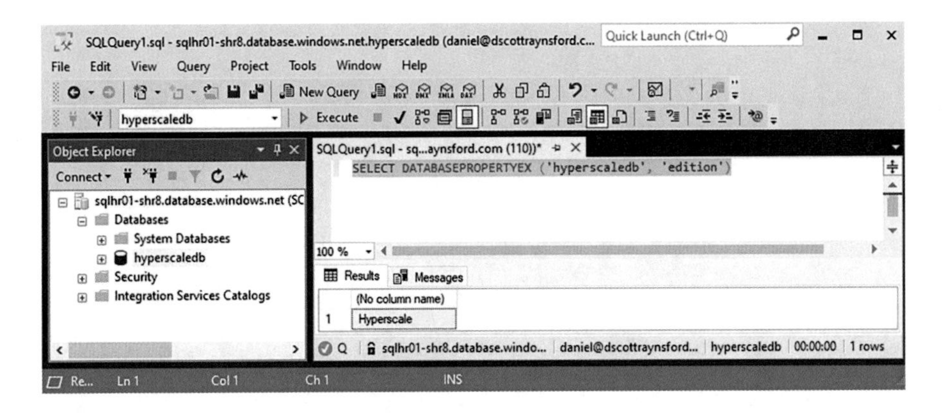

***Figure 5-10.** Checking that the database is Hyperscale*

10. Close SSMS and sign out of the management VM.

Now that we have confirmed we can connect to the Hyperscale database from within a management VM connected to the private network, we can continue to configure the Hyperscale database to meet the requirements defined in Chapter 3.

Summary

In this chapter, we learned how to enable the management of the logical server and Hyperscale database when it is accessible only from within an Azure VNet on a private endpoint. We enabled this by doing the following:

- Creating subnets for the Azure Bastion and the management virtual machine in the virtual network that the logical server is connected to
- Deploying the management virtual machine and connecting it to the virtual network
- Creating an Azure Bastion instance in the virtual network
- Connecting to the management virtual machine using the Azure Bastion, installing SQL management tools, and then connecting them to the logical server over the private endpoint

In the next chapter, we will configure transparent data encryption to use our own encryption key.

CHAPTER 6

Configuring Transparent Data Encryption to Bring Your Own Key

In the previous two chapters, we deployed a Hyperscale database into a virtual network using a private endpoint and learned how to manage it using a management VM and Azure Bastion. In this chapter, we will continue to satisfy the nonfunctional requirements laid out in Chapter 4, by configuring transparent data encryption to use a key managed by our organization.

There are many reasons why we might choose to manage our own encryption key, but some of the most common reasons are the following:

- To meet internal or external compliance requirements
- To store the encryption keys in a hardware security module (HSM), which may also be a compliance requirement
- To centralize and control encryption key management
- To monitor the access and use of the encryption keys

But, whatever your reasons for needing to manage your own encryption key, this chapter will show you how.

Z. Barać and D. Scott-Raynsford, *Azure SQL Hyperscale Revealed*,
https://doi.org/10.1007/978-1-4842-9225-9_6

Enabling Customer-Managed Key Transparent Data Encryption

Many users of Azure SQL Database will not need to manage their own encryption keys, but understanding the process of enabling this demonstrates the use of a managed identity on a Hyperscale database to interact with other services in Azure. Enabling managed identities on Azure services that support them is an important technique that increases your Azure security posture and should be used wherever possible.

As part of the requirements for this Hyperscale environment, we needed to provide transparent data encryption (TDE) with a customer-managed encryption key, which is also known as the TDE protector. The TDE protector must be stored in an Azure Key Vault or Azure Key Vault Managed HSM. The Hyperscale database in the logical server will need reliable access to the TDE protector in the Key Vault to stay online.

Important It is important to understand what happens if the TDE protector is inaccessible or even lost. It is also critical to ensure you're keeping a secure backup of your TDE protector key, because if it is lost, the data in the database, the log files, and any backups are irretrievably lost. You should also ensure you keep backups of your key after it is rotated. For more information, see `https://docs.microsoft.com/azure/azure-sql/database/transparent-data-encryption-byok-overview`.

The Key Vault you choose to use to store the key must have soft-delete and purge protection features enabled. This is to prevent data loss due to accidental deletion of the Key Vault or key. The Key Vault and the Hyperscale database must belong to the same Azure Active Directory tenant, because cross-tenant authorization is not supported.

Tip Even if the soft-delete and purge protection features are enabled on the Key Vault, it is still recommended that you back up the key. You should also put a delete lock on the Key Vault. For more information on backing up a key, see `https://docs.microsoft.com/azure/key-vault/general/backup`.

The steps in this chapter assume that your Key Vault is configured to use Azure role-based access control rather than the older vault access policy. If you are using a Key Vault that is configured to use vault access policy, then you can still enable customer-managed TDE, but you'll need to assign access policies to the Key Vault.

To enable a customer-managed TDE, we will need to perform the following tasks:

- Create a user-assigned managed identity that will be assigned to the logical server. The logical server will use this identity to access the key in the Azure Key Vault.
- Grant yourself the Key Vault Crypto Officer role-based access control (RBAC) role on the Azure Key Vault so that you can generate or import the key to be used as the TDE protector.
- Generate or import a key in the Azure Key Vault in the primary region `sqlhr-<suffix>-kv` to be used as the *TDE protector*. There isn't a Key Vault in the failover region because it has a built-in geo replica. However, best practice for a production environment is to create a failover Key Vault and back up the key to it.
- Grant the user-assigned managed identity the Key Vault Crypto Service Encryption User RBAC role on the key in the Azure Key Vault. We can grant this role to the whole Key Vault if we want, but it is better to limit the permissions scope where possible.
- Assign the user-assigned managed identity to the logical server and set it as the primary identity. The logical server will be able to use this identity to access resources in Azure, such as the Azure Key Vault.

Note We are choosing to use a user-assigned managed identity rather than a system-assigned managed identity so that it can be used by both the primary logical server and the failover logical server. If we used a system-assigned managed identity, then each logical server would have a different one, and we would have to grant permissions on the Key Vault for each. Using a user-assigned managed identity also decouples the life cycle of the identity from the logical server, allowing them to be deployed and managed by separate teams, helping to enable separation of duties.

- Select the key we generate in the Key Vault to use for transparent data encryption in the logical server and enable customer-managed TDE.

Now we will step through each of these tasks in detail.

Creating a User-Assigned Managed Identity

The first task is to create the user-assigned managed identity that the logical server will use to access the key in the Azure Key Vault. To create the user-assigned managed identity, we perform these steps:

1. In the Azure Portal, start by adding a new resource.

2. In the "Create a resource" search box, enter **user assigned managed identity** and press Enter.

3. Click Create on the User Assigned Managed Identity marketplace entry and select User Assigned Managed Identity.

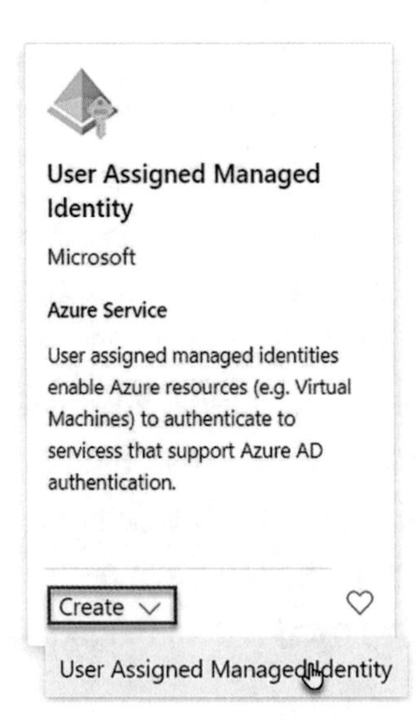

4. Set the resource group to the primary region resource group, `sqlhr01-<suffix>-rg`. Set the region to your primary region. In this case, it is East US. Enter a name for the identity, such as **sqlhr01-<suffix>-umi**.

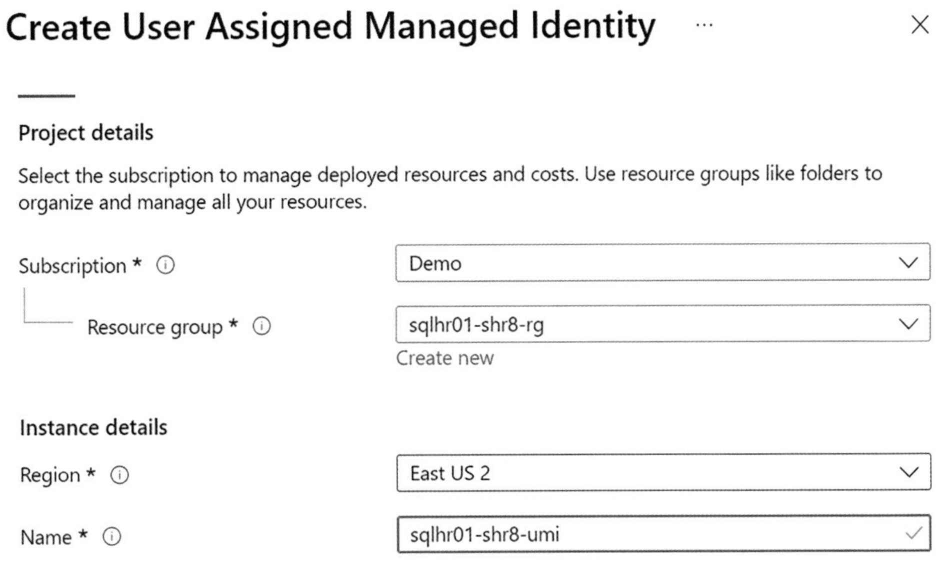

Create User Assigned Managed Identity ... ×

Project details

Select the subscription to manage deployed resources and costs. Use resource groups like folders to organize and manage all your resources.

Subscription * ⓘ Demo

Resource group * ⓘ sqlhr01-shr8-rg

Create new

Instance details

Region * ⓘ East US 2

Name * ⓘ sqlhr01-shr8-umi

5. Click Next.

6. You might want to add an `Environment` tag and set it to `SQL Hyperscale Revealed demo`. Then click "Review + create."

7. Click Create.

After a few moments, the deployment will start and should take only a few seconds to create the user-assigned managed identity.

Granting the Key Vault Crypto Officer Role to a User

Next, we will need to grant ourselves the Key Vault Crypto Officer RBAC role on the Key Vault. This will allow us to generate or import the encryption key.

Tip Once the configuration of the TDE protector has been completed on both the primary and failover logical servers, you will not require this RBAC role, except if you must perform some other tasks on the key. Therefore, you should remove this RBAC role from your account once the TDE configuration process is complete and the key has been assigned to the logical server. It is strongly recommended to use Azure AD Privileged Identity Management to perform a temporary elevation of your user account to this RBAC role, which will provide even greater assurance and protection to your key and Key Vault. See `https://aka.ms/aadpim` for more information on this using role elevation.

To grant the RBAC role, take the following steps:

1. Open the Azure Portal in your web browser.
2. Navigate to the Azure Key Vault `sqlhr-<suffix>-kv`.
3. Select the "Access control (IAM)" blade.
4. Click Add and select "Add role assignment."
5. Select Key Vault Crypto Officer and click Next.
6. Set "Assign access" to "User, group, or service principal" and click "Select members" and add your user account. Figure 6-1 shows how the new RBAC role should be configured. Click Next.

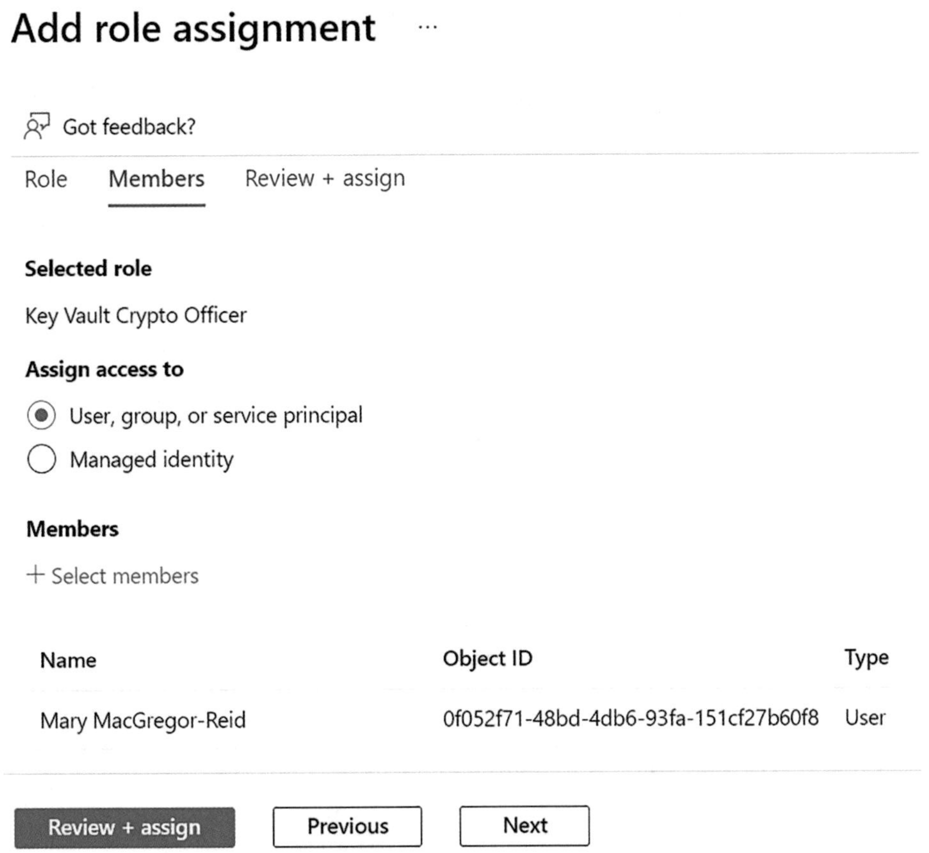

Add role assignment ...

Got feedback?

Role Members Review + assign

Selected role

Key Vault Crypto Officer

Assign access to

(●) User, group, or service principal

( ) Managed identity

Members

+ Select members

Name	Object ID	Type
Mary MacGregor-Reid	0f052f71-48bd-4db6-93fa-151cf27b60f8	User

Review + assign Previous Next

***Figure 6-1.** Assigning the Key Vault Crypto Officer role*

7. Click "Review + assign."
8. After a few seconds, an "Added role assignment message" will appear.

Now that we have permissions on the Key Vault, we can go ahead and generate or import the key.

Generating a Key in the Azure Key Vault

There are two ways that the key that will be used as the TDE protector can be added to the Key Vault. Either it can be generated by the Key Vault or it can be generated externally and imported. Allowing the key to be generated by the Key Vault will mean that you do not have to handle it, which reduces the risk that the key will become compromised. However, you may need to copy the key to another Key Vault or some other location as a backup, so it is acceptable to create the key on another machine and import it. In our case, we are going to generate it in the Key Vault.

To generate the key in the Key Vault, follow these steps:

1. In the Azure Portal, navigate to the Key Vault `sqlhr-<suffix>-kv`.
2. Select the Keys blade.
3. Click Generate/Import to generate the new key.
4. Enter the name for the key. Set it something to indicate its purpose. For example, we will use `sqlhr-<suffix>-tdeprotector`.
5. Set the key type to *RSA*. EC is not supported.
6. Specify a key size of 2048 or 3072.
7. If you choose to specify an activation date, it must be in the past. If you set the expiration date, it must be in the future. It is not necessary to specify these values.
8. Make sure Enabled is set to Yes. Disabling the key will prevent it from being used by the logical server.
9. Using an automatic key rotation policy is supported by Hyperscale. Key rotation takes only a few seconds to complete and doesn't add load to your database or logical server. This is optional. Figure 6-2 shows the "Create a key" page with the key rotation policy shown.

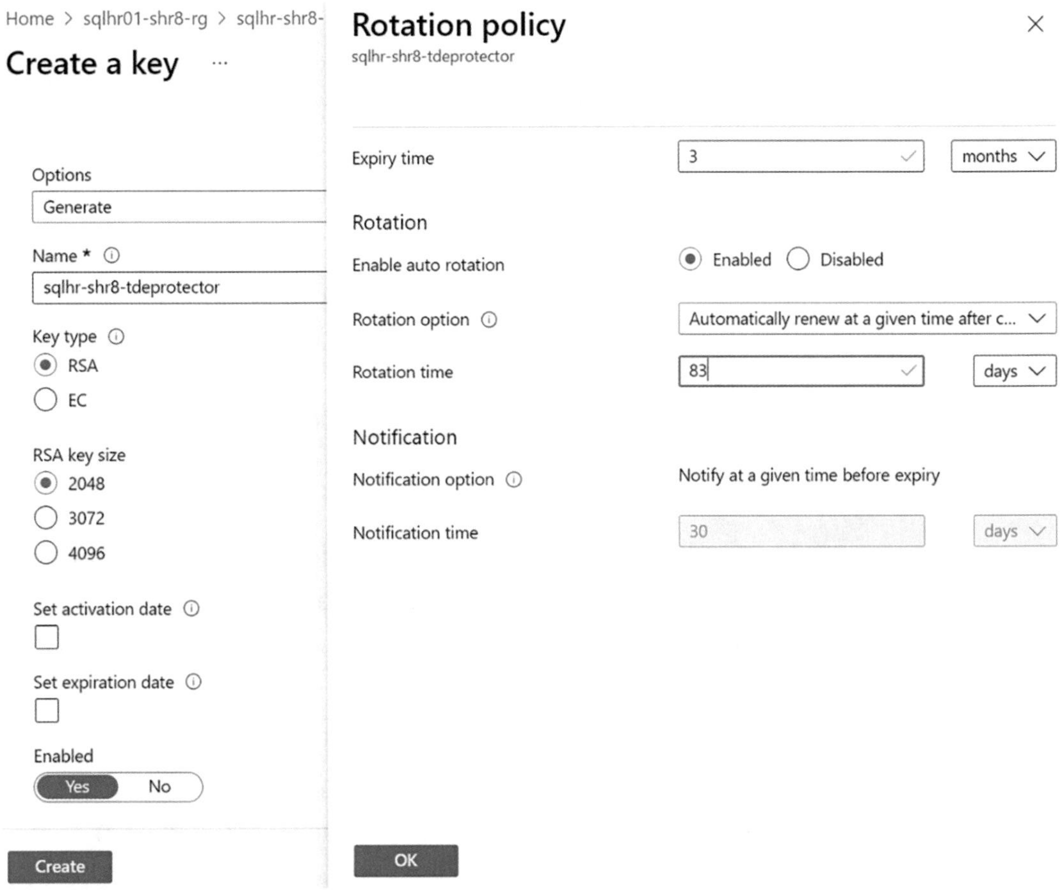

Figure 6-2. Settings for generating a new TDE protector key

10. Click Create to generate the key.
11. After the new key has been generated, click it.
12. Click Download Backup to download a copy of this key and securely store it. The backup is encrypted and can be restored only to a Key Vault in this subscription.

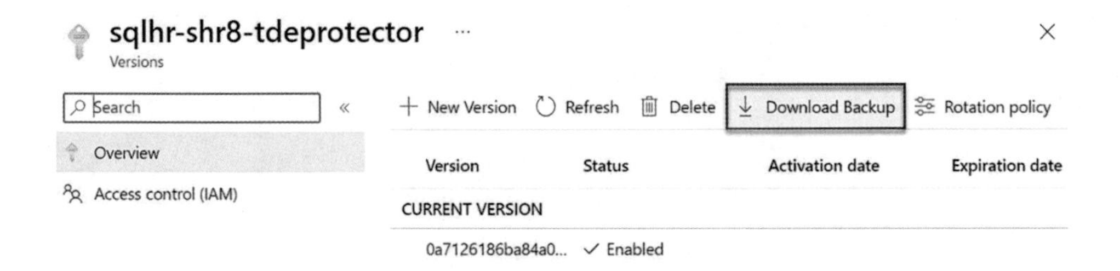

Now that the key has been generated, we will need to grant access to it by the user-assigned managed identity.

Granting Access to the Key by the User-Assigned Managed Identity

The user-assigned managed identity that will be used by the logical server will need to have the Key Vault Crypto Service Encryption User RBAC role granted on the key or the Key Vault. The best practice is always to limit access as much as possible, so we will grant the role on the key only.

Note These instructions assume that the Key Vault is configured in Azure role-based access control mode. You can use vault access policy mode if you want, but you will need to create a policy to grant the identity `get`, `wrapKey`, and `unwrapKey` permissions. It is recommended to use Azure role-based access control mode if possible.

To grant the RBAC role on the key, follow these steps:

1. In the Azure Portal, navigate to the Key Vault `sqlhr-<suffix>-kv`.
2. Select the Keys blade.
3. Click the key that we generated to use as the TDE protector. This will open the overview for the key.
4. Select the Access Control (IAM) blade.
5. Click Add and select "Add role assignment."

6. Select the Key Vault Crypto Service Encryption User role and click Next.

7. Set "Assign access to" to "Managed identity" and click "Select members" and add the user-assigned managed identity we created earlier. Figure 6-3 shows how the new RBAC role should be configured. Click Next.

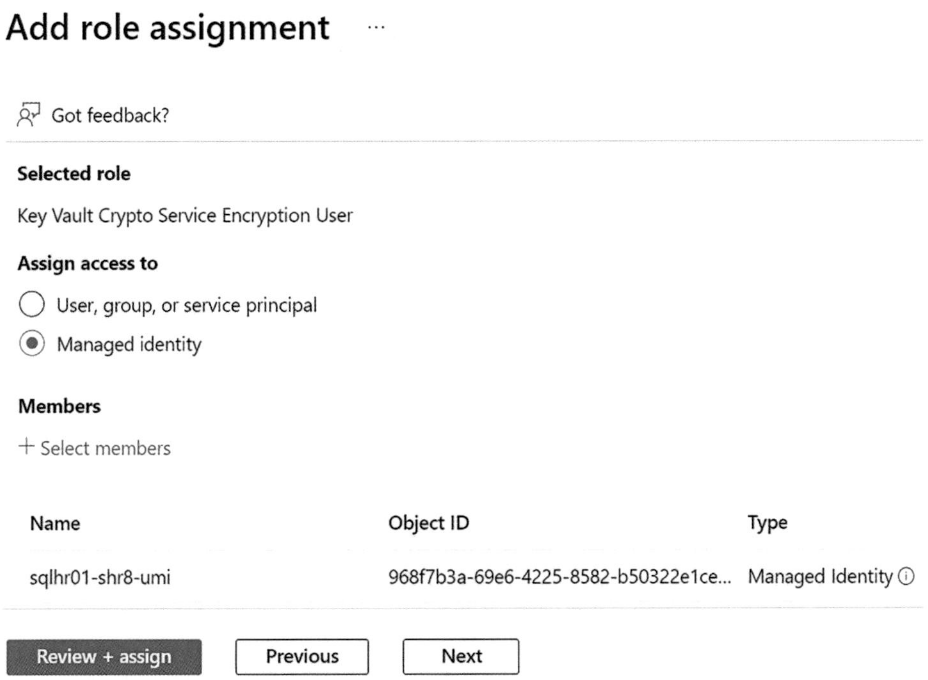

Add role assignment ...

Got feedback?

Selected role

Key Vault Crypto Service Encryption User

Assign access to

User, group, or service principal

Managed identity

Members

+ Select members

Name	Object ID	Type
sqlhr01-shr8-umi	968f7b3a-69e6-4225-8582-b50322e1ce...	Managed Identity

Review + assign Previous Next

Figure 6-3. *Assigning access to the key to the managed identity*

8. Click "Review + assign."

9. After a few seconds, an "Added role assignment message" will appear.

The user-assigned managed identity now has the appropriate permissions to this key.

Note Any Azure service that is assigned to use this managed identity will be able to perform the get, wrap key, and unwrap key operations using this key. Although the permissions granted do not permit any destructive or export activities, it is important to be careful to use the identity only with services that should access it.

Assigning the User-Assigned Managed Identity to the Logical Server

Now that the user-assigned managed identity has appropriate permissions to the key, we can assign it to the logical server. To do this, follow these steps:

1. In the Azure Portal, navigate to the logical server (the SQL Server resource) in the primary region, `sqlhr01-<suffix>`.
2. Select the Identity blade.
3. Click the Add button in the "User assigned managed identity" section.
4. Select the user-assigned managed identity that was created earlier, `sqlhr01-<suffix>-umi`, and click Add.
5. Set the primary identity to `sqlhr01-<suffix>-umi`. Figure 6-4 shows how the identity settings should look after you've configured them.

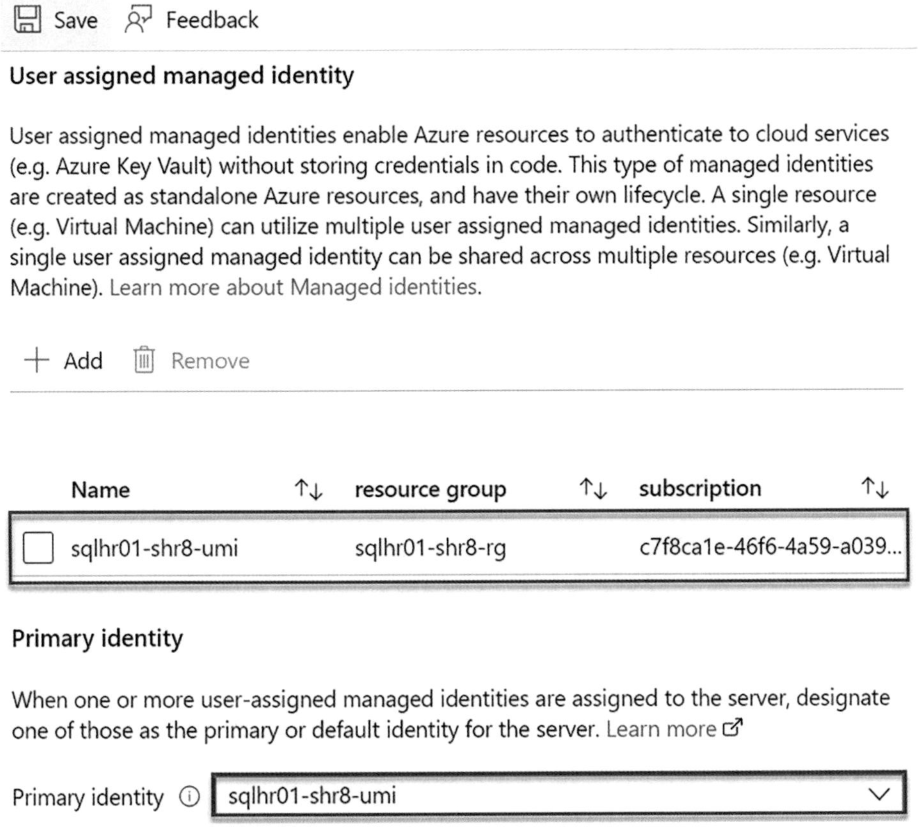

Save Feedback

User assigned managed identity

User assigned managed identities enable Azure resources to authenticate to cloud services (e.g. Azure Key Vault) without storing credentials in code. This type of managed identities are created as standalone Azure resources, and have their own lifecycle. A single resource (e.g. Virtual Machine) can utilize multiple user assigned managed identities. Similarly, a single user assigned managed identity can be shared across multiple resources (e.g. Virtual Machine). Learn more about Managed identities.

+ Add Remove

Name	resource group	subscription
sqlhr01-shr8-umi	sqlhr01-shr8-rg	c7f8ca1e-46f6-4a59-a039...

Primary identity

When one or more user-assigned managed identities are assigned to the server, designate one of those as the primary or default identity for the server. Learn more

Primary identity sqlhr01-shr8-umi

Figure 6-4. *The identity configuration on the logical server*

6. Click Save to save the identity settings.
7. A message confirming that the identity settings have been updated will be displayed after a few seconds.

Once the settings have been updated, we can now perform the final task, which is to enable the customer-managed transparent data encryption.

Enabling Customer-Managed TDE

To enable the customer-managed transparent data encryption, we will need to tell the logical server which key to use in a specified Key Vault. To do this, follow these steps:

1. In the Azure Portal, navigate to the logical server (the SQL Server resource) in the primary region, sqlhr01-<suffix>. You should already be here if you've performed the previous steps.
2. Select the "Transparent data encryption" blade.
3. Set TDE to customer-managed key and specify the key selection method as "Select a key."
4. Click the "Change key" button. The "Select a key" form will be shown.
5. Specify key store type as "Key vault."
6. Specify the Key Vault with our key. In our case, it's sqlhr-<suffix>-kv.
7. Set the key to the key we generated in the Key Vault. It was called sqlhr-shr8-tdeprotector.
8. We don't need to specify a version of the key to use; just use the current version.
9. Click Select to confirm this is the key we will be using.
10. Select the "Make this key the default TDE protector" box. Figure 6-5 shows the transparent data encryption configuration after we've enabled a customer-managed key.
11. If you enabled automatic key rotation policy on the TDE protector key in the Key Vault or if you're planning on manually rotating the key, select the "Auto-rotate key". Figure 6-5 shows how to configure the customer-managed TDE key to retrieve it from a Key Vault.

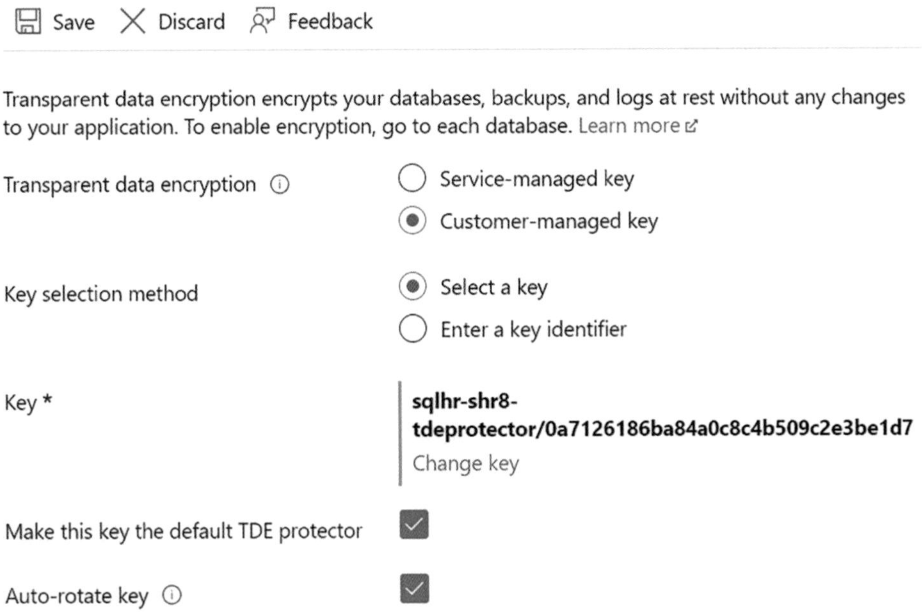

Save Discard Feedback

Transparent data encryption encrypts your databases, backups, and logs at rest without any changes to your application. To enable encryption, go to each database. Learn more

Transparent data encryption: Service-managed key / Customer-managed key (selected)

Key selection method: Select a key (selected) / Enter a key identifier

Key *: sqlhr-shr8-tdeprotector/0a7126186ba84a0c8c4b509c2e3be1d7 — Change key

Make this key the default TDE protector: ☑

Auto-rotate key: ☑

Figure 6-5. *The transparent data encryption configuration*

12. Click Save.
13. After a few seconds, we will receive confirmation that the settings have changed.

We have now successfully configured our Hyperscale database to use a key that we manage for TDE. This may help us meet certain compliance requirements or centralize the management and monitor the use of TDE protectors, but it does increase the work we need to do to deploy and operate the service.

Summary

By default, all data is encrypted at rest in Azure SQL Database instances, including Hyperscale. This includes the data files, the log files, and the backups. These are normally encrypted by a key, called the TDE protector, that the service manages. In this chapter, we learned how to configure a customer-managed key to use as the TDE protector for encryption.

To enable the use of a customer-managed key for transparent data encryption, we covered how to do the following:

- Create a user-assigned managed identity for the logical server to access the Azure Key Vault containing the key protector
- Create a key in an Azure Key Vault to be used as the key protector
- Add the Key Vault Crypto Service Encryption User role to the user-assigned managed identity for the key protector
- Assign the user-assigned managed identity to the logical server
- Enable customer-managed key encryption on the logical server

In the next chapter, we will look at disaster recovery by adding a geo-replication.

CHAPTER 7

Enabling Geo-replication for Disaster Recovery

In the previous few chapters, we defined a comprehensive set of nonfunctional requirements for a Hyperscale database environment and then set about deploying it. This included creating the starting environment, deploying the Hyperscale database into the virtual network using a private endpoint, and enabling transparent data encryption with a customer-managed key. We also took a look at how DBAs can manage the Hyperscale database from within the virtual network.

In this chapter, we will demonstrate how to configure the environment to continue to function in case the primary region is inaccessible. To meet this requirement, we will enable a geo replica of our primary Hyperscale database in a region we call the *failover region* (also sometimes called the *secondary region*). The failover region will often be the Azure region that is paired with the primary region, but in some cases it may need to be a different region.

There are several reasons for not selecting the Azure paired region for your failover region.

- There is no paired region. For example, Brazil South does not currently have a paired region so is usually paired with a US region.
- The paired region does not support the features you require. For example, if you require Hyperscale high-availability replicas with availability zone support and the paired region is not availability zone enabled.

Important It is important to consider data residency requirements if you are using a paired region that is in a different geography.

Z. Barać and D. Scott-Raynsford, *Azure SQL Hyperscale Revealed*,
https://doi.org/10.1007/978-1-4842-9225-9_7

Deploying a Hyperscale Geo Replica

A *geo replica* is an asynchronously replicated copy of the primary database that is usually deployed for the purposes of protecting against a regional outage. It can also be used as an active read region. When a geo replica is deployed, another logical server is created in a region that is chosen to be the failover region. This region is usually the one that is paired with the primary region. The Hyperscale database in the primary region is then configured to replicate to a Hyperscale database with the same name in the secondary region.

Tip Hyperscale also supports named replicas. These are like high-availability replicas in that they share the same page servers as the primary replica. This means there is no need to synchronize data to the named replica but also means they can be placed only in the same region as the primary replica. They can be connected to the same logical server as the primary replica but will need to have a different database name. Alternatively, they can be connected to a different logical server in the primary region, in which case they can have the same database name. The main purpose of these is to enable read scale-out and to prevent read interruption if the primary replica is scaled up or down.

To configure a geo replica, we will need to perform the following tasks:

1. Create a new logical server in the failover region.
2. Connect the logical server in the failover region to the virtual network in that region.
3. If transparent data encryption on the logical server in the primary region is configured to use a customer-managed key, then it must be accessible using the managed identity assigned to the failover logical server.

Important If the logical server in the failover region will use a different Key Vault to store the TDE protector than that of the logical server in the primary region, then the key will need to be copied into the failover Key Vault. A process to keep the key synchronized between the primary and the failover Key Vault will need to be set up.

---

4. Begin geo-replication of the primary Hyperscale database on the failover logical server.

We will now review these tasks in detail.

Creating a Logical Server in the Failover Region

The first step is to create another logical server in the failover region and connect it to the virtual network that is also in that region. However, unlike with the steps for the primary logical server where we created the Hyperscale database at the same time, for the failover region we are going to create only the logical server. We will enable the replica once the failover logical server has been created.

To create a logical server without a database in the failover region, follow these steps:

1. Open your browser and navigate to `https://portal.azure.com`.

2. Start by creating a new resource.

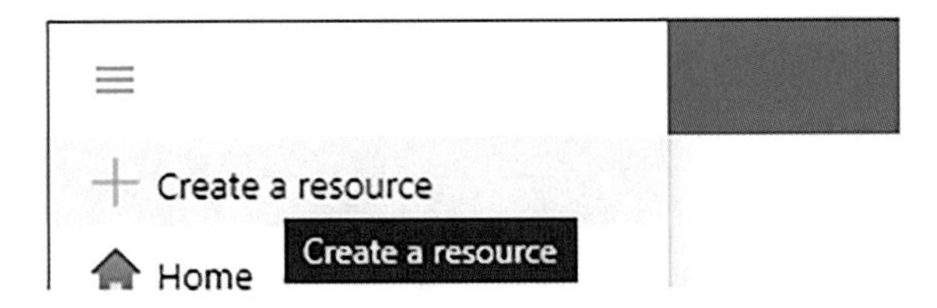

3. Enter **logical server** in the "Create a resource" search box and select the "SQL server (logical server)" entry.

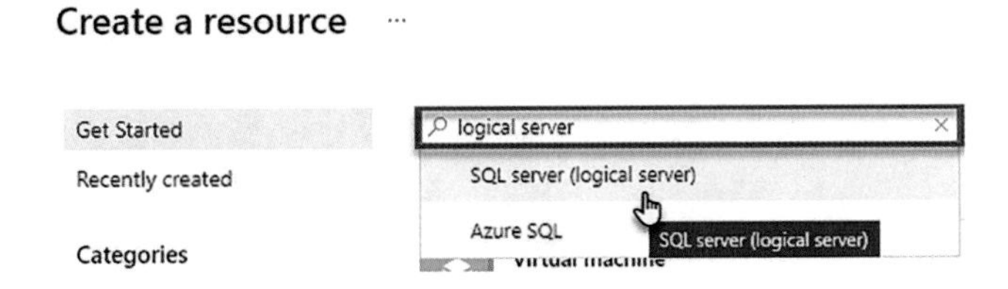

4. The "SQL server (logical server) marketplace" entry will be displayed. Click Create to begin the creation of the resource.

5. Set the resource group to the one created in the failover region. It should be `sqlhr02-<suffix>-rg`. Enter **sqlhr02-<suffix>** for the server's name. The location should be set to the region we

selected for our failover region, which was West US 3 in our case. Configure the Authentication method and set the Azure AD admin just as we did for the primary logical server. Figure 7-1 shows an example of how the configuration of the failover logical server will look.

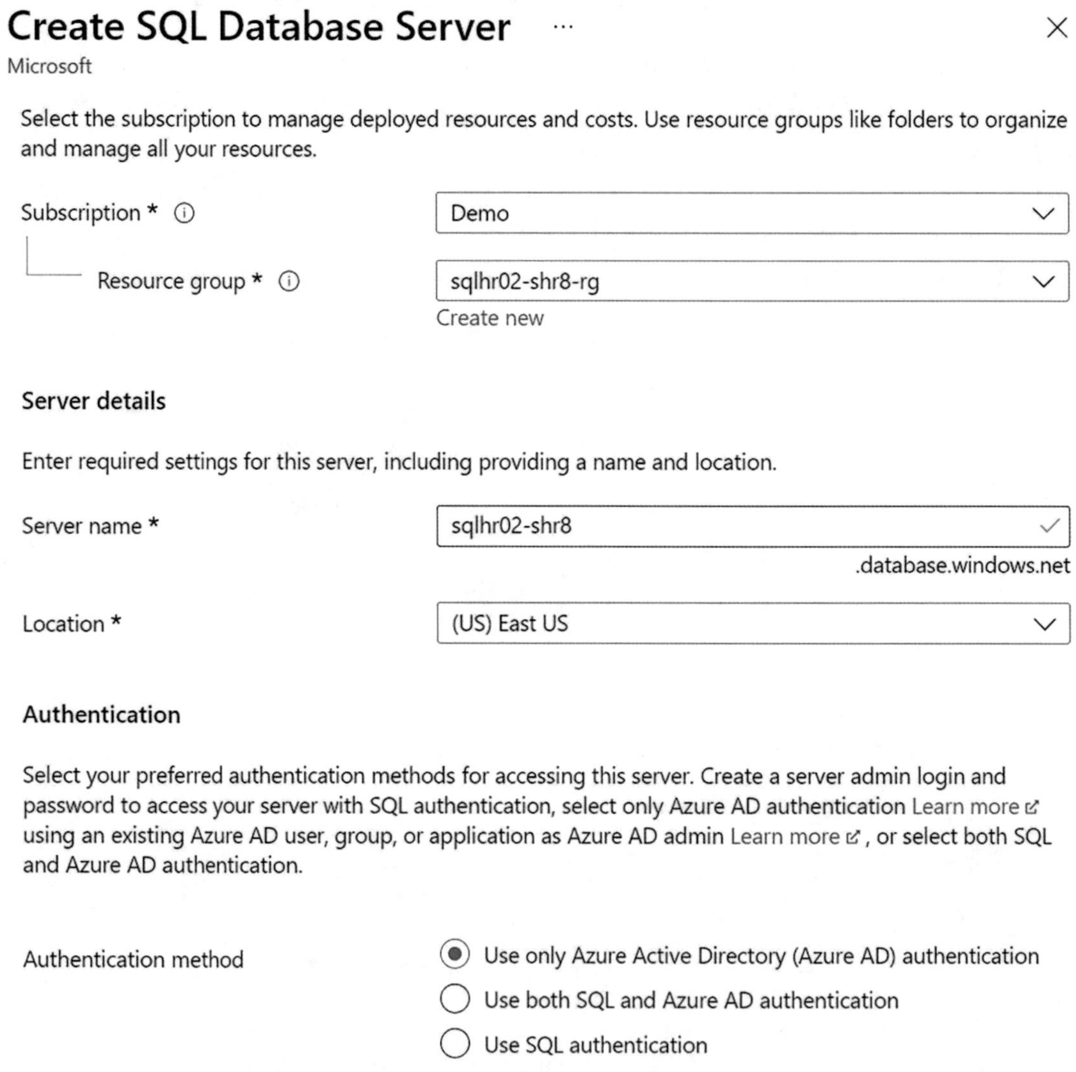

Create SQL Database Server ...

Microsoft

Select the subscription to manage deployed resources and costs. Use resource groups like folders to organize and manage all your resources.

Subscription * Demo

Resource group * sqlhr02-shr8-rg

Create new

Server details

Enter required settings for this server, including providing a name and location.

Server name * sqlhr02-shr8

.database.windows.net

Location * (US) East US

Authentication

Select your preferred authentication methods for accessing this server. Create a server admin login and password to access your server with SQL authentication, select only Azure AD authentication Learn more using an existing Azure AD user, group, or application as Azure AD admin Learn more, or select both SQL and Azure AD authentication.

Authentication method

- (selected) Use only Azure Active Directory (Azure AD) authentication
- Use both SQL and Azure AD authentication
- Use SQL authentication

Figure 7-1. *Configuring the failover logical server*

6. Click Next to configure networking on the logical server. At the time of writing, it is not possible to join the logical server to a virtual network during the initial deployment of just a logical server. Instead, we will do this later in this process. So, click Next.

7. You should specify Not Now when asked to Enable Microsoft Defender for SQL, because we are going to do this in the next section of this chapter. However, in a real-world process, we would enable this during deployment by selecting "Start free trial." Click Next.

8. You might add an `Environment` tag with the value `SQL Hyperscale Revealed demo` so that you can more easily track the costs of this demonstration environment. Click Next.

9. Review the final settings for the logical server and click Create.

After a few minutes, the new logical server in the failover region will be deployed, and we can connect it to the virtual network.

Connecting a Logical Server to a Virtual Network with Private Link

When we deployed the Hyperscale database in the primary region, we included the deployment of the logical server. This method of deployment allowed the logical server to be integrated into the virtual network at deployment time. However, the Azure Portal user interface for only deploying a logical server does not currently provide a method of connecting the virtual network at the initial deployment stage. Therefore, we must make one extra configuration step to connect the logical server to the virtual network.

Tip This is the same process we would use to connect any logical server to a virtual network, so it is useful to be familiar with it. It is also similar to connecting any supported Azure service to a virtual network using Azure Private Link.

To connect an existing logical server to a virtual network, follow these steps:

1. In the Azure Portal, navigate to the logical server (the SQL Server resource) in the failover region, `sqlhr02-<suffix>`.

2. Select the Networking blade.
3. On the "Public access" tab, set "Public network access" to disable it as we do not want this server to be accessible on its public IP address.
4. Click Save.
5. Select the "Private access" tab and click "Create a private endpoint." This will begin the process of creating a private endpoint using Azure Private Link.
6. Set the resource group to the resource group where we will put the private endpoint. In this example, it should be the failover resource group, `sqlhr02-<suffix>-rg`.
7. Enter a name for the private endpoint instance. For this example, `sqlhr02-<suffix>-pe` will be used. The network interface name should be set to indicate that it belongs to this private endpoint, for example, `sqlhr02-<suffix>-pe-nic`.
8. Set the region to the location of the virtual network that you want to connect the private endpoint to. In this case, it will be the failover region, West US 3. Figure 7-2 shows how the basic settings for the private endpoint should be configured.

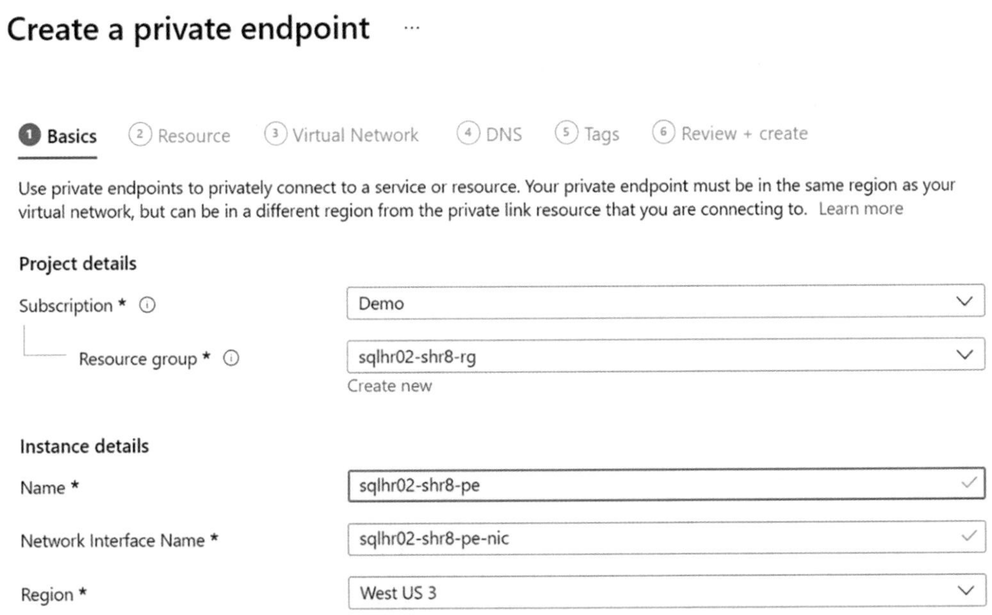

Create a private endpoint ...

1 Basics 2 Resource 3 Virtual Network 4 DNS 5 Tags 6 Review + create

Use private endpoints to privately connect to a service or resource. Your private endpoint must be in the same region as your virtual network, but can be in a different region from the private link resource that you are connecting to. Learn more

Project details

Subscription * : Demo

Resource group * : sqlhr02-shr8-rg

Create new

Instance details

Name * : sqlhr02-shr8-pe

Network Interface Name * : sqlhr02-shr8-pe-nic

Region * : West US 3

***Figure 7-2.** Basic settings for creating a new private endpoint*

9. Click Next.

10. Set the target subresource to sqlServer.

Note Azure Private Link will automatically select the logical server we started creating this private endpoint from. If we had chosen to create a private endpoint from elsewhere in the portal, we would need to specify the target resource manually.

---

11. Click Next.

12. Set the virtual network to the subnet in the virtual network you want to connect the logical server to. In this case, it is the data_subnet in the failover virtual network sqlhr02-<suffix>-vnet.

13. Configure the private IP address so that it will be dynamically allocated.

14. We can ignore the application security group as it is not required for the logical server in this example. Figure 7-3 shows the virtual network configuration for the private endpoint. Click Next.

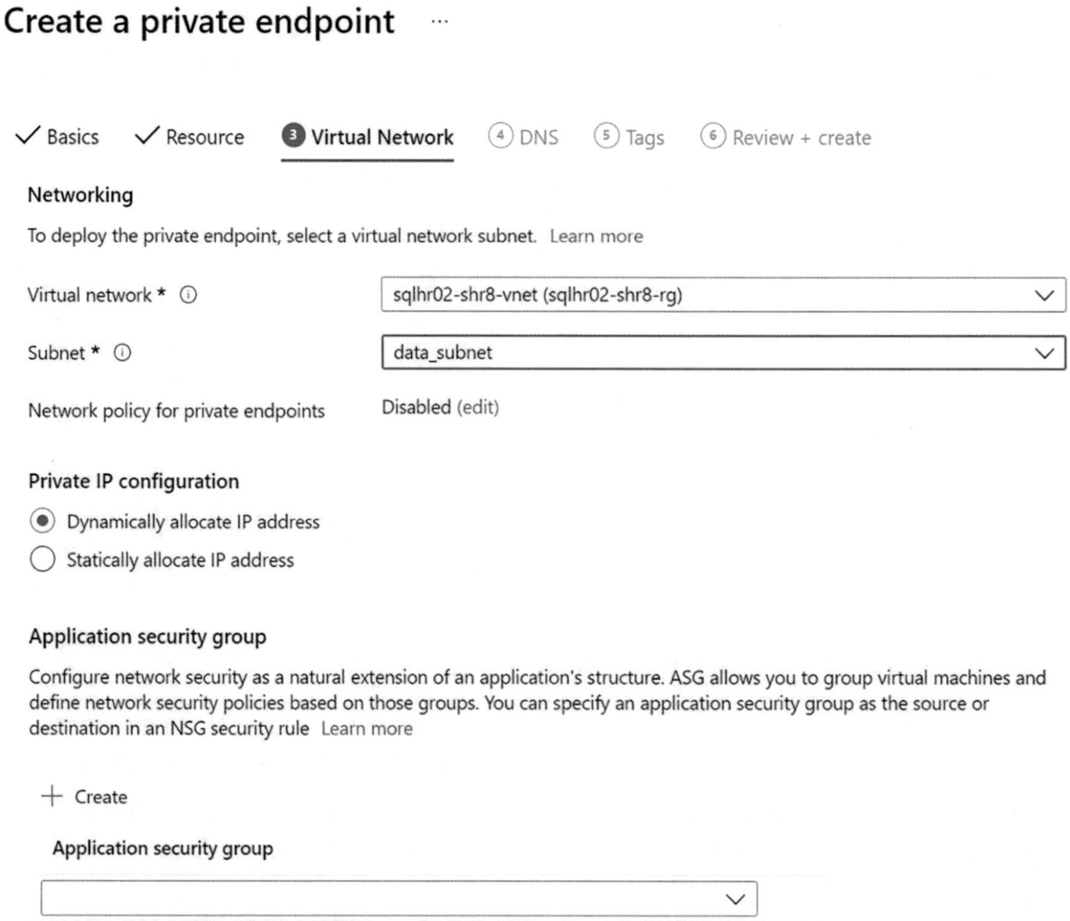

Figure 7-3. *The virtual network to connect the private endpoint to*

15. Ensure "Integration with private DNS zone" is set to Yes so that a private DNS zone is created within the resource group containing the private link. This will also associate the DNS zone with the virtual network. In this case, it should be in the failover resource group `sqlhr02-<suffix>-rg`. You may notice that a new private DNS zone called `privatelink.database.windows.net` is created within the same resource group. If one with that name already exists in the resource group, then a new one won't be created.

16. Enter any tags to assign to these resources and click Next.
17. Click Create.
18. After a few seconds, the deployment will start and should be completed within a minute or two. Once it is completed, you will see three new resources deployed to your resource group. Figure 7-4 shows the resources that make up the private endpoint for the logical server.

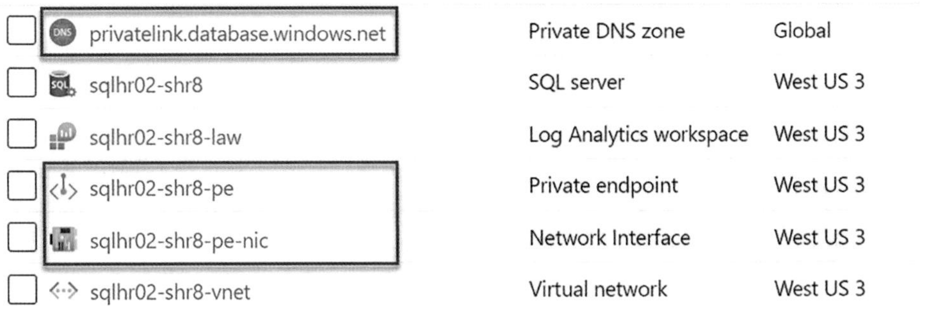

Name	Type	Location
privatelink.database.windows.net	Private DNS zone	Global
sqlhr02-shr8	SQL server	West US 3
sqlhr02-shr8-law	Log Analytics workspace	West US 3
sqlhr02-shr8-pe	Private endpoint	West US 3
sqlhr02-shr8-pe-nic	Network Interface	West US 3
sqlhr02-shr8-vnet	Virtual network	West US 3

Figure 7-4. *Resources deployed as part of the private endpoint*

19. Navigating back to the “Private access” tab of the Networking blade on the logical server will show that it is now connected to the private endpoint that is connected to the virtual network. Figure 7-5 shows the private endpoint connection for the logical server.

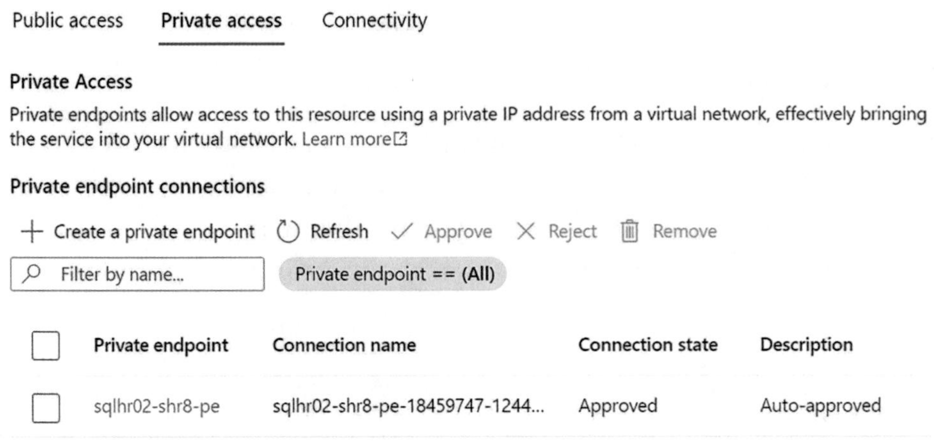

Public access | Private access | Connectivity

Private Access

Private endpoints allow access to this resource using a private IP address from a virtual network, effectively bringing the service into your virtual network. Learn more

Private endpoint connections

+ Create a private endpoint | Refresh | Approve | Reject | Remove

Filter by name... | Private endpoint == (All)

Private endpoint	Connection name	Connection state	Description
sqlhr02-shr8-pe	sqlhr02-shr8-pe-18459747-1244...	Approved	Auto-approved

Figure 7-5. *The logical server is connected to the private endpoint.*

The next step to configure this logical server is to enable the customer-managed key for transparent data encryption.

Enabling the Customer-Managed Key TDE

Both the primary and the failover logical servers must have the appropriate access to the same TDE protector key in the Key Vault. The permissions required are `get`, `wrapKey`, and `unwrapKey` and are granted by the Key Vault Crypto Service Encryption User role.

Important Both the primary and failover logical servers must be accessing the key from the same Key Vault. They cannot use a different Key Vault with the same key material.

To simplify the permissions on the Key Vault, we can reuse the user-assigned managed identity to allow the logical server to retrieve the key. This identity already has the correct permissions for our needs as they were granted earlier. If we had chosen to use a different user-assigned managed identity or a system-assigned managed identity, then we would need to grant the identity permissions to the key in the Key Vault.

This is performed using the same process as was used on the primary logical server.

1. In the failover logical server, assign the user-assigned managed identity we created previously, `sqlhr-<suffix>-umi`, and set it as the primary identity. Figure 7-6 shows the identity assigned to the failover server.

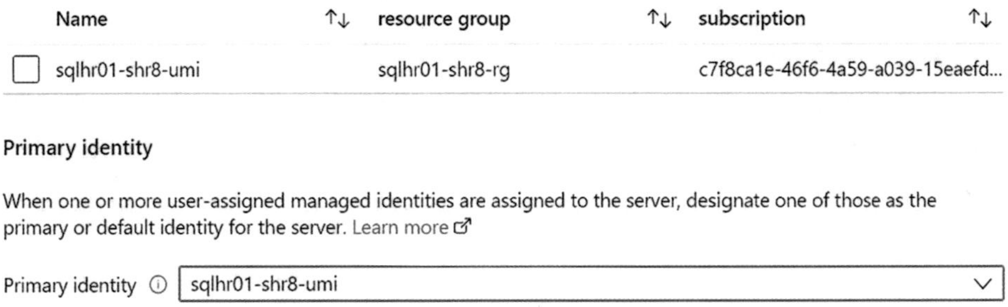

Name	resource group	subscription
sqlhr01-shr8-umi	sqlhr01-shr8-rg	c7f8ca1e-46f6-4a59-a039-15eaefd...

Primary identity

When one or more user-assigned managed identities are assigned to the server, designate one of those as the primary or default identity for the server. Learn more

Primary identity sqlhr01-shr8-umi

***Figure 7-6.** The managed identity for the failover logical server*

2. Configure the transparent data encryption by selecting the key that we created in the Azure Key Vault. Figure 7-7 shows the TDE configuration on the logical server.

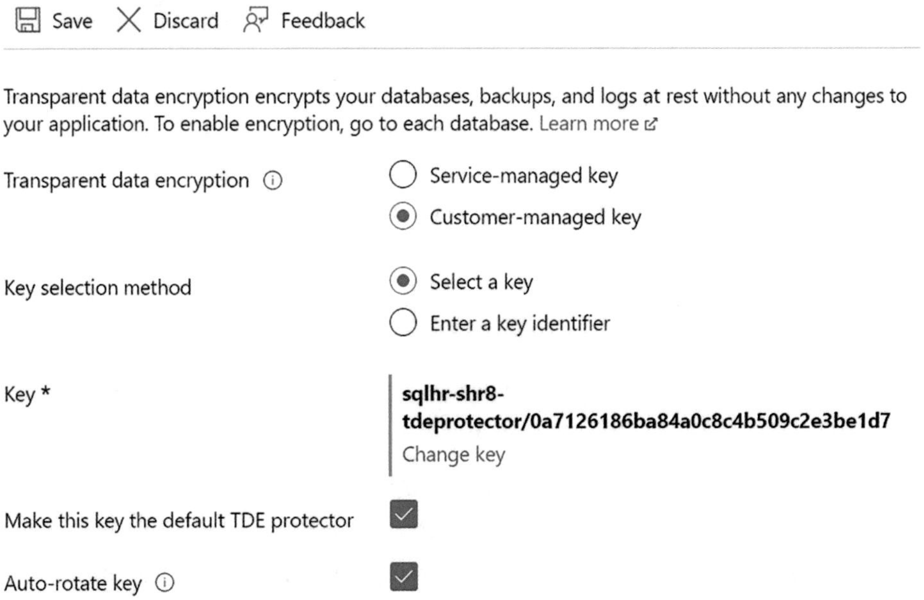

Save Discard Feedback

Transparent data encryption encrypts your databases, backups, and logs at rest without any changes to your application. To enable encryption, go to each database. Learn more

Transparent data encryption: Service-managed key / Customer-managed key

Key selection method: Select a key / Enter a key identifier

Key *: **sqlhr-shr8-tdeprotector/0a7126186ba84a0c8c4b509c2e3be1d7** Change key

Make this key the default TDE protector

Auto-rotate key

***Figure 7-7.** Configure TDE on the failover logical server*

Important You will need to have the Key Vault Crypto Officer RBAC role assigned to your account to perform this task. You may have removed this role previously as a security measure or implemented a privileged identity management elevation process to protect it. If you do not have this role, you will receive the error "This operation is not allowed by RBAC."

The failover logical server is now configured to use the same TDE protector key from the same Key Vault as the primary logical server. This is important so that the data from the primary logical server can be decrypted and used by the failover logical server.

Enabling Geo-replication of a Hyperscale Database

Now that the failover logical server is deployed and ready to be used as a replication target, we can set up the primary Hyperscale database to start replicating to it.

To do this, follow these steps:

1. In the Azure Portal, navigate to the primary Hyperscale database resource `sqlhr01-<suffix>/hyperscaledb`.
2. Select the Replicas blade.
3. Click "Create replica."
4. Set the server to the failover logical server that was just deployed, `sqlhr02-<suffix>`. In our case, it was in West US 3.
5. If we need to, we can change the configuration of the failover database by clicking "Configure database." We can only adjust the number of vCores, the number of high availability secondary replicas, and whether the replicas should be zone redundant, as long as our failover region supports this. It is not possible to change the service tier or hardware configuration for the failover replica.

> **Important** It is strongly recommended that your replica is configured with the same compute size as the primary replica. If it is not, in the case of a regional failover to this database, there may not be enough capacity to serve the workload.

6. Configure the backup storage redundancy as geo-zone-redundant backup storage so that in the case of a regional failover, your backups are replicated back to the primary zone. However, you may decide this is not required, so selecting a different backup storage redundancy for your failover is reasonable.

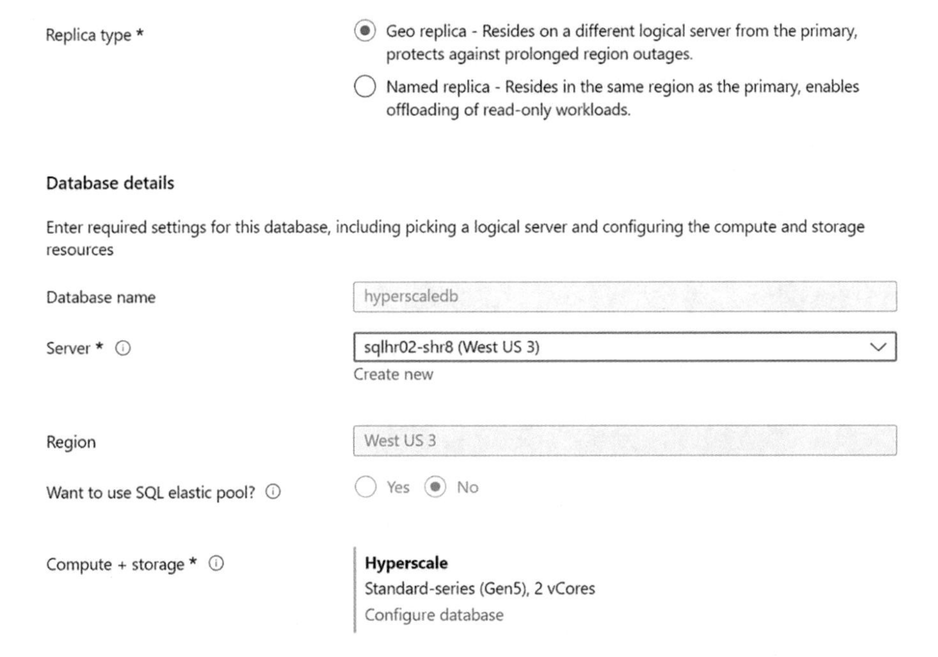

7. Click Next.
8. Click Create to begin creating the replica.
9. After the deployment is completed, the replica database will be shown in the Replicas blade in both the primary and failover Hyperscale databases. Figure 7-8 shows the Replicas blade with the primary and the geo replica displayed.

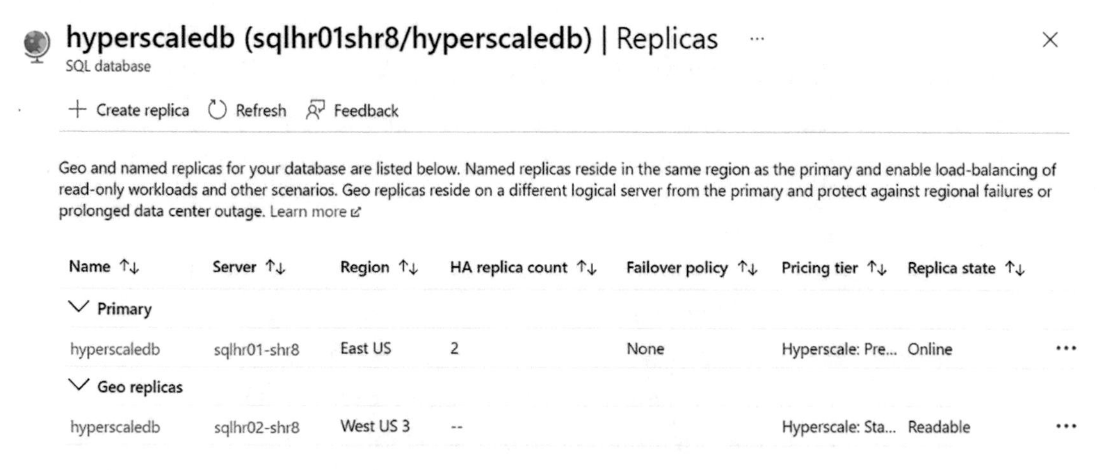

Figure 7-8. *The Replicas blade shows all replicas of a database*

Your failover replica is now available, and you can use it as a read replica or execute a failover to make it your primary database. We will cover business continuity and failovers in more detail in a later chapter.

Summary

In this chapter, we showed how to configure a geo replica of our Hyperscale database in another region to protect against regional outages. To do this, we demonstrated how to do the following:

- We created a new logical server in the failover region with the same authentication settings as the logical server in the primary region.
- We connected the new logical server to the virtual network in the failover region using a private endpoint.
- We configured the failover logical server to use the same user-assigned managed identity as the primary logical server so that it can access the same TDE protector key in the Azure Key Vault.
- We created the geo replica from the primary Hyperscale database to the logical server in the failover region.

In the next chapter, we will finish configuring our Hyperscale environment to meet our nonfunctional requirements by enabling security features with Microsoft Defender for SQL and sending diagnostic and audit logging to a central logging and monitoring service, Azure Log Analytics.

CHAPTER 8

Configuring Security Features and Enabling Diagnostic and Audit Logs

In Chapter 7, we configured a geo replica of our primary Hyperscale database in another Azure region to protect our workload against a failure of the primary region. In this chapter, we will perform the final tasks that we need to do to improve the security and monitoring of our Hyperscale database. This will help our environment align to the practices outlined in the security and operational excellence pillars in the Azure Well-Architected Framework (WAF).

Tip For more information on the Azure Well-Architected Framework, including self-assessment tools and checklists, see `https://aka.ms/waf`.

In this chapter, we will perform the following tasks to improve the security and monitoring of our Hyperscale database:

- Enable Microsoft Defender for SQL to perform automated vulnerability assessments and identify and mitigate vulnerabilities.
- Configure audit and diagnostic logs to be sent to an Azure Log Analytics workspace for storage, analytics, dashboarding, and alerting.

Z. Barać and D. Scott-Raynsford, *Azure SQL Hyperscale Revealed*,
https://doi.org/10.1007/978-1-4842-9225-9_8

Once these remaining tasks have been completed, our Hyperscale database will be ready to handle our production workloads.

Enabling Microsoft Defender for SQL

Microsoft Defender for SQL is part of the Microsoft Defender for Cloud suite of security products. It provides functionality for identifying and mitigating potential database vulnerabilities, as well as detecting anomalous activities that could indicate a threat to your database. The SQL vulnerability assessment feature performs continuous database scanning that can discover, track, and help remediate potential database vulnerabilities.

You can enable Microsoft Defender for SQL either for all logical servers in a subscription or on a per-logical-server basis. Enabling it on a database will enable it on all databases attached to the logical server. There is a per-logical-server monthly cost associated with enabling Microsoft Defender for SQL, so you should take this into account.

Tip In general, it is simplest and most effective to enable Microsoft Defender for SQL at the subscription level. This will automatically enable it on any logical server created in the subscription and ensure the databases in the logical servers are monitored by default. For more information, see `https://docs.microsoft.com/azure/defender-for-cloud/enable-enhanced-security`.

When you enable Microsoft Defender for SQL, an Azure Storage account is created in the same resource group as your logical server. This storage account is used to store the SQL vulnerability assessments that are created. You can configure a different storage account to be used, for example if you want to store all vulnerability assessments for multiple logical servers in a central location.

The simplest way to enable Microsoft Defender for SQL is to enable it on the logical server. To do this, follow these steps:

1. In the Azure Portal, navigate to the logical server in the primary region, `sqlhr01-<suffix>`.
2. Click the Microsoft Defender for Cloud blade.
3. Click Enable Microsoft Defender for SQL.

4. It may take a minute for Microsoft Defender for SQL to be enabled if you haven't already got it enabled on a logical server as it will create an Azure Storage account.

5. To configure Microsoft Defender for SQL settings, click the Configure button.

6. We can change the storage account to store the vulnerability assessments in, enable periodic scans, and configure sending of scan reports to an email address or subscription owners and admins. Figure 8-1 shows the Vulnerability Assessment settings in Microsoft Defender for SQL.

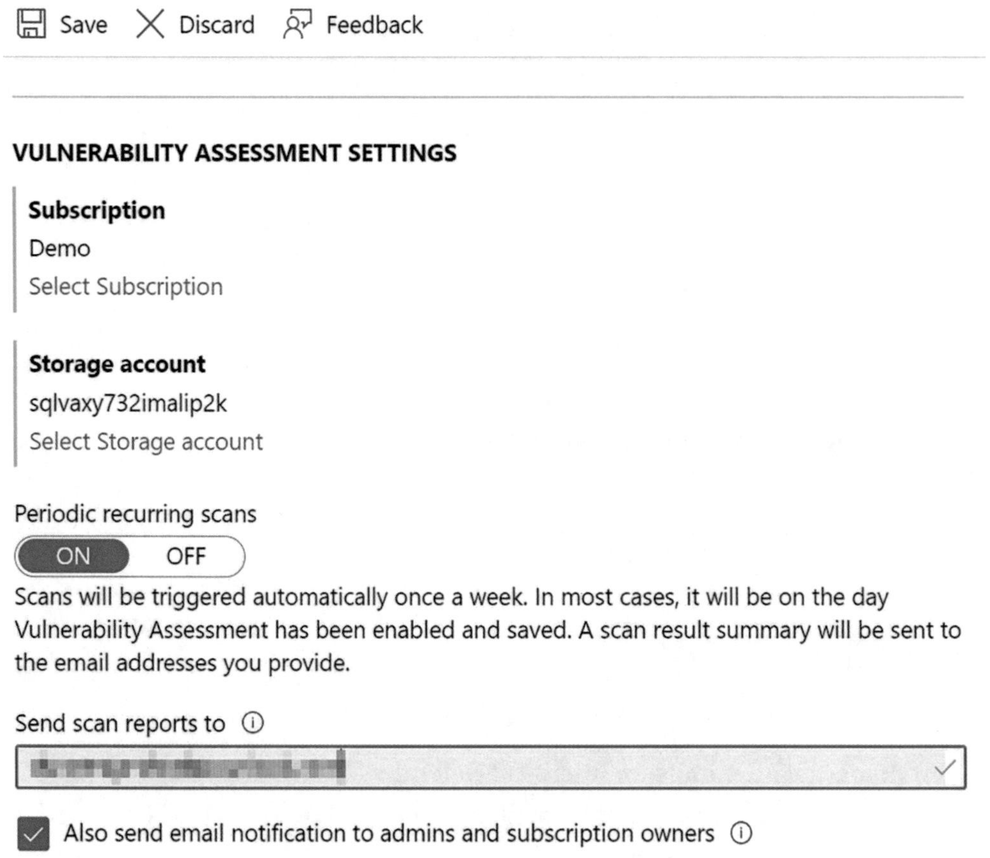

Save Discard Feedback

VULNERABILITY ASSESSMENT SETTINGS

Subscription
Demo
Select Subscription

Storage account
sqlvaxy732imalip2k
Select Storage account

Periodic recurring scans
ON OFF
Scans will be triggered automatically once a week. In most cases, it will be on the day Vulnerability Assessment has been enabled and saved. A scan result summary will be sent to the email addresses you provide.

Send scan reports to

Also send email notification to admins and subscription owners

Figure 8-1. *Configure Microsoft Defender for SQL settings*

7. Click Save.

Microsoft Defender is now set up to monitor the logical server and any databases it contains. It will periodically perform vulnerability assessments and store them in the storage account that was specified.

Storing Diagnostic and Audit Logs

Like most Azure resources, a Hyperscale database will generate diagnostic logs. The diagnostic logs will provide monitoring and diagnostic information for the service. As well as the diagnostic logs, the Hyperscale database will also produce audit logs. It is recommended for production databases to store this information in a central location so that it can be easily available for monitoring, reporting, and analysis. This may also be required to meet compliance requirements.

It is possible to configure the Hyperscale database and the logical server to send this information to multiple destination types.

- *An Azure Storage account*: The diagnostic logs and/or audit logs can be stored as files in an Azure Storage account. This is a cheap way to persist this information and archive it to meet compliance requirements. However, analyzing and reporting on the data in an Azure Storage account will require additional services to be deployed.
- *A Log Analytics workspace*: Sending diagnostic logs and audit information to a Log Analytics workspace allows you to easily analyze and produce reports and dashboards using the data. It can also then be exported out to an archive location. Log Analytics allows you to combine the diagnostics and audit data with data from other Azure services to produce powerful reporting, monitoring, and dashboarding.
- *An event hub*: You would use this to stream the diagnostic and audit logs as events to another monitoring service.
- *A partner solution*: You can stream the logs into a third-party solution that is integrated into Azure. This is available only for diagnostic logs.

You can configure diagnostic logs and audit logs to be sent to a combination of destination types, but for this example, we are going to send them to a Log Analytics workspace.

Sending Audit Logs to a Log Analytics Workspace

An Azure Log Analytics workspace is a unique log data storage environment and querying engine for use by Azure Monitor and other Azure services. Sending logs to an Azure Log Analytics Workspace allows you to combine it with other sources of log data and analyze it using the advanced Kusto Query Language (KQL) to produce reports, dashboards, workbooks, and automated alerting and notifications.

Tip For more information on the Kusto Query Language (KQL), see `https://aka.ms/kql`.

There are two ways you can configure the sending of audit logs for a Hyperscale database.

- *The logical server*: This will result in all databases associated with this logical server having audit logs sent to the destination.
- *The database*: Configuring the auditing settings on the database will enable audit information for just this database to be sent to the destination. Configuring the settings on the database does not override the settings configured for the logical server.

Important You should not enable auditing on both the logical server and a database contained in the logical server unless they are being sent to different destinations. For example, you should *not* configure auditing for the logical server and a database to be sent to the same Log Analytics workspace as this will result in audit duplicate records being sent. However, you can send them to different Log Analytics workspaces.

To configure the audit logs for the logical server to the Log Analytics workspace, perform the following:

1. In the Azure Portal, navigate to the logical server in the primary region, `sqlhr01-<suffix>`.
2. Click the Auditing blade.
3. Select Enable Azure SQL Auditing and select Log Analytics.
4. Set the subscription to the one that contains your Log Analytics workspace.
5. Select the Log Analytics workspace for the *primary region*.
6. You may also choose to enable auditing of any support operations the Microsoft support teams perform.

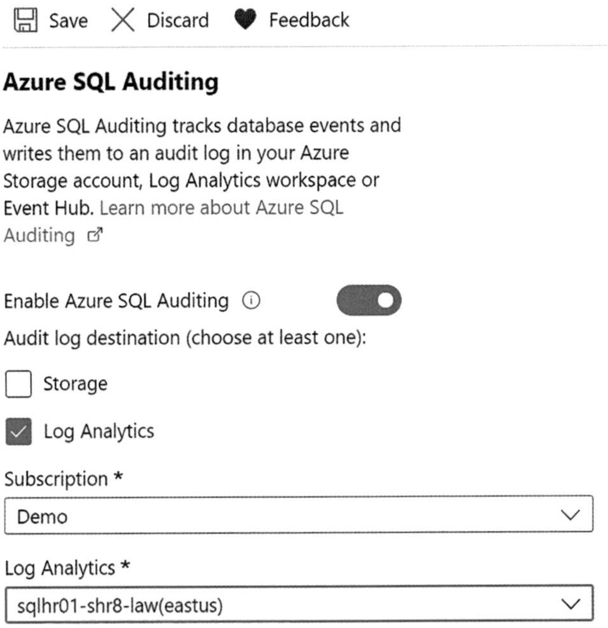

Save Discard Feedback

Azure SQL Auditing

Azure SQL Auditing tracks database events and writes them to an audit log in your Azure Storage account, Log Analytics workspace or Event Hub. Learn more about Azure SQL Auditing

Enable Azure SQL Auditing

Audit log destination (choose at least one):

Storage

Log Analytics

Subscription *

Demo

Log Analytics *

sqlhr01-shr8-law(eastus)

Figure 8-2. *Sending audit logs to Log Analytics*

7. Click the Save button. After a few seconds you will receive a confirmation message that your settings have been saved successfully.

This will also create an SQLAuditing solution in the resource group, associated with the Log Analytics workspace. This solution provides some audit-specific reporting and dashboard functionality to help your logical server.

It is also good practice to send the audit logs for the logical server in the failover region to the Log Analytics workspace in the failover region. So, you should perform the previous tasks on the replica in the failover region.

Sending Database Diagnostic Logs to Log Analytics

In most production environments, it will be a requirement to collect diagnostic logs from the database and store them in a centralized and well-managed location. This will enable reporting and analysis of events occurring within our Hyperscale environment. In our case, we would like to use a Log Analytics workspace for this. To configure this, perform these steps:

1. In the Azure Portal, navigate to the Hyperscale database in the primary region, `sqlhr01-<suffix>/hyperscaledb`.
2. Select the Diagnostic Settings blade.
3. Click the "Add diagnostic" setting.
4. Set the diagnostic settings name to a value indicating the destination and data to be sent. For example, set it to `Send all logs to sqlhr01-shr8-law`.
5. Select the "Send to Log Analytics workspace" box and select the subscription and Log Analytics workspace from your primary region.
6. Select the Logs categories you want to send to Log Analytics.
7. You can also send performance metrics to Log Analytics, but in our case, we are not required to do this. If you enable this, metrics will be converted to log records and sent to the Log Analytics workspace.

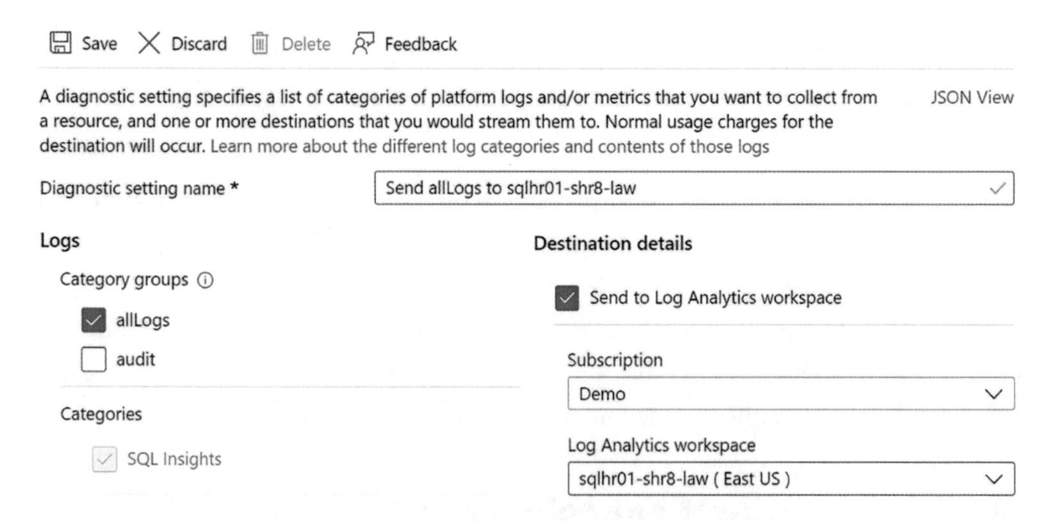

Figure 8-3. *Sending diagnostic logs to Log Analytics*

8. Click the Save button. After a few seconds, you will receive a confirmation message that your settings have been saved successfully.

You can configure up to five diagnostic log settings to send different combinations of logs and metrics to different destinations. It is also a good idea to configure your failover replica to send diagnostic logs to the Log Analytics workspace in the failover region.

Summary

With these final changes to our logical server and Hyperscale database, we have successfully created an environment that meets the requirements defined in Chapter 4 by enabling the following:

- Private endpoint connections to make the database accessible only from within our virtual networks
- The Azure Active Directory identity as the only identity source for our Hyperscale database
- High availability within an availability zone-enabled Azure region by deploying three replicas
- Disaster recovery by replicating our Hyperscale database into our chosen failover region
- Backup redundancy to ensure our backups are available in our primary region even in the case of a zone being unavailable
- Managed identity to be used by the Hyperscale database to manage and communicate with other Azure services
- A customer-managed encryption key to be stored in our Azure Key Vault and used to encrypt the database files and backups
- Microsoft Defender for SQL to produce regular vulnerability assessments for our databases and provide automated threat detection
- Audit and diagnostic logs to be sent to an Azure Log Analytics workspace to provide a central location for monitoring, alerting, analysis, and reporting

Over the next three chapters, we will demonstrate how to deploy the same environment, but we'll be using infrastructure as code with Azure PowerShell, Azure Command-Line Interface (CLI), and Azure Bicep.

CHAPTER 9

Deploying Azure SQL DB Hyperscale Using PowerShell

Over the previous five chapters, we demonstrated how to design a Hyperscale environment that satisfied a set of requirements and then deploy it using the Azure Portal. Although using the portal is a great way to learn about the various components of a Hyperscale database and how they fit together, it is strongly recommended to move to using infrastructure as code (IaC) to deploy your environments.

In this chapter, we'll look at using PowerShell and the Azure PowerShell modules to deploy the resources that make up the environment. We will deploy the same environment we have used in the previous five chapters, but instead, we'll use only the Azure PowerShell module to create and configure the resources.

Tip In this chapter, we'll be sharing individual PowerShell code snippets as well as a complete script that can be used to deploy the environment. These will all be available in the `ch9` directory in the files accompanying this book and in the GitHub repository: `https://github.com/Apress/Azure-SQL-Hyperscale-Revealed`.

Before we get started, we will provide a short primer on infrastructure as code, covering the different approaches and some of the many benefits. However, the finer details and behaviors of different IaC methodologies are beyond the scope of this book. It is just intended to demonstrate how to deploy a more typical production Hyperscale environment using some of the common IaC approaches used with Azure.

Z. Barać and D. Scott-Raynsford, *Azure SQL Hyperscale Revealed*,
https://doi.org/10.1007/978-1-4842-9225-9_9

Note There are many other IaC technologies available that we haven't included in this book, simply for brevity. Some other common IaC technologies used with Azure are Hashicorp Terraform (`https://www.terraform.io`) and Pulumi (`https://www.pulumi.com`), but there are certainly others. If your preferred technology is not demonstrated, it is just to keep the length of the book down, not because it is any less important.

Introduction to Infrastructure as Code

Infrastructure as code is a generic term that is used whenever we configure our cloud infrastructure and services using code instead of manually via the Azure Portal. Infrastructure as code is usually one or more files that when used will deploy the infrastructure defined within them.

Note Although the term *infrastructure* has traditionally been used to refer to virtual machines, storage, and networks, when we are talking about infrastructure in the IaC context, it will refer to any resource that can be deployed into Azure.

Defining your Azure resources with IaC provides you with many benefits.

- *Repeatability*: Your environment can be redeployed the same way each time. You can use the same scripts or templates to deploy multiple copies of the same environment for different purposes, such as development, testing, and production environments.
- *Centralized management*: You can store your IaC files in a centralized location, enabling the definitions to be used and contributed to by multiple teams or a centralized platform team. For example, you might create a definition that can deploy a standard Hyperscale database that meets your organization's standard requirements.

- *Maintainability*: Storing your environment as code allows you to use a source control system, such as Git, to store your infrastructure definition. This allows you to determine what has changed between versions, implement controls over who can change it, and enable automated pipelines to deploy changes.
- *Testability*: Declaring your infrastructure in IaC allows you to adopt DevOps processes to verify changes to your environment definitions. For example, any time a change is proposed to an IaC file, it can be run through a series of tests to ensure it meets your organization's compliance, security, and quality controls.

There are many other benefits to implementing IaC, and your organization might implement IaC methodologies that leverage some or all these benefits. But, fundamentally, IaC is about storing your infrastructure definitions in reusable files.

Imperative vs. Declarative Infrastructure as Code

There are two common IaC methodologies in use today.

- *Imperative*: For this IaC methodology, the infrastructure is defined as a set of imperative steps that must be executed in order to get to the end state of the infrastructure. This process may or may not take into account the current state of the environment, for example, if the environment is already partially deployed or needs to be changed. To implement imperative scripts that can take the environment from any starting state to the desired state can be prohibitively complex. For this reason, declarative IaC is not recommended for anything but the most trivial infrastructure. For Azure infrastructure, imperative IaC is usually written as PowerShell scripts using the Azure PowerShell modules or shell scripts using the Azure CLI.
- *Declarative*: In this IaC methodology, the desired state of the infrastructure is usually defined in a domain-specific language (DSL) that is unique to the chosen IaC technology. The IaC files are then provided to an engine, which is responsible for taking the current state of the environment to the desired state.

In Azure, some of the common declarative technologies are Azure Resource Manager (ARM) templates, Azure Bicep (which are transpiled to ARM templates), and Hashicorp Terraform (using the Hashicorp Configuration Language). When using ARM templates or Azure Bicep, the engine responsible for ensuring the infrastructure is in the desired state is Azure itself using the Azure Resource Manager. However, with Terraform the state engine is external to Azure and provided as part of Hashicorp Terraform and Terraform Azure Resource Provider and relies on maintaining its own internal view of the current state of the Azure environment.

Note Pulumi is also a declarative methodology but uses object-oriented code to define the desired state of the infrastructure, rather than a domain-specific language. So, Pulumi is also considered a declarative methodology.

A discussion about the pros and cons of the different methodologies is beyond the scope of this book, but it is recommended to move to a declarative methodology whenever possible.

The remainder of this chapter will be dedicated to demonstrating the deployment of our Hyperscale environment using PowerShell and the Azure PowerShell modules. In Chapter 10, we will show the deployment of the same Hyperscale environment using Bash and the Azure CLI. Rounding out the IaC samples, Chapter 11 will show a declarative IaC methodology using Azure Bicep to deploy the Hyperscale environment.

Deploying Hyperscale Using Azure PowerShell

In this section, we will break down the deployment process of the SQL Hyperscale environment defined in Chapter 4 into individual steps, highlighting each of the PowerShell commands required to deploy the Azure resources that make up the environment.

The Azure PowerShell Modules

The first thing that is required to deploy and configure a Hyperscale environment using PowerShell is the Azure PowerShell modules. These modules provide the commands that can be used to interact with most of the Azure services. The Azure PowerShell commands are split across different PowerShell modules, each providing commands relevant to a specific Azure service. However, if you install the main Azure PowerShell module `Az`, then all Azure PowerShell modules will be installed.

There are many ways you can install the Azure PowerShell modules into your own PowerShell environment, but if you're using Azure Cloud Shell, they are already pre-installed and connected to your Azure subscriptions for you.

Tip If you want to install the Azure PowerShell modules into your own environment, see `https://learn.microsoft.com/en-us/powershell/azure/install-az-ps`. These modules are compatible with Windows, macOS, and Linux operating systems, using either PowerShell or Windows PowerShell.

For the remainder of this chapter, we'll be using the Azure Cloud Shell. At the time of writing this book, the current version of the main Azure PowerShell module is `9.2.0`, which is the version installed into the Azure Cloud Shell. You should ensure you've got at least this version installed if you're running the commands and scripts in this chapter.

To check the version of main Azure PowerShell module installed, run the following PowerShell command:

```
Get-Module -Name Az -ListAvailable
```

Figure 9-1 shows the result of running the command to get the main Azure PowerShell module installed in the Azure Cloud Shell.

```
PS /home> Get-Module -Name Az -ListAvailable

    Directory: /usr/local/share/powershell/Modules

ModuleType Version    PreRelease Name                                PSEdition
---------- -------    ---------- ----                                ---------
Script     9.2.0                 Az                                  Core,Desk
```

Figure 9-1. *Get the installed Azure PowerShell main module*

To check all the dependent Azure PowerShell modules, run the following PowerShell command:

```
Get-Module -Name Az.* -ListAvailable
```

Figure 9-2 shows some of the Azure PowerShell modules installed with the main Azure PowerShell module.

```
PS /home> Get-Module -Name Az.* -ListAvailable

    Directory: /home/      /.local/share/powershell/Modules

ModuleType Version    PreRelease Name                                PSEdition
---------- -------    ---------- ----                                ---------
Script     0.7.7                 Az.ResourceGraph                    Core,Desk

    Directory: /usr/local/share/powershell/Modules

ModuleType Version    PreRelease Name                                PSEdition
---------- -------    ---------- ----                                ---------
Script     2.10.4                Az.Accounts                         Core,Desk
Script     2.0.0                 Az.Advisor                          Core,Desk
```

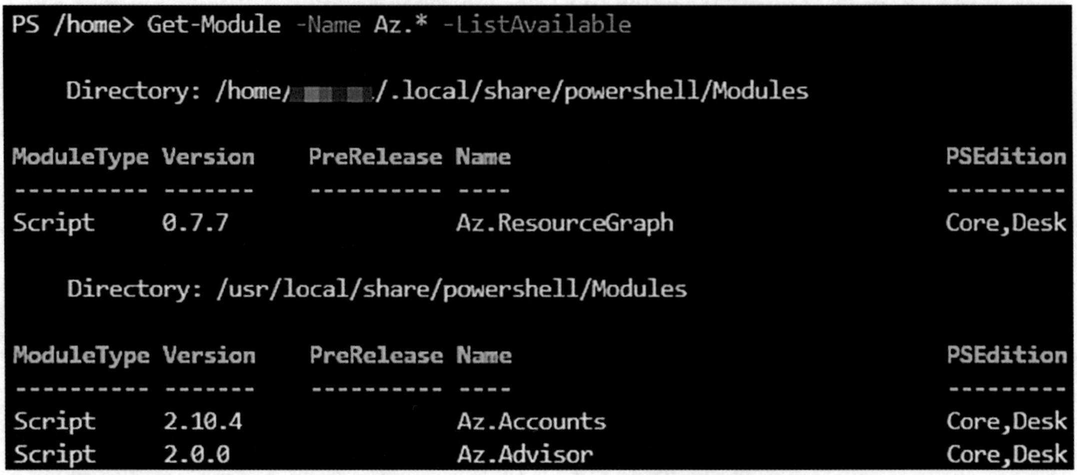

***Figure 9-2.** List of all the installed Azure PowerShell modules*

Now that we have an environment with the Azure PowerShell modules installed and connected, we can begin the process of deploying the environment.

Deploying the Starting Environment

Because we're deploying the same Hyperscale architecture we defined in Chapter 4, we are also going to deploy the same starting environment. We'll use the Starting Environment deployment script that can be found in the `ch4` directory in the files provided for this book; it is called `New-SQLHyperscaleRevealedStartingEnvironment.ps1`.

For a refresher on how to deploy the starting environment using this script, refer to the "Deploying the Starting Environment Using the Azure Cloud Shell" section in Chapter 4.

The Complete Deployment PowerShell Script

Throughout this chapter we will be individually demonstrating each Azure PowerShell command that will be used to deploy and configure the Hyperscale environment and associated services. However, if you'd like to deploy the entire environment using a single go, we've provided a complete PowerShell script. This script is called `Install-SQLHyperscaleRevealedHyperscaleEnvironment.ps1`, and you can find this script in the `ch9` directory of the files provided for this book.

The script accepts a few parameters that allow you to customize the environment for your needs.

- `PrimaryRegion`: This is the name of the Azure region to use as the primary region. This must match the value you specified when you created the starting environment.
- `FailoverRegion`: This is the name of the Azure region to use as the failover region. This must match the value you specified when you created the starting environment.
- `ResourceNameSuffix`: This is the suffix to append into the resource group and resource names. This must match the value you specified when you created the starting environment.
- `AadUserPrincipalName`: This is the Azure Active Directory user principal name (UPN) of the user account running this script. This is usually the user principal name of your account.

Note The UPN is required when running the script in the Azure Cloud Shell because of the way the Azure Cloud Shell executes commands using a service principal, instead of your user account. In this case, the script can't automatically detect your user account to assign appropriate permissions to the Azure Key Vault during the creation of the TDE protector.

To run the `Install-SQLHyperscaleRevealedHyperscaleEnvironment.ps1` script in the Azure Cloud Shell, follow these steps:

1. Open the Azure Cloud Shell.
2. Ensure PowerShell is selected in the shell selector at the top of the Cloud Shell.
3. Clone the GitHub repository that contains the source code for this book by running the following command:

```
git clone https://github.com/Apress/Azure-SQL-Hyperscale-
Revealed.git
```

4. Before running the script, it is important to ensure the correct Azure subscription is selected by running the following command:

```
Select-AzSubscription -Subscription '<subscription name>'
```

5. Run the script by entering the following:

```
./Azure-SQL-Hyperscale-Revealed/ch9/Install-SQLHyperscale
RevealedHyperscaleEnvironment.ps1 -ResourceNameSuffix '<your
resource name suffix>' -AadUserPrincipalName '<your Azure AD
account UPN>'
```

Because of the number of resources being created, the deployment process may take 30 minutes or more.

Azure PowerShell Commands in Detail

This section will show the Azure PowerShell commands that are required to create the logical servers and databases. It will also show supporting commands that are required to configure the virtual network private link integration and customer-managed key encryption.

The commands used will be shown in some detail, but as with many PowerShell commands, they can be used in many different ways, so you can customize them for your needs. You should refer to the specific documentation for each command to better understand how it should be used.

Creating Helper Variables

The first thing we'll do is create some variables that we can use throughout the process by many of the commands. You can customize these as you see fit, but the `PrimaryRegion`, `FailoverRegion`, and `ResourceNameSuffix` must match the values you used to create the starting environment. The `AadUserPrincipalName` must be the Azure Active Directory user principal name (UPN) of the user account running this script.

```
$PrimaryRegion = 'East US'
$FailoverRegion = 'West US 3'
$ResourceNameSuffix = '<your resource name suffix>'
```

```
$AadUserPrincipalName = '<your Azure AD account UPN>'
$tags = @{ Environment = 'SQL Hyperscale Revealed demo' }
$baseResourcePrefix = 'sqlhr'
$primaryRegionPrefix = "$($baseResourcePrefix)01"
$failoverRegionPrefix = "$($baseResourcePrefix)02"
$primaryRegionResourceGroupName = "$primaryRegionPrefix-
$resourceNameSuffix-rg"
$failoverRegionResourceGroupName = "$failoverRegionPrefix-
$resourceNameSuffix-rg"
$subscriptionId = (Get-AzContext).Subscription.Id
$userId = (Get-AzAdUser -UserPrincipalName $AadUserPrincipalName).Id
$privateZone = 'privatelink.database.windows.net'
```

Figure 9-3 shows the creation of the helper variables in the Azure Cloud Shell.

Note These variables will need to be re-created if your Azure Cloud Shell closes between running commands.

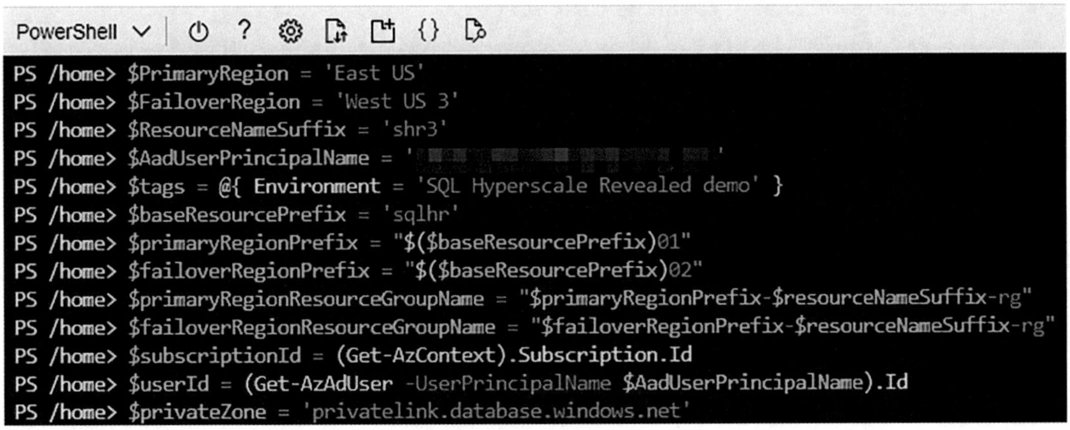

```
PowerShell
PS /home> $PrimaryRegion = 'East US'
PS /home> $FailoverRegion = 'West US 3'
PS /home> $ResourceNameSuffix = 'shr3'
PS /home> $AadUserPrincipalName = '[illegible]'
PS /home> $tags = @{ Environment = 'SQL Hyperscale Revealed demo' }
PS /home> $baseResourcePrefix = 'sqlhr'
PS /home> $primaryRegionPrefix = "$($baseResourcePrefix)01"
PS /home> $failoverRegionPrefix = "$($baseResourcePrefix)02"
PS /home> $primaryRegionResourceGroupName = "$primaryRegionPrefix-$resourceNameSuffix-rg"
PS /home> $failoverRegionResourceGroupName = "$failoverRegionPrefix-$resourceNameSuffix-rg"
PS /home> $subscriptionId = (Get-AzContext).Subscription.Id
PS /home> $userId = (Get-AzAdUser -UserPrincipalName $AadUserPrincipalName).Id
PS /home> $privateZone = 'privatelink.database.windows.net'
```

Figure 9-3. *Creating example helper variables*

Create the User-Assigned Managed Identity for the Hyperscale Database

To allow the logical servers in both the primary and failover regions to access the Key Vault where the TDE protector key will be stored, a user-assigned managed identity needs to be created. To do this, we'll use the `New-AzUserAssignedIdentity` command.

This PowerShell snippet demonstrates how to create the new user-assigned managed identity in the primary region resource group. It then assigns the resource ID of the identity to a variable so it can be used when creating the logical server.

```
$newAzUserAssignedIdentity_parameters = @{
    Name = "$baseResourcePrefix-$resourceNameSuffix-umi"
    ResourceGroupName = $primaryRegionResourceGroupName
    Location = $primaryRegion
    Tag = $tags
}
New-AzUserAssignedIdentity @newAzUserAssignedIdentity_parameters

$userAssignedManagedIdentityId = "/subscriptions/$subscriptionId" + `
    "/resourcegroups/$primaryRegionResourceGroupName" + `
    "/providers/Microsoft.ManagedIdentity" + `
    "/userAssignedIdentities/$baseResourcePrefix-$resourceNameSuffix-umi"
```

Tip The PowerShell snippet uses a technique called *parameter splatting*. It is a useful PowerShell scripting technique that enables assigning parameters to an object that is provided to the command.

Figure 9-4 shows the expected output from the user-assigned managed identity being created.

```
PS /home> $newAzUserAssignedIdentity_parameters = @{
>>     Name = "$baseResourcePrefix-$resourceNameSuffix-umi"
>>     ResourceGroupName = $primaryRegionResourceGroupName
>>     Location = $primaryRegion
>>     Tag = $tags
>> }
PS /home> New-AzUserAssignedIdentity @newAzUserAssignedIdentity_parameters

Location Name           ResourceGroupName
-------- ----           -----------------
eastus   sqlhr-shr3-umi sqlhr01-shr3-rg

PS /home>
PS /home> $userAssignedManagedIdentityId = "/subscriptions/$subscriptionId" + `
>>     "/resourcegroups/$primaryRegionResourceGroupName" + `
>>     "/providers/Microsoft.ManagedIdentity" + `
>>     "/userAssignedIdentities/$baseResourcePrefix-$resourceNameSuffix-umi"
```

Figure 9-4. *Creating the user-assigned managed identity*

Prepare the TDE Protector Key in the Key Vault

The next step is to grant the user account that is being used to run the commands access to the Key Vault to create the TDE protector key. This is done using the `New-AzRoleAssignment` command. We are granting the role to the user account that was specified in the `$userId` helper variable.

```
$newAzRoleAssignment_parameters = @{
    ObjectId = $userId
    RoleDefinitionName = 'Key Vault Crypto Officer'
    Scope = "/subscriptions/$subscriptionId" + `
        "/resourcegroups/$primaryRegionResourceGroupName" + `
        "/providers/Microsoft.KeyVault" + `
        "/vaults/$baseResourcePrefix-$resourceNameSuffix-kv"
}
New-AzRoleAssignment @newAzRoleAssignment_parameters
```

In Figure 9-5 we show example output from assigning the Key Vault Crypto Officer role to the Key Vault.

```
PS /home> New-AzRoleAssignment @newAzRoleAssignment_parameters

RoleAssignmentName : 5e7e3c78-090d-463d-acb5-65411684627a
RoleAssignmentId   : /subscriptions/
Scope              : /subscriptions/
DisplayName        :
SignInName         :
RoleDefinitionName : Key Vault Crypto Officer
RoleDefinitionId   : 14b46e9e-c2b7-41b4-b07b-48a6ebf60603
ObjectId           : 64f0702a-f001-4d3b-bb93-eccf57ec7356
ObjectType         : User
CanDelegate        : False
```

Figure 9-5. *Assigning the Key Vault Crypto Officer role to the Key Vault*

Once the role has been assigned, the user account can be used to create the TDE protector key in the Key Vault using the `Add-AzKeyVaultKey` command and then assign the identity to a variable (for use when creating the logical server).

```
$addAzKeyVaultKey_parameters = @{
    KeyName = "$baseResourcePrefix-$resourceNameSuffix-tdeprotector"
    VaultName = "$baseResourcePrefix-$resourceNameSuffix-kv"
    KeyType = 'RSA'
    Size = 2048
    Destination = 'Software'
    Tag = $tags
}
Add-AzKeyVaultKey @addAzKeyVaultKey_parameters
$tdeProtectorKeyId = (Get-AzKeyVaultKey `
    -KeyName "$baseResourcePrefix-$resourceNameSuffix-tdeprotector" `
    -VaultName "$baseResourcePrefix-$resourceNameSuffix-kv").Id
```

Figure 9-6 shows the output from creating the TDE protector.

```
PS /home> Add-AzKeyVaultKey @addAzKeyVaultKey_parameters

Vault/HSM Name : sqlhr-shr3-kv
Name           : sqlhr-shr3-tdeprotector
Key Type       : RSA
Key Size       : 2048
Curve Name     :
Version        : 285a96481d5048adb2e9370411ab9566
Id             : https://sqlhr-shr3-kv.vault.azure.net:443/keys/sqlhr-shr3-tde
protector/285a96481d5048adb2e9370411ab9566
Enabled        : True
Expires        :
Not Before     :
Created        : 12/9/2022 5:13:10 AM
Updated        : 12/9/2022 5:13:10 AM
Recovery Level : Recoverable
Release Policy :
Tags           : Name        Value
                 Environment SQL Hyperscale Revealed demo
```

Figure 9-6. *Creating the TDE protector key in the Key Vault*

The service principal that is created as part of the user-assigned managed identity then needs to be granted the Key Vault Crypto Service Encryption User role to the key that was just created. The `New-AzRoleAssignment` command is again used for this:

```
$servicePrincipalId = (Get-AzADServicePrincipal -DisplayName
"$baseResourcePrefix-$resourceNameSuffix-umi").Id
$tdeProtectorKeyResourceId = "/subscriptions/$subscriptionId" + `
    "/resourcegroups/$primaryRegionResourceGroupName" + `
    "/providers/Microsoft.KeyVault" + `
    "/vaults/$baseResourcePrefix-$resourceNameSuffix-kv" + `
    "/keys/$baseResourcePrefix-$resourceNameSuffix-tdeprotector"
$newAzRoleAssignment_parameters = @{
    ObjectId = $servicePrincipalId
    RoleDefinitionName = 'Key Vault Crypto Service Encryption User'
    Scope = $tdeProtectorKeyResourceId
}
New-AzRoleAssignment @newAzRoleAssignment_parameters
```

Figure 9-7 shows sample output from assigning the service principal for a role on the TDE protector key.

```
PS /home> New-AzRoleAssignment @newAzRoleAssignment_parameters

RoleAssignmentName : 96079a3d-9654-4ca4-aed7-0b4e47aecbd7
RoleAssignmentId   : /subscriptions/
                     roviders/Microsoft.Authorization/roleAssig
Scope              : /subscriptions/
DisplayName        : sqlhr-shr3-umi
SignInName         :
RoleDefinitionName : Key Vault Crypto Service Encryption User
RoleDefinitionId   : e147488a-f6f5-4113-8e2d-b22465e65bf6
ObjectId           : d2c33674-db8a-44a9-a0c9-d6272e3d4685
ObjectType         : ServicePrincipal
CanDelegate        : False
```

Figure 9-7. *Granting access to the TDE protector key to the service principal*

The TDE protector key in the Key Vault is now ready to be accessed by the new logical servers for encryption.

Create the Logical Server in the Primary Region

Now that the Key Vault is configured, we can create the logical server. This is done using the New-AzSqlServer command.

```
$sqlAdministratorsGroupId = (Get-AzADGroup -DisplayName 'SQL
Administrators').Id
$newAzSqlServer_parameters = @{
    ServerName = "$primaryRegionPrefix-$resourceNameSuffix"
    ResourceGroupName = $primaryRegionResourceGroupName
    Location = $primaryRegion
    ServerVersion = '12.0'
    PublicNetworkAccess = 'Disabled'
    EnableActiveDirectoryOnlyAuthentication = $true
    ExternalAdminName = 'SQL Administrators'
    ExternalAdminSID = $sqlAdministratorsGroupId
```

```
    AssignIdentity = $true
    IdentityType = 'UserAssigned'
    UserAssignedIdentityId = $userAssignedManagedIdentityId
    PrimaryUserAssignedIdentityId = $userAssignedManagedIdentityId
    KeyId = $tdeProtectorKeyId
    Tag = $tags
}
New-AzSqlServer @newAzSqlServer_parameters
```

The first line looks up the object ID of the SQL Administrators group in Azure Active Directory, which is required to be passed to the `ExternalAdminSID` parameter. The `EnableActiveDirectoryOnlyAuthentication` and `ExternalAdminName` parameters are also required for the logical server to use Azure AD for authentication. The `KeyId` is set to the ID of the TDE protector key, and the `UserAssignedIdentityId` and `PrimaryUserAssignedIdentityId` are set to the resource ID of the user-assigned managed identity that was granted access to the TDE protector key in the Key Vault.

Figure 9-8 shows part of the output from creating the logical server.

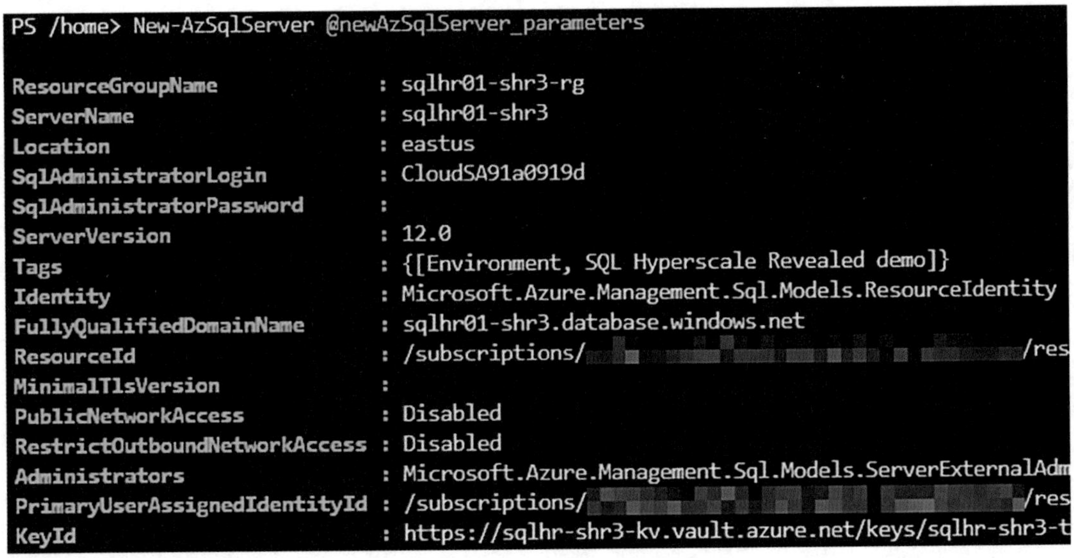

```
PS /home> New-AzSqlServer @newAzSqlServer_parameters

ResourceGroupName             : sqlhr01-shr3-rg
ServerName                    : sqlhr01-shr3
Location                      : eastus
SqlAdministratorLogin         : CloudSA91a0919d
SqlAdministratorPassword      :
ServerVersion                 : 12.0
Tags                          : {[Environment, SQL Hyperscale Revealed demo]}
Identity                      : Microsoft.Azure.Management.Sql.Models.ResourceIdentity
FullyQualifiedDomainName      : sqlhr01-shr3.database.windows.net
ResourceId                    : /subscriptions/                              /res
MinimalTlsVersion             :
PublicNetworkAccess           : Disabled
RestrictOutboundNetworkAccess : Disabled
Administrators                : Microsoft.Azure.Management.Sql.Models.ServerExternalAdm
PrimaryUserAssignedIdentityId : /subscriptions/                              /res
KeyId                         : https://sqlhr-shr3-kv.vault.azure.net/keys/sqlhr-shr3-t
```

Figure 9-8. *Creating the primary region logical server*

Connect the Primary Logical Server to the Virtual Network

The logical server has been created in the primary region, but it is not connected to the virtual network. To make the logical server available on a private endpoint in the virtual network, we need to run a number of commands.

The first command is to create the private link service connection for the logical server using the `New-AzPrivateLinkServiceConnection` command.

```
$sqlServerResourceId = (Get-AzSqlServer `
    -ServerName "$primaryRegionPrefix-$resourceNameSuffix" `
    -ResourceGroupName $primaryRegionResourceGroupName).ResourceId
$newAzPrivateLinkServiceConnection_parameters = @{
    Name = "$primaryRegionPrefix-$resourceNameSuffix-pl"
    PrivateLinkServiceId = $sqlServerResourceId
    GroupId = 'SqlServer'
}
$privateLinkServiceConnection = New-AzPrivateLinkServiceConnection
@newAzPrivateLinkServiceConnection_parameters
```

The next step is to use the service connection for the logical server to create a private endpoint in the virtual network subnet using the `New-AzPrivateEndpoint` command. This command requires us to provide the subnet object of the subnet we're connecting the private endpoint to.

```
$vnet = Get-AzVirtualNetwork `
    -Name "$primaryRegionPrefix-$resourceNameSuffix-vnet" `
    -ResourceGroupName $primaryRegionResourceGroupName
$subnet = Get-AzVirtualNetworkSubnetConfig `
    -VirtualNetwork $vnet `
    -Name 'data_subnet'
$newAzPrivateEndpoint_parameters = @{
    Name = "$primaryRegionPrefix-$resourceNameSuffix-pe"
    ResourceGroupName = $primaryRegionResourceGroupName
    Location = $primaryRegion
    Subnet = $subnet
    PrivateLinkServiceConnection = $privateLinkServiceConnection
```

```
    Tag = $tags
}
New-AzPrivateEndpoint @newAzPrivateEndpoint_parameters
```

Figure 9-9 shows part of the output from creating the new private endpoint connected to the subnet.

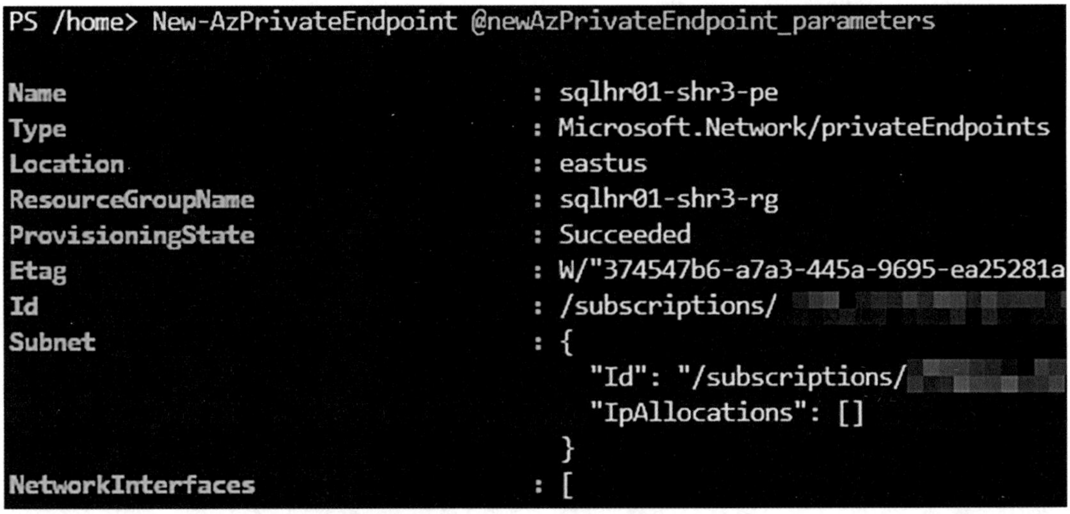

```
PS /home> New-AzPrivateEndpoint @newAzPrivateEndpoint_parameters

Name                          : sqlhr01-shr3-pe
Type                          : Microsoft.Network/privateEndpoints
Location                      : eastus
ResourceGroupName             : sqlhr01-shr3-rg
ProvisioningState             : Succeeded
Etag                          : W/"374547b6-a7a3-445a-9695-ea25281a
Id                            : /subscriptions/
Subnet                        : {
                                  "Id": "/subscriptions/
                                  "IpAllocations": []
                                }
NetworkInterfaces             : [
```

***Figure 9-9.** Creating a private endpoint in the subnet*

Once the private endpoint is created and connected, we need to configure the DNS entry for it so that it can be resolved correctly from within the virtual network. To do this, we first create a new private DNS zone with the `New-AzPrivateDnsZone` command.

```
$newAzPrivateDnsZone_parameters = @{
    Name = $privateZone
    ResourceGroupName = $primaryRegionResourceGroupName
}
$privateDnsZone = New-AzPrivateDnsZone @newAzPrivateDnsZone_parameters
```

This command won't produce any output but will create the `privatelink.database.windows.net` DNS zone in the primary region resource group. The `$privateDnsZone` variable will be set to the private zone object and will be used later when creating the DNS zone record.

The DNS zone then needs to be connected to the virtual network using the New-AzPrivateDnsVirtualNetworkLink command. We specify the virtual network ID and the DNS zone name to connect these two resources.

```
$newAzPrivateDnsVirtualNetworkLink_parameters = @{
    Name = "$primaryRegionPrefix-$resourceNameSuffix-dnslink"
    ResourceGroupName = $primaryRegionResourceGroupName
    ZoneName = $privateZone
    VirtualNetworkId = $vnet.Id
    Tag = $tags
}
New-AzPrivateDnsVirtualNetworkLink @newAzPrivateDnsVirtualNetworkLink_
parameters
```

Figure 9-10 shows the output from connecting the DNS zone to the virtual network.

```
PS /home> New-AzPrivateDnsVirtualNetworkLink @newAzPrivateDnsVirtualNetw

Name                      : sqlhr01-shr3-dnslink
ResourceId                : /subscriptions/
ResourceGroupName         : sqlhr01-shr3-rg
ZoneName                  : privatelink.database.windows.net
VirtualNetworkId          : /subscriptions/
Location                  :
Etag                      : "02008d27-0000-0100-0000-6392cef00000"
Tags                      : {[Environment, SQL Hyperscale Revealed demo]}
RegistrationEnabled       : False
VirtualNetworkLinkState   : Completed
ProvisioningState         : Succeeded
```

Figure 9-10. *Connecting the DNS zone to the virtual network*

The final step is to create the DNS record for the logical server in the DNS zone using the New-AzPrivateDnsZoneConfig and New-AzPrivateDnsZoneGroup commands.

```
$privateDnsZoneConfig = New-AzPrivateDnsZoneConfig `
    -Name $privateZone `
    -PrivateDnsZoneId $privateDnsZone.ResourceId
$newAzPrivateDnsZoneGroup_parameters = @{
    Name = "$primaryRegionPrefix-$resourceNameSuffix-zonegroup"
```

```
    ResourceGroupName = $primaryRegionResourceGroupName
    PrivateEndpointName = "$primaryRegionPrefix-$resourceNameSuffix-pe"
    PrivateDnsZoneConfig = $privateDnsZoneConfig
}
New-AzPrivateDnsZoneGroup @newAzPrivateDnsZoneGroup_parameters
```

Figure 9-11 shows partial output from creating the DNS record.

```
PS /home> New-AzPrivateDnsZoneGroup @newAzPrivateDnsZoneGroup_para

Name                 : sqlhr01-shr3-zonegroup
Id                   : /subscriptions/
ProvisioningState    : Succeeded
PrivateDnsZoneConfigs : [
                          {
                            "Name": "privatelink.database.windows.
                            "PrivateDnsZoneId": "/subscriptions/c
                            "RecordSets": [
                              {
                                "RecordType": "A",
                                "RecordSetName": "sqlhr01-shr3",
                                "Fqdn": "sqlhr01-shr3.privatelink.
                                "ProvisioningState": "Succeeded",
```

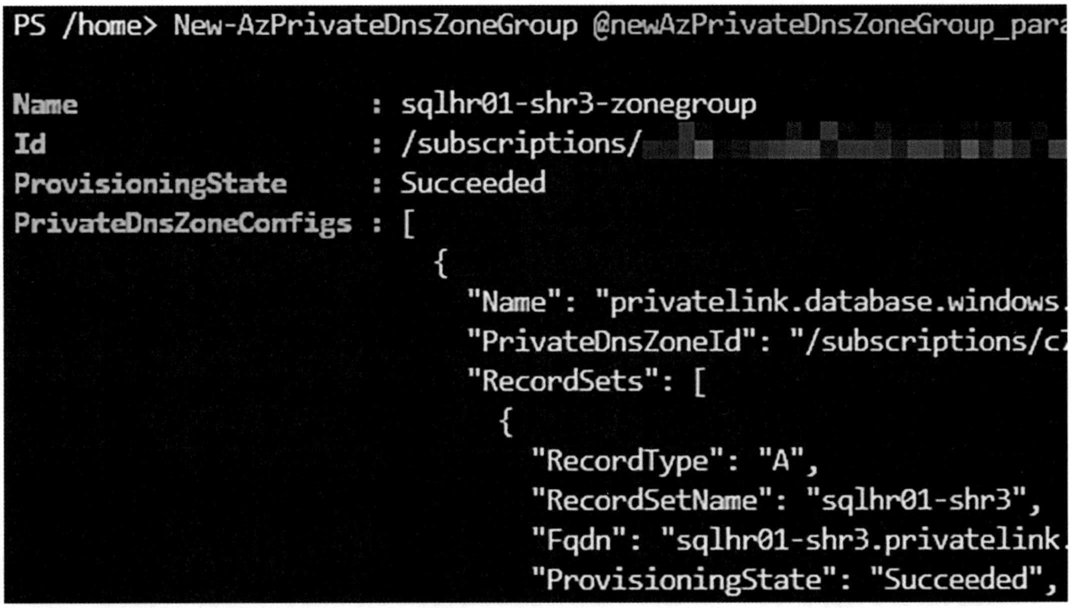

Figure 9-11. *Creating the DNS record for the logical server*

The logical server is now connected to the virtual network with a private endpoint and is not accessible on a public IP address.

Create the Hyperscale Database in the Primary Region

We are now ready to create the Hyperscale database in the logical server. This is done by using the `New-AzSqlDatabase` command. We need to ensure we configure the following parameters:

- `DatabaseName`: This is the name of the database we're creating.
- `ServerName`: This is the name of the logical server that the database will be attached to.

- ResourceGroupName: This is the name of the resource group that contains the logical server and where the database will be put.
- Edition: This must be set to Hyperscale.
- Vcore: This is the number of vCores to assign to the Hyperscale database.
- ComputeGeneration: This must be set to Gen5 or above.
- ComputeModel: This should be set to Provisioned.
- HighAvailabilityReplicaCount: This is the number of high-availability replicas to deploy.
- ZoneRedundant: This should be set to $true to ensure the high-availability replicas are spread across the availability zones.
- BackupStorageRedundancy: This should be set to GeoZone to enable geographically zone-redundant backups to be created.

The command will look like this:

```
$newAzSqlDatabase_parameters = @{
    DatabaseName = 'hyperscaledb'
    ServerName = "$primaryRegionPrefix-$resourceNameSuffix"
    ResourceGroupName = $primaryRegionResourceGroupName
    Edition = 'Hyperscale'
    Vcore = 2
    ComputeGeneration = 'Gen5'
    ComputeModel = 'Provisioned'
    HighAvailabilityReplicaCount = 2
    ZoneRedundant = $true
    BackupStorageRedundancy = 'GeoZone'
    Tags = $tags
}
New-AzSqlDatabase @newAzSqlDatabase_parameters
```

Figure 9-12 shows part of the output from creating the Hyperscale database using the New-AzSqlDatabase command. The full output is truncated for brevity.

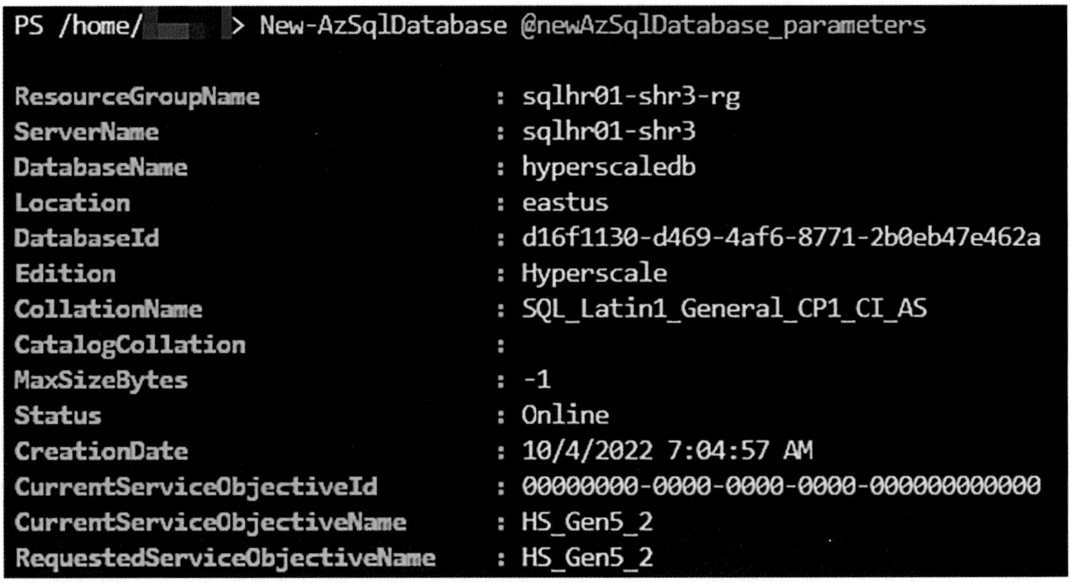

```
PS /home/      > New-AzSqlDatabase @newAzSqlDatabase_parameters

ResourceGroupName                 : sqlhr01-shr3-rg
ServerName                        : sqlhr01-shr3
DatabaseName                      : hyperscaledb
Location                          : eastus
DatabaseId                        : d16f1130-d469-4af6-8771-2b0eb47e462a
Edition                           : Hyperscale
CollationName                     : SQL_Latin1_General_CP1_CI_AS
CatalogCollation                  :
MaxSizeBytes                      : -1
Status                            : Online
CreationDate                      : 10/4/2022 7:04:57 AM
CurrentServiceObjectiveId         : 00000000-0000-0000-0000-000000000000
CurrentServiceObjectiveName       : HS_Gen5_2
RequestedServiceObjectiveName     : HS_Gen5_2
```

Figure 9-12. *Creating the Hyperscale database*

Configure Diagnostic and Audit Logs to Be Sent to a Log Analytics Workspace

The final tasks to perform on the environment in the primary region is to configure audit logs for the logical server and diagnostic logs for the database to be sent to the Log Analytics workspace.

Audit logs are configured to be sent to the Log Analytics workspace by running the `Set-AzSqlServerAudit` command, providing the resource ID of the Log Analytics workspace in the `WorkspaceResourceId` parameter.

```
$logAnalyticsWorkspaceResourceId = "/subscriptions/$subscriptionId" + `
    "/resourcegroups/$primaryRegionResourceGroupName" + `
    "/providers/microsoft.operationalinsights" + `
    "/workspaces/$primaryRegionPrefix-$resourceNameSuffix-law"
$setAzSqlServerAudit_Parameters = @{
    ServerName = "$primaryRegionPrefix-$resourceNameSuffix"
    ResourceGroupName = $primaryRegionResourceGroupName
    WorkspaceResourceId = $logAnalyticsWorkspaceResourceId
    LogAnalyticsTargetState = 'Enabled'
}
Set-AzSqlServerAudit @setAzSqlServerAudit_Parameters
```

The command does not return any output.

We can then configure diagnostic logs to be sent to the Log Analytics workspace from the database by using the `New-AzDiagnosticSetting` command. The `WorkspaceId` parameter will be set to the resource ID of the Log Analytics workspace. The `Category` parameter should be set to an array of diagnostic log categories to send.

We are using the `Get-AzDiagnosticSettingCategory` command to get a list of all the categories of diagnostic logs that are emitted from an Azure SQL Database instance and then looping over them to assemble the `$log` variable, which gets passed to the `Log` parameter. There are many other ways to assemble this object, but this is a fairly compact method and will always contain all supported diagnostic categories.

```
$logAnalyticsWorkspaceResourceId = "/subscriptions/$subscriptionId" + `
    "/resourcegroups/$primaryRegionResourceGroupName" + `
    "/providers/microsoft.operationalinsights" + `
    "/workspaces/$primaryRegionPrefix-$resourceNameSuffix-law"
$databaseResourceId = (Get-AzSqlDatabase `
    -ServerName "$primaryRegionPrefix-$resourceNameSuffix" `
    -ResourceGroupName $primaryRegionResourceGroupName `
    -DatabaseName 'hyperscaledb').ResourceId

# Get the Diagnostic Settings category names for the Hyperscale database
$log = @()
$categories = Get-AzDiagnosticSettingCategory -ResourceId
$databaseResourceId |
    Where-Object -FilterScript { $_.CategoryType -eq 'Logs' }
$categories | ForEach-Object -Process {
    $log += New-AzDiagnosticSettingLogSettingsObject -Enabled $true
    -Category $_.Name -RetentionPolicyDay 7 -RetentionPolicyEnabled $true
}
$newAzDiagnosticSetting_parameters = @{
    ResourceId = $databaseResourceId
    Name = "Send all logs to $primaryRegionPrefix-$resourceNameSuffix-law"
    WorkspaceId = $logAnalyticsWorkspaceResourceId
    Log = $log
}
New-AzDiagnosticSetting @newAzDiagnosticSetting_parameters
```

Figure 9-13 shows partial output from configuring the diagnostic settings on the database.

```
PS /home> New-AzDiagnosticSetting @newAzDiagnosticSetting_parameters

Name
----
Send all logs to sqlhr01-shr3-law
```

Figure 9-13. *Creating the diagnostic log settings*

The deployment and configuration of the Hyperscale database in the primary region is now complete. We can now deploy the failover region.

Create the Logical Server in the Failover Region and Connect It to the VNet

Deploying the failover logical server is the same process and code as we used in the primary region. We'll summarize the process again here for clarity but will omit the code for brevity.

Note You can still find the code for this in the `Install-SQLHyperscaleRevealedHyperscaleEnvironment.ps1` file in the `ch9` directory provided in the source code for this book.

---

1. Create the logical server using the `New-AzSqlServer` command, this time setting the `ServerName`, `ResourceGroup`, and `Location` parameters to the values appropriate for the failover region. The same user-assigned managed identity as the primary logical server is used to access the TDE protector key in the primary region Key Vault. The object ID of the SQL Administrators group is also passed to the command.
2. Create the private endpoint service connection for the failover logical server using the `New-AzPrivateLinkServiceConnection` command.

3. Connect the private endpoint service connection to the failover virtual network as a private endpoint using the `New-AzPrivateEndpoint` command.
4. Create the private DNS zone to enable name resolution of the private endpoint for the logical server using the `New-AzPrivateDnsZone` command.
5. Connect the private DNS zone to the virtual network using the `New-AzPrivateDnsVirtualNetworkLink` command.
6. Create the DNS zone group in the private endpoint using the `New-AzPrivateDnsZoneGroup` command.

The failover logical server will now be connected to the virtual network, configured to use customer-managed encryption and to use the SQL Administrators Azure AD group.

Create the Replica Hyperscale Database in the Failover Region

It is now possible to create the Hyperscale geo replica database in the logical server in the failover region. We do this by using the `New-AzSqlDatabaseSecondary` command, passing the following parameters:

- `DatabaseName`: This is the name of the primary database to replicate.
- `ServerName`: This is the name of the primary logical server that contains the database to replicate.
- `ResourceGroupName`: This is the name of the resource group that contains the primary logical server and Hyperscale database that will be replicated.
- `PartnerDatabaseName`: This is the name of the failover database. This should usually be the same as the primary database, in which case it can be omitted.
- `PartnerServerName`: This is the name of the failover logical server where the failover database will be created.
- `PartnerResourceGroupName`: This is the resource group containing the failover logical server.

- `SecondaryType`: This must be set to `Geo` to create a geo replica.
- `SecondaryVcore`: This is the number of vCores to assign to the replica Hyperscale database. This should have the same number of vCores as the primary database.
- `SecondaryComputeGeneration`: This must be set to `Gen5` or above and should be the same generation as the primary Hyperscale database.
- `HighAvailabilityReplicaCount`: This should be the same as the primary region, to enable high availability, but in this demonstration, we are configuring it to 1 to reduce cost.
- `ZoneRedundant`: This should be set to `$true` to ensure the high-availability replicas are spread across the availability zones. However, because the demonstration environment has only a single replica, the setting is irrelevant, so we will set it to `$false`.
- `AllowConnections`: This configures whether the replica accepts both read and read/write connections or just read. This will be set to `All` to accept both read and read/write connections.

The command will look like this:

```
$newAzSqlDatabaseSecondary = @{
    DatabaseName = 'hyperscaledb'
    ServerName = "$primaryRegionPrefix-$resourceNameSuffix"
    ResourceGroupName = $primaryRegionResourceGroupName
    PartnerServerName = "$failoverRegionPrefix-$resourceNameSuffix"
    PartnerResourceGroupName = $failoverRegionResourceGroupName
    SecondaryType = 'Geo'
    SecondaryVCore = 2
    SecondaryComputeGeneration = 'Gen5'
    HighAvailabilityReplicaCount = 1
    ZoneRedundant = $false
    AllowConnections = 'All'
}
New-AzSqlDatabaseSecondary @newAzSqlDatabaseSecondary
```

Figure 9-14 shows the output from running the command to create the geo replica.

```
PS /home/         > New-AzSqlDatabaseSecondary @newAzSqlDatabaseSecondary

LinkId                       : 95acc2d6-8e66-40cd-b445-01883a951e7f
ResourceGroupName            : sqlhr01-shr3-rg
ServerName                   : sqlhr01-shr3
DatabaseName                 : hyperscaledb
Role                         : Primary
Location                     : East US
PartnerResourceGroupName     : sqlhr02-shr3-rg
PartnerServerName            : sqlhr02-shr3
PartnerDatabaseName          : hyperscaledb
PartnerRole                  : Secondary
PartnerLocation              : West US 3
AllowConnections             : All
ReplicationState             : CATCH_UP
PercentComplete              : 100
StartTime                    : 10/5/2022 7:52:38 AM
```

Figure 9-14. *Creating the geo replica using PowerShell*

Configure Diagnostic and Audit Logs

The next tasks are to configure the audit logs and diagnostic logs for the failover components to be sent the Log Analytics workspace in the failover region. This is done in the same way as the primary region using the `Set-AzSqlServerAudit` and `New-AzDiagnosticSetting` commands. These commands will be omitted for brevity.

Note You can still find the code for this in the `Install-SQLHyperscaleRevealedHyperscaleEnvironment.ps1` file in the `ch9` directory provided in the source code for this book.

Remove the Key Vault Crypto Officer Role from the Key Vault

The final task is to remove the Key Vault Crypto Officer role from the Key Vault for the account running the script. This is simply a security hygiene task that can be performed by running the `Remove-AzRoleAssignment` command.

```
$roleAssignmentScope = "/subscriptions/$subscriptionId" + `
    "/resourcegroups/$primaryRegionResourceGroupName" + `
    "/providers/Microsoft.KeyVault" + `
    "/vaults/$baseResourcePrefix-$resourceNameSuffix-kv"
$removeAzRoleAssignment_parameters = @{
    ObjectId = $userId
    RoleDefinitionName = 'Key Vault Crypto Officer'
    Scope = $roleAssignmentScope
}
Remove-AzRoleAssignment @removeAzRoleAssignment_parameters
```

This command will report a simple success message on deletion of the role:

```
Succesfully removed role assignment for AD object '64f0702a-f001-4d3b-bb93-
eccf57ec7356' on scope '/subscriptions/<subscription id>/resourcegroups/
sqlhr01-shr3-rg/providers/Microsoft.KeyVault/vaults/sqlhr-shr3-kv' with
role definition 'Key Vault Crypto Officer'
```

The Hyperscale architecture that was defined in Chapter 4's Figure 4-1 has now been deployed and is ready to use.

Summary

In this chapter, we introduced the concept of infrastructure as code and briefly looked at the difference between imperative and declarative methods. We demonstrated the imperative infrastructure as code method of deploying a Hyperscale database environment using PowerShell and the Azure PowerShell modules. We also described some of the key Azure PowerShell commands required to deploy the resources.

In the next chapter, we will deploy a similar environment using imperative infrastructure as code, except using Bash and Azure CLI.

CHAPTER 10

Deploying Azure SQL DB Hyperscale Using Bash and Azure CLI

The previous chapter introduced the concept of infrastructure as code (IaC) as well as briefly describing the difference between implicit and declarative IaC models. The chapter also stepped through the process of deploying the example SQL Hyperscale environment using Azure PowerShell, as well as providing a closer look at some of the commands themselves.

In this chapter, we are going to replicate the same process we used in the previous chapter, deploying an identical SQL Hyperscale environment that was defined in Chapter 4, except we'll be using shell script (Bash specifically) and the Azure CLI instead of Azure PowerShell. We will break down the important Azure CLI commands required to deploy the environment as well as provide a complete Bash deployment script you can use for reference.

Tip In this chapter we'll be sharing several scripts. These will all be available in the `ch10` folder in the source files accompanying this book and in this GitHub repository: `https://github.com/Apress/Azure-SQL-Hyperscale-Revealed`.

Deploying Hyperscale Using the Azure CLI

In this section, we will break down the deployment process into the logical steps required to deploy the SQL Hyperscale environment, including connecting it to the virtual network, configuring the customer-managed encryption, and setting up diagnostic logging and audit.

Z. Barać and D. Scott-Raynsford, *Azure SQL Hyperscale Revealed*,
https://doi.org/10.1007/978-1-4842-9225-9_10

The Azure CLI

The Azure CLI is a command-line tool, written in Python, that can run on most operating systems that support the Python 3.7 (or newer) environment. Once the Azure CLI command-line tool is available on a machine, it can be run with the `az` command.

Supported Azure CLI Environments

The Azure CLI can also be used in PowerShell on Windows, Linux, or macOS. Because PowerShell Core 6.0 and PowerShell 7.0 can also be used on Linux and macOS, it is also possible to run Azure PowerShell on those operating systems. However, the Azure PowerShell modules can't be run directly within a shell script or Windows CMD, they can be run only within PowerShell. Table 10-1 provides a comparison of the different scripting languages available on common operating systems and whether Azure PowerShell or Azure CLI is supported.

Table 10-1. *Supported Environments for Running Azure CLI and Azure PowerShell*

Operating System	Azure PowerShell	Azure CLI
Windows Server 2016 and above or Windows 7 and above	Windows PowerShell 5, PowerShell Core 6, or PowerShell 7, and above	Windows PowerShell 5, PowerShell Core 6, PowerShell 7 and above, or CMD
Linux, multiple distributions, and versions supported.	PowerShell Core 6 or PowerShell 7 and above	Shell script, PowerShell Core 6, or PowerShell 7 and above
macOS 10.9 and above	PowerShell Core 6 or PowerShell 7 and above	Shell script, PowerShell Core 6, or PowerShell 7 and above

Installing the Azure CLI

There are multiple methods that can be used to install the Azure CLI onto your machine, depending on which operating system you're using. You can choose to install the Azure CLI into your own environment using OS-specific package management systems or you can run a Docker container with the Azure CLI commands installed.

Tip If you want to review instructions for all the different methods of installing the Azure CLI into your own environment, see `https://learn.microsoft.com/cli/azure/install-azure-cli`.

If you're using Azure Cloud Shell with either Bash or PowerShell, then these commands are already pre-installed for you. The Azure CLI commands in the Azure Cloud Shell will also be already connected to your Azure account when you open it. For the remainder of this chapter, we're going to be using the Azure CLI in the Azure Cloud Shell.

The current version of the Azure CLI that is available at the time of writing this book is v2.41.0. The Azure CLI is updated frequently, so you should ensure that you keep up-to-date to access the latest features and bug fixes.

Note It is possible that a new feature or bug fix is released for Azure CLI that breaks your existing scripts, so you should either ensure your automation *pins* a specific version or you're testing and updating your infrastructure as code regularly. Breaking changes are usually indicated by a major version number release and may require you to change your code.

Deploying the Starting Environment

As we will be deploying a Hyperscale architecture like what was outlined in Chapter 4, we will need to create the starting environment resources again. We'll use the Starting Environment deployment Bash script that can be found in the `ch4` directory in the files provided for this book and is called `newsqlhyperscalerevealedstartingenvrionment.sh`.

This Bash script is like the PowerShell version we used in the "Deploying the Starting Environment using Azure Cloud Shell" section of Chapter 4, except it is written in Bash script. It accepts the same basic parameters.

- `--primary-region`: This is the name of the Azure region to use as the primary region. If you choose to use a different Azure region, you should ensure it is availability zone enabled.

- `--failover-region`: This is the name of the Azure region to use as the failover region. If you don't specify this parameter, it will default to `West US 3`.
- `--resource-name-suffix`: This is the suffix to append into the resource group and resource names. If you want to use your own four-character code for the suffix instead of a randomly generated one, then omit this parameter and specify the `--use-random-resource-name-suffix` switch instead.

To deploy the starting environment, we are going to use Azure Cloud Shell. To create the starting environment, follow these steps:

1. Open an Azure Cloud Shell.
2. Ensure Bash is selected in the shell selector at the top of the Cloud Shell, as shown in Figure 10-1.

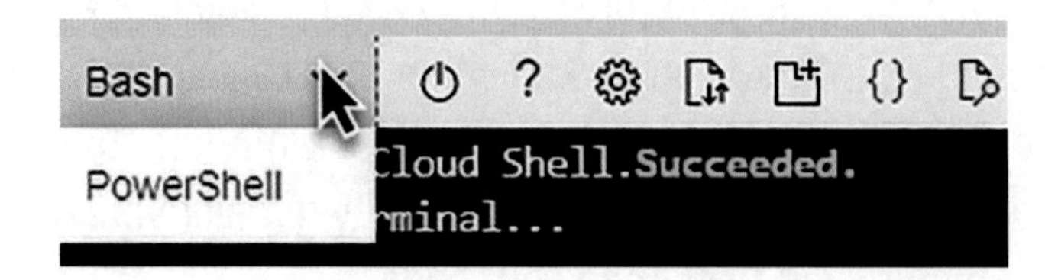

Figure 10-1. *Selecting Bash as the shell to use*

3. Clone the GitHub repository that contains the source code for this book by running the following command:

```
git clone https://github.com/Apress/Azure-SQL-Hyperscale-
Revealed.git
```

4. Before running the script, it is important to ensure the correct Azure subscription is selected by running the following command:

```
az account set --subscription '<subscription name>'
```

5. Deploy the starting environment by running the following command:

```
./Azure-SQL-Hyperscale-Revealed/ch4/newsqlhyperscalerevealed
startingenvrionment.sh --resource-name-suffix '<your
resource name suffix>'
```

It will take a few minutes to deploy the resources but will produce output like Figure 10-2, showing JSON output from the deployment.

```
@Azure:~$ SQLHyperscaleRevealed/ch4/newsqlhyperscalerevealedstartingenvrionment.sh --resource-name-suffix 'shr5'
{
  "id": "/subscriptions/                /providers/Microsoft.Resources/deployments/sql-hyperscale-
  "location": "eastus",
  "name": "sql-hyperscale-revealed-starting-env--202210300428",
  "properties": {
    "correlationId": "02ad2ea6-7080-4b4d-8703-8f024e879303",
    "debugSetting": null,
    "dependencies": [
```

Figure 10-2. *The output from running the script*

The starting environment is now ready for the remaining Hyperscale resources to be deployed into it, using Azure CLI commands.

The Complete Deployment Bash Script

The remainder of this chapter will demonstrate each Azure CLI command that is needed to deploy and configure the Hyperscale environment and associated services. However, you may prefer to see a complete Bash script that deploys the entire environment. A complete Bash script is provided for you in the `ch10` directory of the files provided for this book and is called `installsqlhyperscalerevealedhyperscalenevironment.sh`.

The script contains accepts a few parameters that allow you to customize the environment for your needs:

- `--primary-region`: This is the name of the Azure region to use as the primary region. This must match the value you specified when you created the starting environment.
- `--failover-region`: This is the name of the Azure region to use as the failover region. This must match the value you specified when you created the starting environment.
- `--resource-name-suffix`: This is the suffix to append into the resource group and resource names. This must match the value you specified when you created the starting environment.
- `--aad-user-principal-name:` This is the Azure Active Directory user principal name (UPN) of the user account running this script. This is usually the user principal name of your account.

Note The UPN is required when running the script in Azure Cloud Shell because of the way Azure Cloud Shell executes commands using a service principal, instead of your user account. In this case, the script can't automatically detect your user account to assign appropriate permissions to the Azure Key Vault during the creation of the TDE protector.

To run the `installsqlhyperscalerevealedhyperscalenevironment.sh` script in Azure Cloud Shell, follow these steps:

1. Open the Azure Cloud Shell.
2. Ensure Bash is selected in the shell selector at the top of the Cloud Shell.
3. Clone the GitHub repository that contains the source code for this book by running the following command:

   ```
   git clone https://github.com/Apress/Azure-SQL-Hyperscale-
   Revealed.git
   ```

4. Before running the script, it is important to ensure the correct Azure subscription is selected by running the following command:

   ```
   az account set --subscription '<subscription name>'
   ```

5. Run the script by entering the following:

   ```
   Azure-SQL-Hyperscale-Revealed/ch10/installsqlhyperscalerevealed
   hyperscalenevironment.sh --resource-name-suffix '<your resource
   name suffix>' --aad-user-principal-name '<your Azure AD
   account UPN>'
   ```

6. Because of the number of resources being created, the deployment process may take 30 minutes or more.

Azure CLI Commands in Detail

In this section, we demonstrate each Azure CLI command that is required to create the logical servers and databases as well as the supporting commands that are required to configure the virtual network private link integration and customer-managed key encryption.

Creating Helper Variables

As with most scripts, it is useful to set some variables that contain values that will be used throughout the remainder of the script. This includes the parameters that will control the names of the resources and the Azure regions to deploy to. You can customize these as you see fit, but the `PrimaryRegion`, `FailoverRegion`, and `ResourceNameSuffix` must match the values you used to create the starting environment.

```
PrimaryRegion='East US'
FailoverRegion='West US 3'
ResourceNameSuffix='<your resource name suffix>'
AadUserPrincipalName='<your Azure AD account UPN>'
Environment='SQL Hyperscale Revealed demo'
baseResourcePrefix='sqlhr'
primaryRegionPrefix=$baseResourcePrefix'01'
failoverRegionPrefix=$baseResourcePrefix'02'
primaryRegionResourceGroupName=$primaryRegionPrefix-$ResourceNameSuffix-rg
failoverRegionResourceGroupName=$failoverRegionPrefix-
$ResourceNameSuffix-rg
subscriptionId="$(az account list --query "[?isDefault].id" -o tsv)"
userId="$(az ad user show --id $AadUserPrincipalName --query 'id' -o tsv)"
privateZone='privatelink.database.windows.net'
sqlAdministratorsGroupSid="$(az ad group show --group 'SQL Administrators'
--query 'id' -o tsv)"
```

Figure 10-3 shows the creation of the helper variables in the Azure Cloud Shell.

Note These variables will need to be re-created if your Azure Cloud Shell closes between running commands.

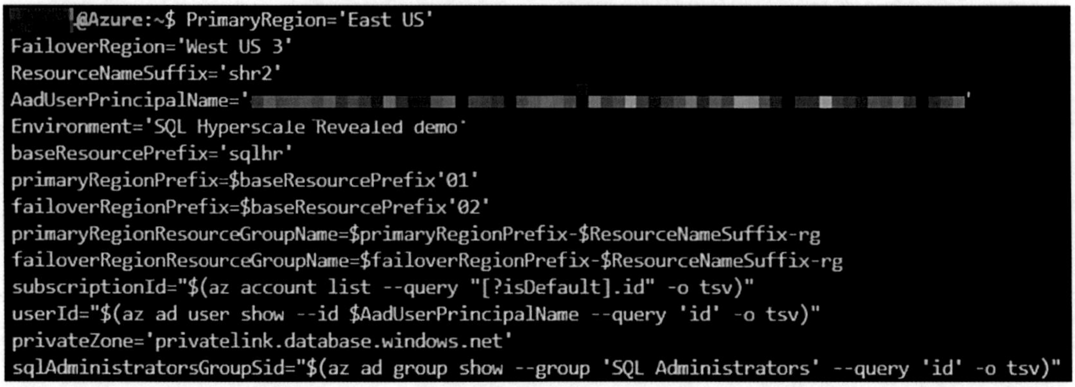

```
@Azure:~$ PrimaryRegion='East US'
FailoverRegion='West US 3'
ResourceNameSuffix='shr2'
AadUserPrincipalName=''
Environment='SQL Hyperscale Revealed demo'
baseResourcePrefix='sqlhr'
primaryRegionPrefix=$baseResourcePrefix'01'
failoverRegionPrefix=$baseResourcePrefix'02'
primaryRegionResourceGroupName=$primaryRegionPrefix-$ResourceNameSuffix-rg
failoverRegionResourceGroupName=$failoverRegionPrefix-$ResourceNameSuffix-rg
subscriptionId="$(az account list --query "[?isDefault].id" -o tsv)"
userId="$(az ad user show --id $AadUserPrincipalName --query 'id' -o tsv)"
privateZone='privatelink.database.windows.net'
sqlAdministratorsGroupSid="$(az ad group show --group 'SQL Administrators' --query 'id' -o tsv)"
```

Figure 10-3. *Creating helper variables*

Create the User-Assigned Managed Identity for the Hyperscale Database

The logical servers need to be able to access the Key Vault where the TDE protector key will be stored, so a user-assigned managed identity needs to be created. This is done by running the `az identity create` command.

This script snippet shows how to create the new user-assigned managed identity in the primary region resource group. The resource ID of the user-assigned managed identity to a variable, so it can be passed to the command used to create the logical server.

```
az identity create \
    --name "$baseResourcePrefix-$ResourceNameSuffix-umi" \
    --resource-group "$primaryRegionResourceGroupName" \
    --location "$primaryRegion" \
    --tags Environment="$Environment"
userAssignedManagedIdentityId="/subscriptions/$subscriptionId"\
"/resourcegroups/$primaryRegionResourceGroupName"\
"/providers/Microsoft.ManagedIdentity"\
"/userAssignedIdentities/$baseResourcePrefix-$ResourceNameSuffix-umi"
```

Figure 10-4 shows the expected output from the user-assigned managed identity being created.

```
@Azure:~$ az identity create \
    --name "$baseResourcePrefix-$ResourceNameSuffix-umi" \
    --resource-group "$primaryRegionResourceGroupName" \
    --location "$primaryRegion" \
    --tags Environment="$Environment"
userAssignedManagedIdentityId="/subscriptions/$subscriptionId"\
"/resourcegroups/$primaryRegionResourceGroupName"\
"/providers/Microsoft.ManagedIdentity"\
"/userAssignedIdentities/$baseResourcePrefix-$ResourceNameSuffix-umi"
{
  "clientId": "[illegible]",
  "id": "/subscriptions/[illegible]/resourcegroups/sqlhr01-shr2-rg/
  "location": "eastus",
  "name": "sqlhr-shr2-umi",
  "principalId": "[illegible]",
  "resourceGroup": "sqlhr01-shr2-rg",
  "tags": {
    "Environment": "SQL Hyperscale Revealed demo"
  },
  "tenantId": "[illegible]",
  "type": "Microsoft.ManagedIdentity/userAssignedIdentities"
}
```

Figure 10-4. *Creating the user-assigned managed identity*

Prepare the TDE Protector Key in the Key Vault

The user account that is being used to run the commands to create the TDE protector key needs to be granted access to the Key Vault. This is done using the `az role assignment create` command. We are granting the role to the user account that was specified in the `userId` helper variable.

```
scope="/subscriptions/$subscriptionId"\
"/resourcegroups/$primaryRegionResourceGroupName"\
"/providers/Microsoft.KeyVault"\
"/vaults/$baseResourcePrefix-$ResourceNameSuffix-kv"
az role assignment create \
    --role 'Key Vault Crypto Officer' \
    --assignee-object-id "$userId" \
    --assignee-principal-type User \
    --scope "$scope"
```

Note If we were running this command in a pipeline such as GitHub Actions or Azure Pipelines, then we'd most likely be setting the assignee to a service principal or federated workload identity, but this is beyond the scope of this book.

Figure 10-5 shows part of the output from assigning the Key Vault Crypto Officer role to the Key Vault.

```
az role assignment create \
    --role 'Key Vault Crypto Officer' \
    --assignee-object-id "$userId" \
    --assignee-principal-type User \
    --scope "$scope"
{
  "canDelegate": null,
  "condition": null,
  "conditionVersion": null,
  "description": null,
  "id": "/subscriptions/                /resourcegroups/sqlhr01-shr2-rg
  "name": "0cf82d95-76d5-408c-b7aa-3a7385038871",
  "principalId": "                ",
  "principalType": "User",
  "resourceGroup": "sqlhr01-shr2-rg",
  "roleDefinitionId": "/subscriptions/                /providers/Micros
  "scope": "/subscriptions/                /resourcegroups/sqlhr01-shr2
  "type": "Microsoft.Authorization/roleAssignments"
}
```

Figure 10-5. *Assigning role to key vault*

Once the role has been assigned, the user account specified in the `userId` variable can be used to create the TDE protector key in the Key Vault using the `az keyvault key create` command. The identity of the TDE protector key is then assigned to a variable to use when creating the logical server.

```
az keyvault key create \
    --name "$baseResourcePrefix-$ResourceNameSuffix-tdeprotector" \
    --vault-name "$baseResourcePrefix-$ResourceNameSuffix-kv" \
    --kty RSA \
    --size 2048 \
    --ops wrapKey unwrapKey \
    --tags Environment="$Environment"
```

```
tdeProtectorKeyId="$(az keyvault key show --name "$baseResourcePrefix-
$ResourceNameSuffix-tdeprotector" --vault-name "$baseResourcePrefix-
$ResourceNameSuffix-kv" --query 'key.kid' -o tsv)"
```

Figure 10-6 shows partial output from creating the TDE protector.

```
@Azure:~$ az keyvault key create \
  --name "$baseResourcePrefix-$ResourceNameSuffix-tdeprotector" \
  --vault-name "$baseResourcePrefix-$ResourceNameSuffix-kv" \
  --kty RSA \
  --size 2048 \
  --ops wrapKey unwrapKey \
  --tags Environment="$Environment"
tdeProtectorKeyId="$(az keyvault key show --name "$baseResourcePrefix-$Resource
{
  "attributes": {
    "created": "2022-11-03T21:42:25+00:00",
    "enabled": true,
    "expires": null,
    "exportable": false,
    "notBefore": null,
    "recoverableDays": 90,
    "recoveryLevel": "Recoverable",
    "updated": "2022-11-03T21:42:25+00:00"
  },
  "key": {
```

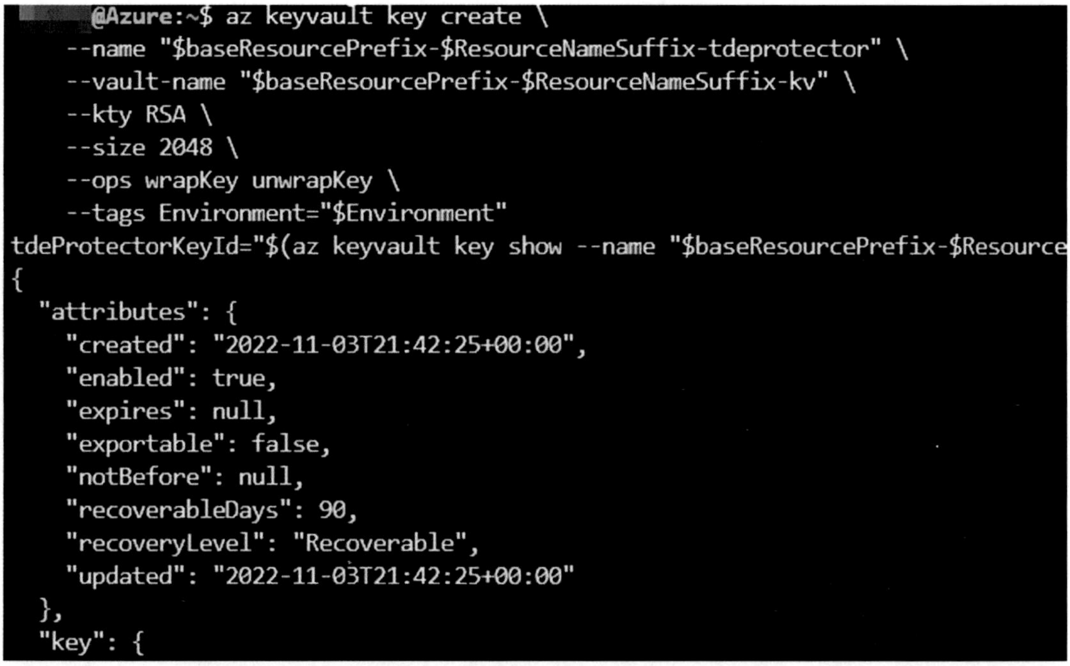

Figure 10-6. *Creating the TDE protector key in the Key Vault*

The `az role assignment create` command is used again to grant the Key Vault Crypto Service Encryption User role to the service principal that is created as part of the user-assigned managed identity. The `az ad sp list` command is used to look up the ID of the service principal.

```
servicePrincipalId="$(az ad sp list --display-name "$baseResourcePrefix-
$ResourceNameSuffix-umi" --query '[0].id' -o tsv)"
scope="/subscriptions/$subscriptionId"\
"/resourcegroups/$primaryRegionResourceGroupName"\
"/providers/Microsoft.KeyVault"\
"/vaults/$baseResourcePrefix-$ResourceNameSuffix-kv"\
"/keys/$baseResourcePrefix-$ResourceNameSuffix-tdeprotector"
az role assignment create \
    --role 'Key Vault Crypto Service Encryption User' \
```

```
    --assignee-object-id "$servicePrincipalId" \
    --assignee-principal-type ServicePrincipal \
    --scope "$scope"
```

Figure 10-7 shows part of the output from assigning the service principal a role on the key.

```
az role assignment create \
    --role 'Key Vault Crypto Service Encryption User' \
    --assignee-object-id "$servicePrincipalId" \
    --assignee-principal-type ServicePrincipal \
    --scope "$scope"
{
  "canDelegate": null,
  "condition": null,
  "conditionVersion": null,
  "description": null,
  "id": "/subscriptions/                    /resourcegroups/sqlhr01-shr2-rg
  "name": "389dd4d2-cbd6-4140-9c8f-704c3158938b",
  "principalId": "                    ",
  "principalType": "ServicePrincipal",
  "resourceGroup": "sqlhr01-shr2-rg",
  "roleDefinitionId": "/subscriptions/                    /providers/Micros
  "scope": "/subscriptions/                    /resourcegroups/sqlhr01-shr2
  "type": "Microsoft.Authorization/roleAssignments"
}
```

Figure 10-7. *Grant access to the key to the service principal*

The logical servers will now be able to access the TDE protector key in the Key Vault once they've been assigned the user-assigned managed identity.

Create the Logical Server in the Primary Region

The logical server in the primary region can now be created. This is done using the `az sql server create` command. The `AssignIdentity`, `IdentityType`, `UserAssignedIdentityId`, and `PrimaryUserAssignedIdentityId` parameters all need to be set to ensure the user-assigned managed identity is assigned and available to the logical server. The `KeyId` parameter needs to set to the full ID of the TDE protector in the Key Vault that was retrieved earlier.

```
az sql server create \
    --name "$primaryRegionPrefix-$ResourceNameSuffix" \
    --resource-group "$primaryRegionResourceGroupName" \
    --location "$primaryRegion" \
```

```
--enable-ad-only-auth \
--assign-identity \
--identity-type UserAssigned \
--user-assigned-identity-id "$userAssignedManagedIdentityId" \
--primary-user-assigned-identity-id "$userAssignedManagedIdentityId" \
--key-id "$tdeProtectorKeyId" \
--external-admin-principal-type Group \
--external-admin-name 'SQL Administrators' \
--external-admin-sid "$sqlAdministratorsGroupSid"
```

Figure 10-8 shows part of the output from creating the logical server.


```
{
  "administratorLogin": "",
  "administratorLoginPassword": null,
  "administrators": {
    "administratorType": "ActiveDirectory",
    "azureAdOnlyAuthentication": true,
    "login": "SQL Administrators",
    "principalType": "Group",
    "sid": "",
    "tenantId": ""
  },
  "federatedClientId": null,
  "fullyQualifiedDomainName": "sqlhr01-shr2.database.windows.
  "id": "/subscriptions/
```


Figure 10-8. *Creating the primary region logical server*

Connect the Primary Logical Server to the Virtual Network

Now that the logical server has been created in the primary region, it can be connected to the virtual network. The process of making the logical server available on a private endpoint in the virtual network requires a number of commands to be run.

The first command creates the private endpoint for the logical server using the `az network private-endpoint create` command. The `--private-connection-resource` parameter needs to be set to the resource ID of the logical server. The `--vnet-name` and `--subnet` parameters are used to specify the virtual network the private endpoint is connected to.

```
sqlServerResourceId="/subscriptions/$subscriptionId"\
"/resourcegroups/$primaryRegionResourceGroupName"\
"/providers/Microsoft.Sql"\
"/servers/$primaryRegionPrefix-$ResourceNameSuffix"
az network private-endpoint create \
    --name "$primaryRegionPrefix-$ResourceNameSuffix-pe" \
    --resource-group "$primaryRegionResourceGroupName" \
    --location "$primaryRegion" \
    --vnet-name "$primaryRegionPrefix-$ResourceNameSuffix-vnet" \
    --subnet "data_subnet" \
    --private-connection-resource-id "$sqlServerResourceId" \
    --group-id sqlServer \
    --connection-name "$primaryRegionPrefix-$ResourceNameSuffix-pl"
```

Once the private endpoint has been created, we can create a private DNS zone that will be used to resolve the name of the logical server within the virtual network. The private DNS zone is created using the `az network private-dns zone create` command. The `--name` is set to the domain name suffix used by databases that is connected to private endpoints. For Azure Cloud, this is privatelink.database.windows.net.

```
az network private-dns zone create \
    --name "$privateZone" \
    --resource-group "$primaryRegionResourceGroupName"
```

Next, the private DNS zone is connected to the virtual network using the `az network private-dns link vnet create` command, specifying the virtual network name in the `--virtual-network` parameter and the private DNS zone name in the `--zone-name` parameter.

```
az network private-dns link vnet create \
    --name "$primaryRegionPrefix-$ResourceNameSuffix-dnslink" \
    --resource-group "$primaryRegionResourceGroupName" \
    --zone-name "$privateZone" \
    --virtual-network "$primaryRegionPrefix-$ResourceNameSuffix-vnet" \
    --registration-enabled false
```

Finally, a DNS zone group for the logical server private endpoint is created in the private DNS zone. This will ensure that the DNS A-record for the logical server is created and kept up-to-date in the private DNS zone. This is done by the `az network private-endpoint dns-zone-group create` command.

```
az network private-endpoint dns-zone-group create \
    --name "$primaryRegionPrefix-$ResourceNameSuffix-zonegroup" \
    --resource-group "$primaryRegionResourceGroupName" \
    --endpoint-name "$primaryRegionPrefix-$ResourceNameSuffix-pe" \
    --private-dns-zone "$privateZone" \
    --zone-name "$privateZone"
```

The logical server is now connected to the virtual network with a private endpoint and is not accessible on a public IP address.

Create the Hyperscale Database in the Primary Region

We can now use the `az sql db create` command to create the Hyperscale database in the logical server. The following parameters need to be included:

- `--name`: This is the name of the database we're creating.
- `--server`: This is the name of the logical server that the database is being created in.
- `--resource-group`: This is the name of the resource group that contains the logical server and where the database will be put.
- `--edition`: This must be set to `Hyperscale`.
- `--capacity`: This is the number of vCores to assign to the Hyperscale database.
- `--family`: This must be set to `Gen5` or above.
- `--compute-model`: This should be set to `Provisioned`.
- `--ha-replicas`: This is the number of high-availability replicas to deploy.

- `--zone-redundant`: This should be included to ensure the high-availability replicas are spread across the availability zones.
- `--backup-storage-redundancy`: This should be set to `GeoZone` to enable geographically zone-redundant backups to be created.

The command will look like this:

```
az sql db create \
    --name 'hyperscaledb' \
    --server "$primaryRegionPrefix-$ResourceNameSuffix" \
    --resource-group "$primaryRegionResourceGroupName" \
    --edition 'Hyperscale' \
    --capacity 2 \
    --family 'Gen5' \
    --compute-model 'Provisioned' \
    --ha-replicas 2 \
    --zone-redundant \
    --backup-storage-redundancy 'GeoZone' \
    --tags $tags
```

Figure 10-9 shows part of the output from creating the Hyperscale database using the `az sql db create` command. The full output is truncated for brevity.

```
@Azure:~$ az sql db create \
    --name 'hyperscaledb' \
    --server "$primaryRegionPrefix-$ResourceNameSuffix" \
    --resource-group "$primaryRegionResourceGroupName" \
    --edition 'Hyperscale' \
    --capacity 2 \
    --family 'Gen5' \
    --compute-model 'Provisioned' \
    --ha-replicas 2 \
    --zone-redundant \
    --backup-storage-redundancy 'GeoZone' \
    --tags $tags
{
  "autoPauseDelay": null,
  "catalogCollation": "SQL_Latin1_General_CP1_CI_AS",
  "collation": "SQL_Latin1_General_CP1_CI_AS",
  "createMode": null,
  "creationDate": "2022-11-07T05:32:19.890000+00:00",
  "currentBackupStorageRedundancy": "GeoZone",
  "currentServiceObjectiveName": "HS_Gen5_2",
  "currentSku": {
    "capacity": 2,
    "family": "Gen5",
    "name": "HS_Gen5",
    "size": null,
    "tier": "Hyperscale"
```

Figure 10-9. *Creating the Hyperscale database*

Configure Diagnostic and Audit Logs to Be Sent to the Log Analytics Workspace

The last tasks we need to do on the environment in the primary region are to configure audit logs for the logical server and diagnostic logs for the database to be sent to the Log Analytics workspace.

Audit logs are configured to be sent to the Log Analytics workspace by running the `az sql server audit-policy update` command, providing the resource ID of the Log Analytics workspace in the `--log-analytics-workspace-resource-id` parameter.

```
logAnalyticsWorkspaceResourceId="/subscriptions/$subscriptionId"\
"/resourcegroups/$primaryRegionResourceGroupName"\
"/providers/microsoft.operationalinsights"\
"/workspaces/$primaryRegionPrefix-$ResourceNameSuffix-law"
az sql server audit-policy update \
    --name "$primaryRegionPrefix-$ResourceNameSuffix" \
    --resource-group "$primaryRegionResourceGroupName" \
    --log-analytics-workspace-resource-id
    "$logAnalyticsWorkspaceResourceId" \
    --log-analytics-target-state Enabled \
    --state Enabled
```

Figure 10-10 shows some partial output from the `az sql server audit-policy update` command.

```
az sql server audit-policy update \
    --name "$primaryRegionPrefix-$ResourceNameSuffix" \
    --resource-group "$primaryRegionResourceGroupName" \
    --log-analytics-workspace-resource-id "$logAnalyticsWorkspaceResourceId" \
    --log-analytics-target-state Enabled \
    --state Enabled
{
  "auditActionsAndGroups": [
    "SUCCESSFUL_DATABASE_AUTHENTICATION_GROUP",
    "FAILED_DATABASE_AUTHENTICATION_GROUP",
    "BATCH_COMPLETED_GROUP"
  ],
  "id": "/subscriptions/                                    /resourceGroups/sqlh
  "isAzureMonitorTargetEnabled": true,
```

Figure 10-10. *Output from configuring logical server audit logs*

The diagnostic logs also need to be sent to the Log Analytics workspace from the database by using the `az monitor diagnostic-settings create` command. The `--workspace` parameter will be set to the resource ID of the Log Analytics workspace. The `--logs` parameter should be set to a JSON array of diagnostic log categories to send to the workspace. Here's an example:

```
logs='[
    {
        "category": "SQLInsights",
```

```
        "enabled": true,
        "retentionPolicy": {
            "enabled": false,
            "days": 0
        }
    },
    {
        "category": "AutomaticTuning",
        "enabled": true,
        "retentionPolicy": {
            "enabled": false,
            "days": 0
        }
    }
]'
logAnalyticsWorkspaceResourceId="/subscriptions/$subscriptionId"\
"/resourcegroups/$primaryRegionResourceGroupName"\
"/providers/microsoft.operationalinsights"\
"/workspaces/$primaryRegionPrefix-$ResourceNameSuffix-law"
databaseResourceId="/subscriptions/$subscriptionId"\
"/resourcegroups/$primaryRegionResourceGroupName"\
"/providers/Microsoft.Sql"\
"/servers/$primaryRegionPrefix-$ResourceNameSuffix"\
"/databases/hyperscaledb"
az monitor diagnostic-settings create \
    --name "Send all logs to $primaryRegionPrefix-
    $ResourceNameSuffix-law" \
    --resource "$databaseResourceId" \
    --logs "$logs" \
    --workspace "$logAnalyticsWorkspaceResourceId"
```

Tip The list of available diagnostic logging categories for a Hyperscale database can be obtained by running the `az monitor diagnostic-settings list` command and setting `--resource` to the resource ID of the Hyperscale database.

Figure 10-11 shows partial output from configuring the diagnostic settings on the database.

```
az monitor diagnostic-settings create \
    --name "Send all logs to $primaryRegionPrefix-$ResourceNameSuffix-law" \
    --resource "$databaseResourceId" \
    --logs "$logs" \
    --workspace "$logAnalyticsWorkspaceResourceId"

{
  "eventHubAuthorizationRuleId": null,
  "eventHubName": null,
  "id": "/subscriptions/                                    /resourcegroups/s
  "identity": null,
  "kind": null,
  "location": null,
  "logAnalyticsDestinationType": null,
  "logs": [
    {
      "category": "SQLInsights",
      "categoryGroup": null,
      "enabled": true,
```

Figure 10-11. *Creating the diagnostic log settings*

The deployment and configuration of the Hyperscale database in the primary region is now complete. We can now deploy the failover region.

Create the Logical Server in the Failover Region and Connect It to the Virtual Network

Creating a failover logical server and database is almost the same process and code as we used in the primary region. The steps are summarized again here for clarity, but the code is omitted for brevity. You can still find the code for this in the `installsqlhyperscalerevealedhyperscalenevironment.sh` file in the `ch10` directory provided in the source code for this book.

1. Create the logical server using the `az sql server create` command, this time setting the `--name`, `--resource-group`, and `--location` parameters to the values appropriate for the failover

region. The failover logical server will use the same user-assigned managed identity as the primary logical server so that it can access the TDE protector key in the primary region's Key Vault.

2. Create the private endpoint in the failover virtual network for the failover logical server using the `az network private-endpoint create` command.
3. Create the private DNS zone to enable name resolution of the private endpoint for the logical server using the `az network private-dns zone create` command.
4. Connect the private DNS zone to the virtual network using the `az network private-dns link vnet create` command.
5. Create the DNS zone group in the private endpoint using the `az network private-endpoint dns-zone-group create` command.

The failover logical server will now be connected to the virtual network, configured to use customer-managed encryption and to use the SQL Administrators Azure AD group.

Create the Replica Hyperscale Database in the Failover Region

A geo replica of the primary Hyperscale database can now be created in the failover region. We do this by using the `az sql db replica create` command, passing the following parameters:

- `--name`: This is the name of the primary database to replicate.
- `--server`: This is the name of the primary logical server that contains the database to replicate.
- `--resource-group`: This is the name of the resource group that contains the primary logical server and Hyperscale database that will be replicated.
- `--partner-database`: This is the name of the failover database. This should usually be the same as the primary database, in which case it can be omitted.

- --partner-server: This is the name of the failover logical server where the failover database will be created.
- --partner-resource-group: This is the resource group containing the failover logical server.
- --secondary-type: This must be set to Geo to create a geo replica.
- --capacity: This is the number of vCores to assign to the replica Hyperscale database. This should have the same number of vCores as the primary database.
- --family: This must be set to Gen5 or above and should be the same generation as the primary Hyperscale database.
- --ha-replicas: This should be the same as the primary region, to enable high availability, but in this demonstration, we are configuring it to 1 to reduce the cost.
- --zone-redundant: This should be set to true to ensure the high-availability replicas are spread across the availability zones. However, because the demonstration environment has only a single replica, the setting will be set to false.
- -read-scale: This configures whether the replica accepts read-only connections. It should be set to Enabled in our case so that we can read from this database.

The command will look like this:

```
az sql db replica create \
    --name "hyperscaledb" \
    --resource-group "$primaryRegionResourceGroupName" \
    --server "$primaryRegionPrefix-$ResourceNameSuffix" \
    --partner-resource-group "$failoverRegionResourceGroupName" \
    --partner-server "$failoverRegionPrefix-$ResourceNameSuffix" \
    --secondary-type Geo \
    --family Gen5 \
    --capacity 2 \
    --zone-redundant false \
    --ha-replicas 1 \
    --read-scale "Enabled"
```

Figure 10-12 shows partial output from running the command to create the geo replica.

```
@Azure:~$ az sql db replica create \
    --name "hyperscaledb" \
    --resource-group "$primaryRegionResourceGroupName" \
    --server "$primaryRegionPrefix-$ResourceNameSuffix" \
    --partner-resource-group "$failoverRegionResourceGroupName" \
    --partner-server "$failoverRegionPrefix-$ResourceNameSuffix" \
    --secondary-type Geo \
    --family Gen5 \
    --capacity 2 \
    --zone-redundant false \
    --ha-replicas 1 \
    --read-scale "Enabled"
{
  "autoPauseDelay": null,
  "catalogCollation": "SQL_Latin1_General_CP1_CI_AS",
  "collation": "SQL_Latin1_General_CP1_CI_AS",
  "createMode": null,
  "creationDate": "2022-11-07T06:26:17.657000+00:00",
  "currentBackupStorageRedundancy": "GeoZone",
  "currentServiceObjectiveName": "HS_Gen5_2",
  "currentSku": {
    "capacity": 2,
    "family": "Gen5",
    "name": "HS_Gen5",
    "size": null,
    "tier": "Hyperscale"
```

Figure 10-12. *Creating the geo replica using the Azure CLI*

Configure Diagnostic and Audit Logs

The next tasks are to configure the audit logs and diagnostic logs for the failover components to be sent to the Log Analytics workspace in the failover region. This is done in the same way as the primary region using the `az sql server audit-policy update` and `az monitor diagnostic-settings create` commands. These commands will be omitted for brevity.

Note You can still find the code for this in the `Install-installsqlhyperscalerevealedhyperscalenevironment.sh` file in the `ch10` directory provided in the source code for this book.

Remove the Key Vault Crypto Officer Role from the Key Vault

The final task is to remove the Key Vault Crypto Officer role from the Key Vault for the account running the script. This is simply a security hygiene task that can be performed by running the `az role assignment delete` command.

```
scope="/subscriptions/$subscriptionId"\
"/resourcegroups/$primaryRegionResourceGroupName"\
"/providers/Microsoft.KeyVault"\
"/vaults/$baseResourcePrefix-$ResourceNameSuffix-kv"
az role assignment delete \
    --assignee "$userId" \
    --role "Key Vault Crypto Officer" \
    --scope "$scope"
```

This command will not produce any output. The Hyperscale architecture that was defined in Chapter 4's Figure 4-1 has now been deployed using the Azure CLI commands and is ready to use.

Summary

In this chapter, we introduced imperative infrastructure as code to deploy a Hyperscale environment using the Azure CLI with a shell script. We looked at the Azure CLI commands to create the logical server as well as connect it to a virtual network. We also showed the Azure CLI commands that are used to configure transparent data encryption using a customer-managed key stored in a Key Vault. The command for creating the Hyperscale database and the geo replica were also demonstrated. Finally, we showed how to configure audit and diagnostic logging to an Azure Log Analytics workspace.

Although imperative infrastructure as code will be familiar to most readers, it is recommended that declarative infrastructure as code is adopted. There are very few situations where declarative IaC doesn't provide significant benefits when compared to imperative IaC. In the next chapter, we will demonstrate the use of declarative IaC using Azure Bicep.

CHAPTER 11

Deploying Azure SQL DB Hyperscale Using Azure Bicep

The previous two chapters deployed a Hyperscale environment using implicit infrastructure as code (IaC) with Azure PowerShell and the Azure CLI (with Bash). However, as we've noted in the previous chapters, declarative IaC provides many benefits so should be used whenever possible.

In this chapter, we are going to deploy the same SQL Hyperscale environment that was defined in Chapter 4, except this time we'll be using Azure Bicep. We will look at a few of the key Azure Bicep resources required to deploy the SQL Hyperscale environment and supporting resources. We will not cover all aspects of Azure Bicep as this would require an entire book on its own.

Note This chapter is not intended to be a primer on Azure Bicep, its syntax, or other declarative methods. Microsoft provides many great free resources to help learn Azure Bicep here: `http://aka.ms/learnbicep` and `https://aka.ms/bicep`.

If you're not familiar with Azure Bicep, it is strongly recommended to at least learn the basics before reading through it. If you don't have the time to learn the Azure Bicep language, then you should feel free to skip this chapter and refer to it at a later date once you've become familiar with the language.

Z. Barać and D. Scott-Raynsford, *Azure SQL Hyperscale Revealed*,
https://doi.org/10.1007/978-1-4842-9225-9_11

As with the previous chapters, the complete set of Azure Bicep files and scripts required to deploy the environment can be found in the `ch11` directory in the files provided for this book. You should review these files to familiarize yourself with the process of deploying a SQL Hyperscale environment with all the "bells and whistles."

About Azure Bicep

Azure Bicep is a declarative IaC domain-specific language (DSL) that is provided by Microsoft to deploy and configure resources in Azure. This language declares what the desired state of the Azure resources should be, rather than the implicit steps required to deploy the resources. This is what makes Azure Bicep a declarative IaC methodology.

Azure Bicep deployments are usually made up of one or more Azure Bicep files (which have a `bicep` extension) and are often parameterized so that they can be easily reused. The Azure Bicep files can then be sent to the Azure Resource Manager (ARM) engine. The ARM engine then determines if the resources are in the desired state or not and will make the necessary changes to put the resources into this state. If everything is already in the desired state, then no changes will be made; this is referred to as *idempotency*.

Note The Azure Bicep files are transparently transpiled into an older declarative syntax called an *ARM template* by the commands that are used to send them to the ARM engine. The transpilation process can be performed manually for diagnostic purposes or if your tools support only the older ARM template syntax.

Deploying Using Azure Bicep

In this section, we will break down and review some of the key resources that are found in Azure Bicep files for deploying Azure SQL Hyperscale databases and related resources.

A Complete Azure Bicep Deployment

As with the previous chapters, we've provided you with the ability to deploy the environment in one go using a single script. The script simply runs the appropriate command that will transpile and send the Azure Bicep files to the ARM engine.

Unlike the previous chapters, however, we don't need to start by deploying a starting environment. Instead, the entire environment, including the starting resources (virtual networks, Key Vault, and Log Analytics workspaces), are deployed using the Azure Bicep files.

Both PowerShell and Bash versions of the script are provided for your convenience. For the PowerShell version, the Azure PowerShell `New-AzDeployment` is used to send the files to the ARM engine. For the Bash version, the Azure CLI `az deployment sub create` command is used. Both versions use the same parameters and the same Azure Bicep files.

Both scripts are provided for you in the `ch11` directory of the files provided for this book. The Bash script is called `installsqlhyperscalerevealedhyperscalenevironment.sh`, and the PowerShell script is called `Install-SQLHyperscaleRevealedHyperscaleEnvironment.ps1`.

Deploying Using PowerShell

The PowerShell script to deploy the environment using Azure Bicep is called `Install-SQLHyperscaleRevealedHyperscaleEnvironment.ps1` and can be found in the `ch11` directory of the files provided for this book.

The script accepts a few parameters that allow you to customize the environment for your needs.

- `-PrimaryRegion`: This is the name of the Azure region to use as the primary region. This must match the value you specified when you created the starting environment.
- `-FailoverRegion`: This is the name of the Azure region to use as the failover region. This must match the value you specified when you created the starting environment.
- `-ResourceNameSuffix`: This is the suffix to append into the resource group and resource names. This must match the value you specified when you created the starting environment.

To run the `Install-SQLHyperscaleRevealedHyperscaleEnvironment.ps1` script in the Azure Cloud Shell, follow these steps:

1. Open the Azure Cloud Shell.
2. Ensure Bash is selected in the shell selector at the top of the Cloud Shell.

3. Clone the GitHub repository that contains the source code for this book by running the following command:

```
git clone https://github.com/Apress/Azure-SQL-Hyperscale-
Revealed.git
```

4. Before running the script, it is important to ensure the correct Azure subscription is selected by running the following command:

```
Select-AzSubscription -Subscription '<subscription name>'
```

5. Run the script by entering the following:

```
./Azure-SQL-Hyperscale-Revealed/ch10/Install-SQLHyperscaleRevealed
HyperscaleEnvironment.ps1 -ResourceNameSuffix
'<your resource name suffix>'
```

Because of the number of resources being created, the deployment process may take 30 minutes or more.

Deploying Using Bash

The Bash script to deploy the environment using Azure Bicep is called `installsqlhyperscalerevealedhyperscalenevironment.sh` and can be found in the `ch11` directory of the files provided for this book.

The script accepts a few parameters that allow you to customize the environment for your needs.

- `--primary-region`: This is the name of the Azure region to use as the primary region. This must match the value you specified when you created the starting environment.
- `--failover-region`: This is the name of the Azure region to use as the failover region. This must match the value you specified when you created the starting environment.
- `--resource-name-suffix`: This is the suffix to append into the resource group and resource names. This must match the value you specified when you created the starting environment.

To run the `installsqlhyperscalerevealedhyperscalenevironment.sh` script in the Azure Cloud Shell, follow these steps:

1. Open the Azure Cloud Shell.
2. Ensure Bash is selected in the shell selector at the top of the Cloud Shell.
3. Clone the GitHub repository that contains the source code for this book by running the following command:

   ```
   git clone https://github.com/Apress/Azure-SQL-Hyperscale-
   Revealed.git
   ```

4. Before running the script, it is important to ensure the correct Azure subscription is selected by running the following command:

   ```
   az account set --subscription '<subscription name>'
   ```

5. Run the script by entering the following:

   ```
   ./Azure-SQL-Hyperscale-Revealed/ch11/installsqlhyperscalerevealedhy
   perscalenevironment.sh --resource-name-suffix '<your resource
   name suffix>'
   ```

Because of the number of resources being created, the deployment process may take 30 minutes or more.

Hyperscale Resources in Azure Bicep

In this section, we will review some of the most important resource types required when deploying an Azure SQL Hyperscale environment. We will note some of the key parameters of the resource type and identify some of the important features to be aware of, but we won't go into detail on every resource and parameter as this would require a book in itself.

Note Some of the resource type versions that are used in this book are in preview. These are subject to change but are often required to be used to support the latest features of the underlying resources.

Creating the User-Assigned Managed Identity

The SQL logical server in the primary and failover regions will need a user-assigned managed identity that it will use to access the TDE protector key in the Azure Key Vault. This user-assigned managed identity resource can be created using the `Microsoft.ManagedIdentity/userAssignedIdentities` resource type.

```
resource userAssignedManagedIdentity 'Microsoft.ManagedIdentity/
userAssignedIdentities@2018-11-30' = {
  name: name
  location: location
}
```

This resource type just needs the `name` and `location` properties to be set.

You can find the complete Azure Bicep module to create the user-assigned managed identity in the `user_assigned_managed_identity.bicep` file found in the `ch11/modules` directory in the files provided for this book.

Configuring the Key Vault for Customer-Managed Encryption

The TDE protector key that will be used by the SQL logical server for encryption can be created within the Azure Key Vault using the `Microsoft.KeyVault/vaults/keys` resource type.

```
resource tdeProtectorKey 'Microsoft.KeyVault/vaults/keys@2022-07-01' = {
  name: '${keyVault.name}/${keyName}'
  properties: {
    kty: 'RSA'
    keySize: 2048
    keyOps: [
      'wrapKey'
      'unwrapKey'
```

```
    ]
  }
}
```

The name needs to be set to the name of the Key Vault concatenated with the name of the key. The kty, keySize, and keyOps properties must be set according to the requirements for Azure SQL customer-managed keys for transparent data encryption. If wrapKey or unwrapKey is not included in the keyOps, then errors will occur when enabling encryption.

The user-assigned managed identity will need to be granted the Key Vault Crypto Service Encryption role on the TDE protector key so that it can access it. This is done by obtaining the role ID for the Key Vault crypto service encryption user using its GUID with the Microsoft.Authorization/roleDefinitions resource type. The role ID and the user-assigned managed identity are provided to the Microsoft.Authorization/roleAssignments resource type to assign the role.

```
var keyVaultCryptoServiceEncryptionUserRoleId = 'e147488a-f6f5-4113-8e2d-
b22465e65bf6'

resource keyVaultCryptoServiceEncryptionRoleDefinition 'Microsoft.
Authorization/roleDefinitions@2022-04-01' existing = {
  scope: subscription()
  name: keyVaultCryptoServiceEncryptionUserRoleId // Key Vault Crypto
Service Encryption User Role
}

resource roleAssignment 'Microsoft.Authorization/
roleAssignments@2022-04-01' = {
  name: guid(subscription().id, userAssignedManagedIdentityPrincipalId,
keyVaultCryptoServiceEncryptionRoleDefinition.id)
  scope: tdeProtectorKey
  properties: {
    roleDefinitionId: keyVaultCryptoServiceEncryptionRoleDefinition.id
    principalId: userAssignedManagedIdentityPrincipalId
    principalType: 'ServicePrincipal'
  }
}
```

This is all that is required to create the TDE protector key. If we compared this to the process for using Azure PowerShell or the Azure CLI, it is much simpler and requires fewer steps.

You can find the complete Azure Bicep module to create the Azure Key Vault and the TDE protector key in the `key_vault_with_tde_protector.bicep` file found in the `ch11/modules` directory in the files provided for this book.

Creating the Azure SQL Logical Server

Creating the Azure SQL logical server requires the `Microsoft.Sql/servers` resource.

```
resource sqlLogicalServer 'Microsoft.Sql/servers@2022-05-01-preview' = {
  name: name
  location: location
  identity: {
    type: 'UserAssigned'
    userAssignedIdentities: {
      '${userAssignedManagedIdentityResourceId}': {}
    }
  }
  properties: {
    administrators: {
      administratorType: 'ActiveDirectory'
      azureADOnlyAuthentication: true
      login: 'SQL Administrators'
      principalType: 'Group'
      sid: sqlAdministratorsGroupId
      tenantId: tenantId
    }
    keyId: tdeProtectorKeyId
    primaryUserAssignedIdentityId: userAssignedManagedIdentityResourceId
    publicNetworkAccess: 'Disabled'
  }
}
```

The identity.userAssignedIdentities and primaryUserAssignedIdentityId needs to be set to the resource ID of the user-assigned managed identity. The administrators properties should have the sid and tenantId attributes set to the group ID of the Azure AD group and the Azure AD tenant ID, respectively. The keyId should be set to the resource ID of the TDE protector key in the Azure Key Vault. The publicNetworkAccess property is set to Disabled to prevent public network access to this logical server as we will be connecting it to a virtual network next.

Connecting the Logical Server to the Virtual Network

The logical server will also need to be connected to the virtual network using a private endpoint. To do this, four resource types are required. The first is the Microsoft.Network/privateEndpoints resource type, which will create the private endpoint for the logical server.

```
resource sqlLogicalServerPrivateEndpoint 'Microsoft.Network/
privateEndpoints@2022-05-01' = {
  name: '${name}-pe'
  location: location
  properties: {
    privateLinkServiceConnections: [
      {
        name: '${name}-pe'
        properties: {
          groupIds: [
            'sqlServer'
          ]
          privateLinkServiceId: sqlLogicalServer.id
        }
      }
    ]
    subnet: {
      id: dataSubnetId
    }
  }
}
```

This will require the privateLinkServiceId to be set to the resource ID of the logical server. The subnet.id will need to be set to the resource ID of the subnet to connect the logical server to.

Next, the private DNS zone needs to be created and then linked to the virtual network to enable name resolution of the logical server in the virtual network. This is done using the Microsoft.Network/privateDnsZones and Microsoft.Network/privateDnsZones/virtualNetworkLinks resource types.

```
var privateZoneName = 'privatelink${az.environment().suffixes.
sqlServerHostname}'

resource privateDnsZone 'Microsoft.Network/privateDnsZones@2020-06-01' = {
  name: privateZoneName
  location: 'global'
}

resource privateDnsZoneLink 'Microsoft.Network/privateDnsZones/
virtualNetworkLinks@2020-06-01' = {
  name:  '${name}-dnslink'
  parent: privateDnsZone
  location: 'global'
  properties: {
    registrationEnabled: false
    virtualNetwork: {
      id: vnetId
    }
  }
}
```

The name of the private DNS zone needs to be assembled by looking up the Azure SQL server hostname for the Azure environment being used, prefixed with privatelink.

The final step of configuring the networking is to link the private endpoint for the SQL logical server to the private DNS zone using the Microsoft.Network/privateEndpoints/privateDnsZoneGroups resource type.

```
resource privateEndpointDnsGroup 'Microsoft.Network/privateEndpoints/
privateDnsZoneGroups@2021-05-01' = {
  name: '${name}-zonegroup'
  parent: sqlLogicalServerPrivateEndpoint
  properties: {
    privateDnsZoneConfigs: [
      {
        name: '${name}-config1'
        properties: {
          privateDnsZoneId: privateDnsZone.id
        }
      }
    ]
  }
}
```

This creates a `privateDnsZoneGroups` child resource under the `Microsoft.Network/privateEndpoints` resource. The `privateDnsZone` needs to be set to the resource ID of the private DNS zone.

Like the Azure Key Vault configuration, this is much simpler when using Azure Bicep than when using Azure PowerShell or the Azure CLI.

Configuring Audit Logs to Be Sent to Log Analytics

Once the logical server has been created, it need to be configured so that both audit logs are sent to a Log Analytics workspace in the same region. To do this, we need to use three resource types.

```
resource masterDatabase 'Microsoft.Sql/servers/databases@2022-05-01-
preview' = {
  name: 'master'
  parent: sqlLogicalServer
  location: location
  properties: {}
}
```

```
resource sqlLogicalServerAuditing 'Microsoft.Insights/
diagnosticSettings@2021-05-01-preview' = {
  name: 'Send all audit to ${logAnalyticsWorkspaceName}'
  scope: masterDatabase
  properties: {
    logs: [
      {
        category: 'SQLSecurityAuditEvents'
        enabled: true
        retentionPolicy: {
          days: 0
          enabled: false
        }
      }
    ]
    workspaceId: logAnalyticsWorkspaceId
  }
}

resource sqlLogicalServerAuditingSettings 'Microsoft.Sql/servers/
auditingSettings@2022-05-01-preview' = {
  name: 'default'
  parent: sqlLogicalServer
  properties: {
    state: 'Enabled'
    isAzureMonitorTargetEnabled: true
  }
}
```

The usual part about this configuration is that we need to create the master database, before we can create the `Microsoft.Insights/diagnosticSettings` extension resource.

Note The master database is a hidden database on each logical server, so we won't be able to see this in the Azure Portal.

The `logs` array of the `Microsoft.Insights/diagnosticSettings` resource needs to be configured with the audit categories that we want to send, and the `workspaceId` needs to be configured with the resource ID of the Log Analytics workspace.

The `Microsoft.Sql/servers/auditingSettings` resource is created as a child resource of the logical server and is simply used to enable audit logs to be sent to Azure Monitor Log Analytics by setting `isAzureMonitorTargetEnabled` to `true` and `state` to `Enabled`.

You can find the complete Azure Bicep module to create the logical server, connect it to the virtual network, and configure audit logging in the `sql_logical_server.bicep` file found in the `ch11/modules` directory in the files provided for this book. The file can be used to create the logical server in the primary and failover regions because it is parameterized.

Creating the Hyperscale Database

The last step to create the resources in the primary region is to deploy the Hyperscale database. This requires the use of the `Microsoft.Sql/servers/databases` resource.

```
resource sqlHyperscaleDatabase 'Microsoft.Sql/servers/databases@2022-05-01-
preview' = {
  name: name
  location: location
  parent: sqlLogicalServer
  sku: {
    name: 'HS_Gen5'
    capacity: 2
    family: 'Gen5'
  }
  properties: {
    highAvailabilityReplicaCount: 2
    readScale: 'Enabled'
    requestedBackupStorageRedundancy: 'GeoZone'
    zoneRedundant: true
  }
}
```

You will note the `parent` is set to the logical server that was created earlier. The `sku` must be configured as follows:

- `name`: `HS_Gen5` indicates that this is using Hyperscale Gen 5 SKU.
- `capacity`: This is the number of vCores to assign to the Hyperscale database.
- `family`: This should be set to `Gen5`.

The `highAvailabilityReplicaCount` parameter should be set to the number of high-availability replicas. `readScale` should be `Enabled` so that the high availability replicas can be read from. To ensure the backups are replicated across availability zones and geographic regions, we will set `requestedBackupStorageRedundancy` to `GeoZone`. Finally, the `zoneRedundant` flag should be set to `true` to ensure the high-availability replicas are spread across availability zones.

Configuring Diagnostic Logs to Be Sent to Log Analytics

Once the Hyperscale database has been defined on the logical server, we can configure the extension resource `Microsoft.Insights/diagnosticSettings` so that diagnostic logs will be sent to the Log Analytics workspace.

```
resource sqlHyperscaleDatabaseDiagnostics 'Microsoft.Insights/
diagnosticSettings@2021-05-01-preview' = {
  name: 'Send all logs to ${logAnalyticsWorkspaceName}'
  scope: sqlHyperscaleDatabase
  properties: {
    logs: [
      {
        category: 'SQLInsights'
        enabled: true
        retentionPolicy: {
          days: 0
          enabled: false
        }
      }
      {
        category: 'Deadlocks'
```

```
        enabled: true
        retentionPolicy: {
          days: 0
          enabled: false
        }
      }
    ]
    workspaceId: logAnalyticsWorkspaceId
  }
}
```

Because this is an extension resource type, the scope needs to be set to the SQL Hyperscale database resource.

Note An extension resource is a resource type that can be attached only to another resource and extends the capabilities or configuration of the resource it is attached to, for example, sending diagnostic logs.

The logs array can be configured with the categories of diagnostic logs to send to the Log Analytics workspace, while workspaceId should be set to the resource ID of the Log Analytics workspace.

You can find the complete Azure Bicep module to create the Hyperscale database and configure diagnostic logging in the sql_hyperscale_database.bicep file found in the ch11/modules directory in the files provided for this book.

Creating the Failover Resources and Replica Hyperscale Database

Because we've defined the resources in parameterized Azure Bicep modules, we can simply reference these again with different parameters to deploy the resources in the failover region. Here's an example:

```
module failoverLogicalServer './modules/sql_logical_server.bicep' = {
  name: 'failoverLogicalServer'
  scope: failoverResourceGroup
  params: {
```

```
    name: '${failoverRegionPrefix}-${resourceNameSuffix}'
    location: failoverRegion
    environment: environment
    tenantId: subscription().tenantId
    userAssignedManagedIdentityResourceId: userAssignedManagedIdentity.
    outputs.userAssignedManagedIdentityResourceId
    tdeProtectorKeyId: keyVault.outputs.tdeProtectorKeyId
    sqlAdministratorsGroupId: sqlAdministratorsGroupId
    vnetId: failoverVirtualNetwork.outputs.vnetId
    dataSubnetId: failoverVirtualNetwork.outputs.dataSubnetId
    logAnalyticsWorkspaceName: failoverLogAnalyticsWorkspace.outputs.
    logAnalyticsWorkspaceName
    logAnalyticsWorkspaceId: failoverLogAnalyticsWorkspace.outputs.
    logAnalyticsWorkspaceId
  }
}
```

However, to deploy the replica database, we will again use the `Microsoft.Sql/servers/databases` resource type, but this time, we'll need to adjust the properties slightly.

```
resource sqlHyperscaleDatabase 'Microsoft.Sql/servers/databases@2022-05-01-
preview' = {
  name: name
  location: location
  parent: sqlLogicalServer
  sku: {
    name: 'HS_Gen5'
    capacity: 2
    family: 'Gen5'
  }
  properties: {
    createMode: 'Secondary'
    readScale: 'Enabled'
    requestedBackupStorageRedundancy: 'GeoZone'
    secondaryType: 'Geo'
```

```
    sourceDatabaseId: sourceDatabaseId
    zoneRedundant: false
  }
}
```

When we're creating a replica, we need to set createMode to Secondary and set sourceDatabaseId to the resource ID of the primary Hyperscale database. We also need to set secondaryType to Geo to create a geo replica. Again, the code here is much simpler to understand than with Azure PowerShell and the Azure CLI.

You can find an example of reusing modules in a single deployment in the sql_hyperscale_revealed_environment.bicep file found in the ch11 directory in the files provided for this book.

Putting It All Together

Now that we have all the main Azure Bicep files and the Azure Bicep modules, we can deploy it using either the Azure PowerShell New-AzDeployment command or the Azure CLI az deployment sub create command. Here's an example:

```
$resourceNameSuffix = '<resource name suffix>'
$sqlAdministratorsGroupId = (Get-AzADGroup -DisplayName 'SQL
Administrators').Id

New-AzDeployment `
    -Name "sql-hyperscale-revealed-env-$resourceNameSuffix-$(Get-Date
    -Format 'yyyyMMddHHmm')" `
    -Location 'West US 3' `
    -TemplateFile ./Azure-SQL-Hyperscale-Revealed/ch11/sql_hyperscale_
    revealed_environment.bicep `
    -TemplateParameterObject @{
        'resourceNameSuffix' = $resourceNameSuffix
        'sqlAdministratorsGroupId' = $sqlAdministratorsGroupId
    }
```

The template parameter resourceNameSuffix is set to a four-character string that will be suffixed into each resource name to ensure it is globally unique. You may also notice that the template parameter sqlAdministratorsGroupId is looked up first to determine the ID of the Azure AD group used for SQL Administrators. This is required to deploy the logical server because currently there is no way of doing this directly in Azure Bicep.

Figure 11-1 shows the sample output produced once the resources have been successfully deployed.

```
Id                      : /subscriptions/
DeploymentName          : sql-hyperscale-revealed-env-
Location                : westus3
ProvisioningState       : Succeeded
Timestamp               : 11/25/2022 7:19:49 AM
Mode                    : Incremental
TemplateLink            :
```

***Figure 11-1.** Sample output from running the New-AzDeployment command*

Because Azure Bicep is idempotent, running the command again will quickly return the same output and will not change any of the resources, unless you've changed the Azure Bicep files or the deployed resources have been changed by a different method.

Summary

In this chapter, we demonstrated the deployment of a complete Hyperscale environment using declarative infrastructure as code with Azure Bicep. The environment was able to be deployed more quickly using the Azure Bicep than with a typical imperative method because many of the resources are able to be deployed in parallel. Parallel deployments with imperative methods are possible but require significantly more code to achieve.

We also showed that Azure Bicep syntax can be much more compact and readable than the equivalent Azure PowerShell or Azure CLI. Many other benefits are also more easily achievable with Azure Bicep, such as best-practice evaluation, easier-to-detect issues, and linting (removing unnecessary code).

There is almost no downside, except for initially learning the new syntax, to using Azure Bicep or another declarative method.

In the next chapter, we will look at the performance of SQL Hyperscale in depth, especially when compared with other Azure SQL offerings.

CHAPTER 12

Testing Hyperscale Database Performance Against Other Azure SQL Deployment Options

In the previous chapter, we talked about how to deploy the SQL Hyperscale environment using Azure Bicep. We explained a few of the key Azure Bicep resources required to deploy the SQL Hyperscale environment and supporting resources.

In this chapter, we are going to do an overall performance comparison between the traditional Azure architecture and Hyperscale architecture. For this purpose, we are going to deploy three different Azure SQL database service tiers.

- Azure SQL Database General Purpose service tier
- Azure SQL Database Business Critical service tier
- Azure SQL Database Hyperscale service tier

All Azure SQL resources will be deployed in the same Azure region and on the same logical Azure SQL Server instance. We will also deploy all of these on Standard-series (Gen5) hardware configuration with eight vCores. Figure 12-1 shows the Azure SQL Database service tiers and hardware we will use for testing.

Z. Barać and D. Scott-Raynsford, *Azure SQL Hyperscale Revealed*,
https://doi.org/10.1007/978-1-4842-9225-9_12

Database ↑↓	Status ↑↓	Pricing tier
tpcc_bc	Online	Business Critical: Standard-series (Gen5), 8 vCores
tpcc_gp	Online	General Purpose: Standard-series (Gen5), 8 vCores
tpcc_hs	Online	Hyperscale: Standard-series (Gen5), 8 vCores

Figure 12-1. *Azure SQL Database service tiers*

There are multiple ways to run benchmark tests against your Azure SQL deployments. One of the available tools for this purpose is HammerDB. It is open source and offers Transaction Processing Performance Council (TPC) standard metrics for SQL in the cloud or in on-premises servers. The Transaction Processing Performance Council establishes standards that are intended to promote the fast, reliable execution of database transactions. TPC benchmarking measures the performance, speed, and reliability of transactions.

Note Founded in 1985, the TPC consists of major hardware and software vendors such as Microsoft, Hewlett-Packard (HP), IBM, and Dell. The TPC specifications are published benchmarks widely accepted by all leading database vendors.

HammerDB

For testing purposes, we are going to use the HammerDB tool. HammerDB is free and open source software and one of the leading benchmarking and load testing tools for the most popular databases such as SQL Server.

Note HammerDB was our preferred tool for these tests and performing a comparison between different Azure SQL Database deployments, including the Hyperscale service tier. However, HammerDB is not an officially endorsed measuring standard, and results may vary.

HammerDB TPROC-C Workload

For these tests, we will use the TPROC-C OLTP workload implemented in HammerDB and derived from TPC-C. The TPROC-C HammerDB workload name means "Transaction Processing Benchmark derived from the TPC-C standard specification." The reason for choosing this open source workload is to run the HammerDB tool so that we can run these tests in the easiest and most efficient way against our supported environment. The HammerDB TPROC-C workload is designed to be reliable and tuned to produce consistently repeatable and accurate results. Therefore, if we rerun the same test with the same hardware configuration with the same parameters, we will get almost identical results, with only marginal variations.

Tip TPROC-C complies with a subset of TPC-C benchmark standards, so the results cannot be compared to published TPC-C results.

The main goal of executing these tests is to gain some certainty as to whether the new database deployment will be able to handle the anticipated workload.

HammerDB Step-by-Step

For these tests we will use HammerDB v4.5 on Windows using the graphical user interface. HammerDB is also available for Linux. Both the Windows and Linux versions (since v3.0) have a command-line version as well. Figure 12-2 shows the current HammerDB version at the time of writing.

Figure 12-2. *Showing the HammerDB version*

For more information on HammerDB, see the official HammerDB web page at `https://www.HammerDB.com`.

Schema Build

The first step of using HammerDB to benchmark your database is to build the schema of the test database. The following steps show how to execute a schema build using the HammerDB tool:

1. Start HammerDB and choose the database management system hosting the database you want to benchmark by double-clicking it in the Benchmark box. We will choose the SQL Server option for DBMS and TPROC-C for our benchmark option, as shown in Figure 12-3.

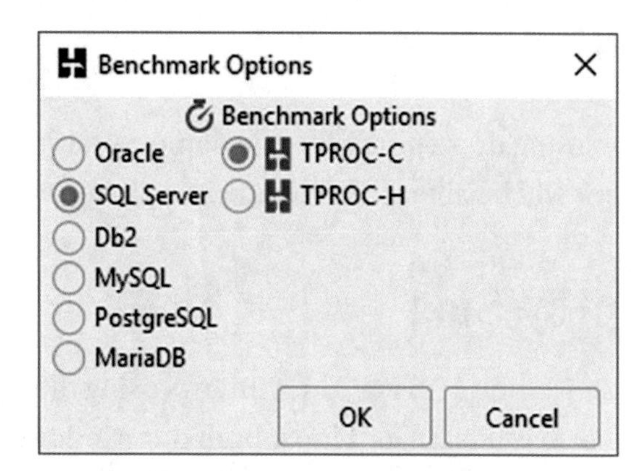

***Figure 12-3.** HammerDB TPROC-C benchmark options*

2. In the Benchmark pane, expand SQL Server and then TPROC-C Branches. See Figure 12-4.

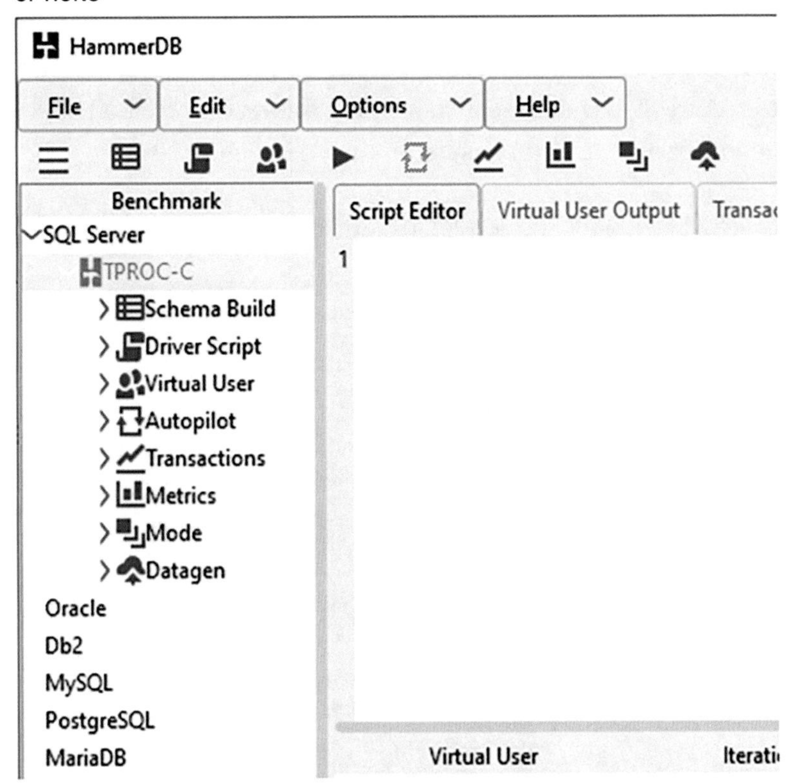

***Figure 12-4.** HammerDB TPROC-C benchmark tree pane*

3. Expand the Schema Build branch and then select Options. See Figure 12-5.

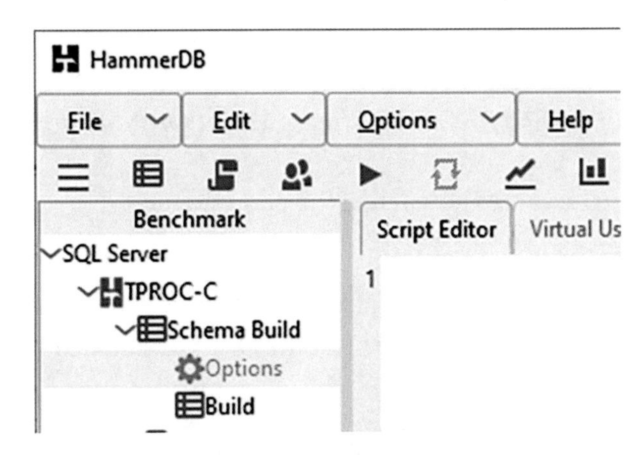

***Figure 12-5.** HammerDB TPROC-C Schema Build option menu*

4. Configure the settings in the Microsoft SQL Server TPROC-C Build Options box to enable the HammerDB benchmark tool to connect to the Azure SQL Database instance. For this test we created 50 warehouses and 32 virtual users, as shown in Figure 12-6.

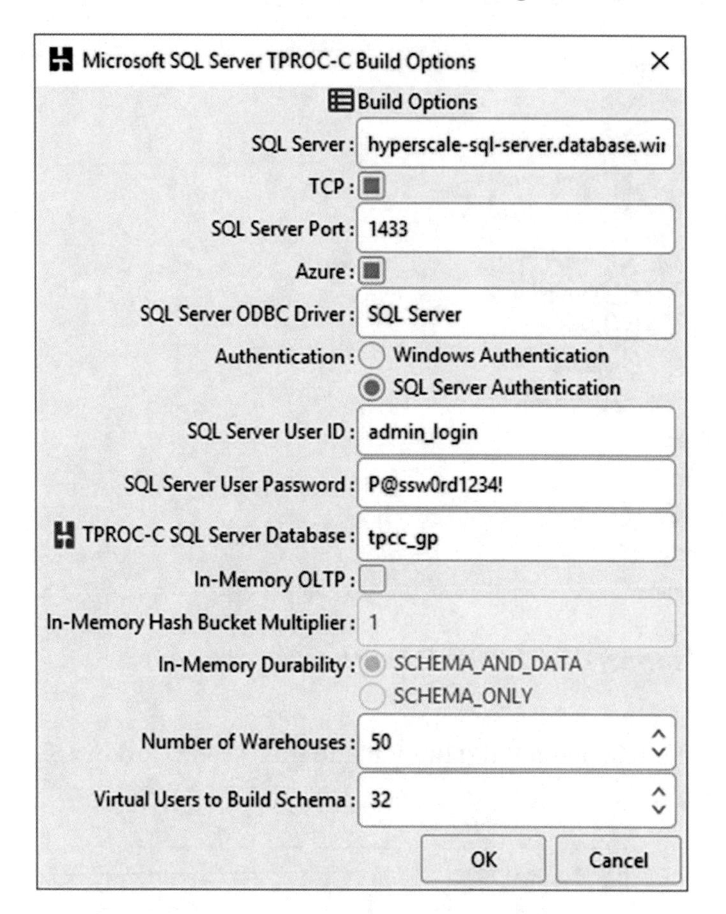

Figure 12-6. *HammerDB TPROC-C Schema Build Options box*

The following are the values for each of the parameters:

- *SQL Server*: Set this to the DNS name of the logical SQL server containing the Azure SQL Database instance we are evaluating. This can be found in the Azure Portal on the Overview blade of the logical SQL server resource.
- *TCP*: Use TCP for communication. Set to true.

- *SQL Server Port*: Since we enabled TCP, the default SQL Server port 1433 will be used to connect to the logical server.
- *Azure*: Select this if box if you are connecting to an Azure SQL Database.
- *SQL Server ODBC Driver*: Choose the available Microsoft SQL ODBC driver you will use to connect to the Azure SQL Server database.
- *Authentication*: Select from Windows or SQL Server Authentication. For this test, we used SQL Authentication.
- *SQL Server User ID*: Specify a SQL user account with access to the logical SQL Server instance and the database being tested.
- *SQL Server User Password*: Specify the SQL user account password.
- *TRPOC-C SQL Server Database*: Use an existing empty Azure SQL Database instance.
- *Number of Warehouses*: Set this to the number of warehouses you will use for your test. For these tests, set it to 50.
- *Virtual Users to Build Schema*: This is the number of virtual users to be created to build the schema. For these tests, set it to 32. This value should be greater than or equal to the number of vCPUs of your Azure SQL Database deployment. This number also cannot be higher than the specified number of warehouses.

5. After setting all the parameters and supplying the credentials that HammerDB will use to connect to your database, HammerDB will be ready to build the schema and populate the data for the database that will be used for benchmarking.

Tip To be able to run the schema build process, the database needs to be created first.

---

6. HammerDB can now build the schema and populate the tables by selecting TPROC-C, then Schema Build, and then Build from the Benchmark box or selecting Build from the toolbar. See Figure 12-7.

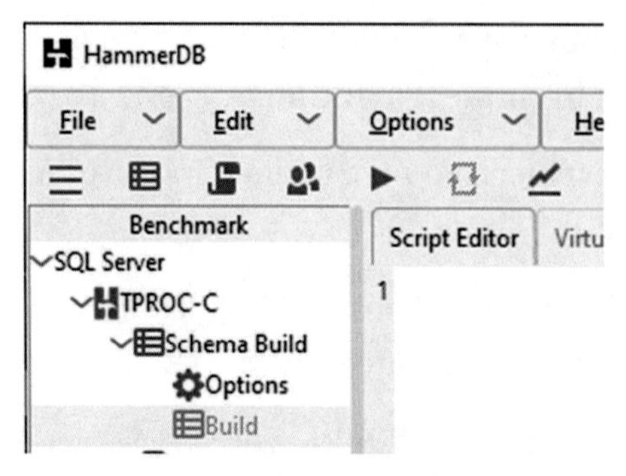

Figure 12-7. *HammerDB TPROC-C Schema build, generating schema and populating tables*

7. To finalize the schema build process, click Yes to confirm the prompt screen to execute the schema build, as shown in Figure 12-8.

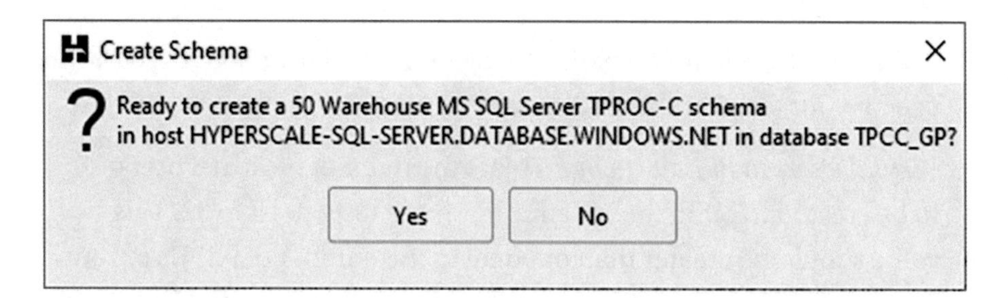

Figure 12-8. *HammerDB TPROC-C schema build, create TPROC-C schema prompt screen*

Depending on the parameters that were entered and the environment that is being tested, this process could take from a few minutes to a few hours. See Figure 12-9.

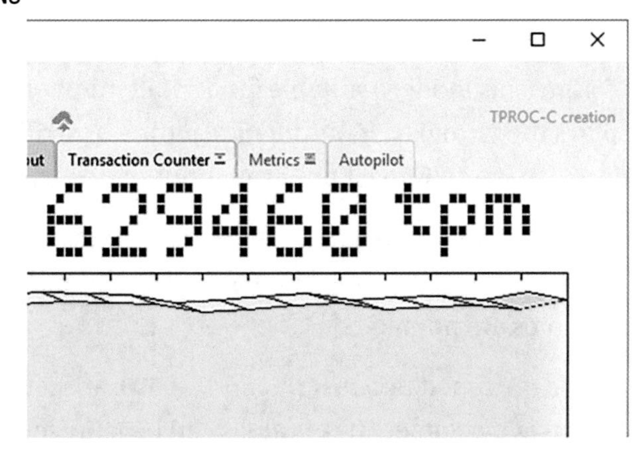

Figure 12-9. *HammerDB TPROC-C creation progress example*

Driver Script Options

Now that the schema has been created in the database that will be tested, we can configure the script that will drive the testing process.

1. In HammerDB, select TPROC-C, then Driver Script, and then Options in the Benchmark box. See Figure 12-10.

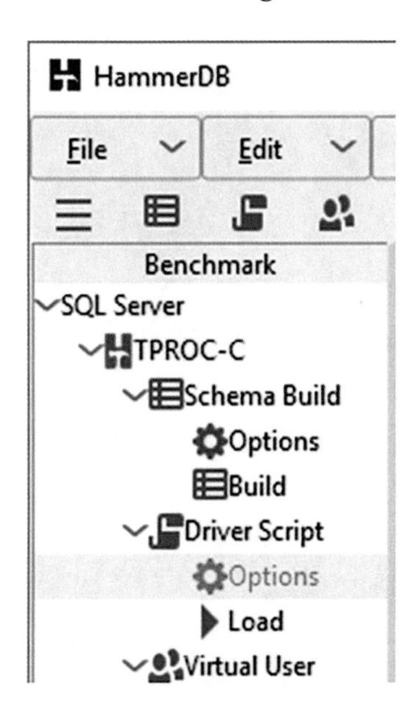

Figure 12-10. *HammerDB Driver Script options*

There are two driver scripts you can choose from.

- *Test Driver Script*: This mode is suitable for a small number of virtual users. It displays the virtual user output in real time by writing to the display. This may have an impact on throughput performance.
- *Timed Driver Script*: This mode is suited to testing with a higher number of virtual users. Use this script once your schema build is done and the tables are populated.

For this test we will use the timed driver script and 1,000 transactions per user to be run against all warehouses available. These values can be changed to suit your test requirements. Figure 12-11 shows an example of setting the HammerDB SQL Server TPROC-C Driver Script Options parameters.

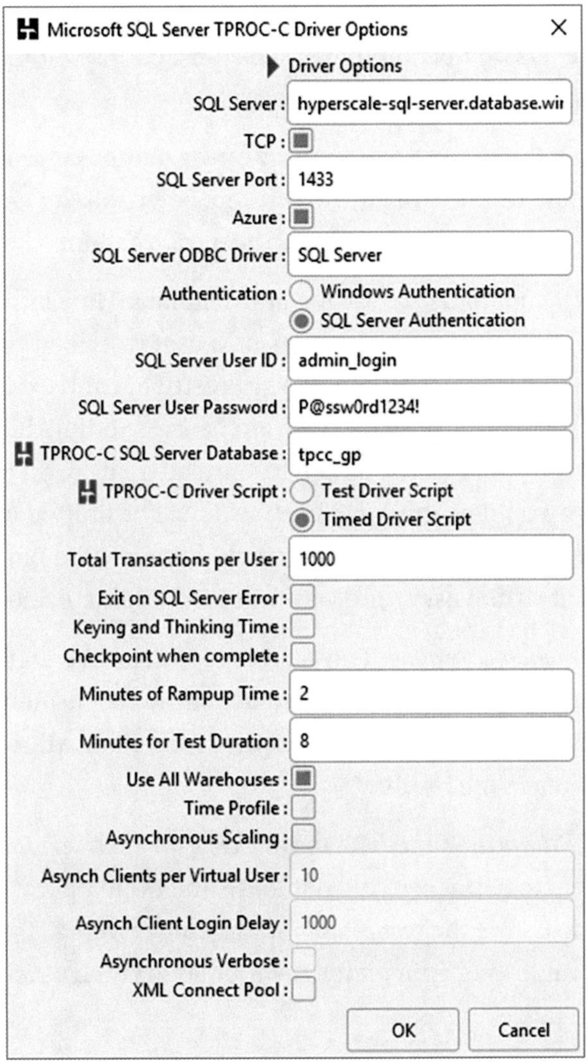

Figure 12-11. *HammerDB SQL Server TPROC-C Driver Script Options box*

For these tests, we will configure the Microsoft SQL Server TPROC-C Driver Options parameters like this:

- *TPROC-C Driver Script*: Set this to Time Driver Script.
- *Total Transactions per User*: The number of total transactions per user determines how long each virtual user will remain active for, which may depend on the DB server performance. If using the Timed

Driver Script option, it should be set to a high enough value so that the virtual users do not complete their transactions before the timed test is complete.

- *Exit on Error*: Set to true to report any errors to the HammerDB console before terminating the execution. Set to false to ignore any errors and continue execution of the next transaction.
- *Keying and Thinking Time*: Keying and Thinking Time is a feature of a TPROC-C test that simulates real-world user workload patterns. If this is set to false, each user will be unrestricted and execute transactions as fast as possible, with no pauses. Setting this value to true will cause pauses to be added to simulate real-world behavior where users would be thinking. Each user will be limited to a few transactions per minute. You will need to have created hundreds or thousands of virtual users and warehouses for valid results.
- *Checkpoint when complete*: This enables the database to trigger a checkpoint to write out the modified data from the in-memory cache to the disk after the workload is complete rather than during the test, which may affect the results.
- *Minutes of Ramp-up Time*: The monitoring for each virtual user will wait for them to connect to the database and build up the database buffer cache before the test results start being counted. This should be set to a value long enough for a workload to reach a steady transaction rate.
- *Minutes for Test Duration*: This is the test duration set in the driver script, which is measured as the time the monitor thread waits between the first timed value and the second one when the test is complete.
- *Use All Warehouses*: If set to true, the virtual users will select a different warehouse for each transaction from the shared list of warehouses to ensure greater I/O activity. If set to false, each virtual user selects a home warehouse at random for the entire test by default.

- *Time Profile*: If this is set to true, it sets client-side time profiling that is configured in the `generic.xml` file. There are two options that can be set.
    - *Xtprof*: This profiles all virtual users and prints their timing outputs at the end of a run.
    - *Etprof*: This profiler times only the first virtual user and prints the output every 10 seconds.
- *Asynchronous Scaling*: Enable the event-driven scaling feature to configure multiple client sessions per virtual user. This requires the Keying and Thinking Time option to also be enabled. This will also test connection pooling by scaling up the number of client sessions that connect to the database.
- *Asynch Client per Virtual User*: Configure the number of sessions that each virtual user will connect to the database and manage. For example, for a test with 5 virtual users with 10 asynchronous clients, there will be 50 active connections to the database.
- *Asynch Client Login Delay*: The delay that each virtual user will have before each asynchronous client is logged in.
- *Asynchronous Verbose*: Set this to true to report asynchronous operations such as the time taken for keying and thinking time.
- *XML Connect Pool*: Use this for simultaneously testing multiple instances of related clustered databases, as defined in the database-specific XML file. Each virtual user (or asynchronous client) will open and hold open all the defined connections.
- *Mode*: This is the operational mode. If set to Local or Primary, then the monitor thread takes snapshots; if set to Replica, no snapshots are taken. If running multiple instances of HammerDB, you can set only one instance to take snapshots.

Virtual User Options

This will allow us to set the number of virtual users that will be used to simulate load on the database.

After configuring the driver script, we can specify the number of virtual users and iterations that will be used to simulate load. In HammerDB, select TPROC-C, then Virtual User, and then Options in the Benchmark box. For this example, we used the parameters shown in Figure 12-12.

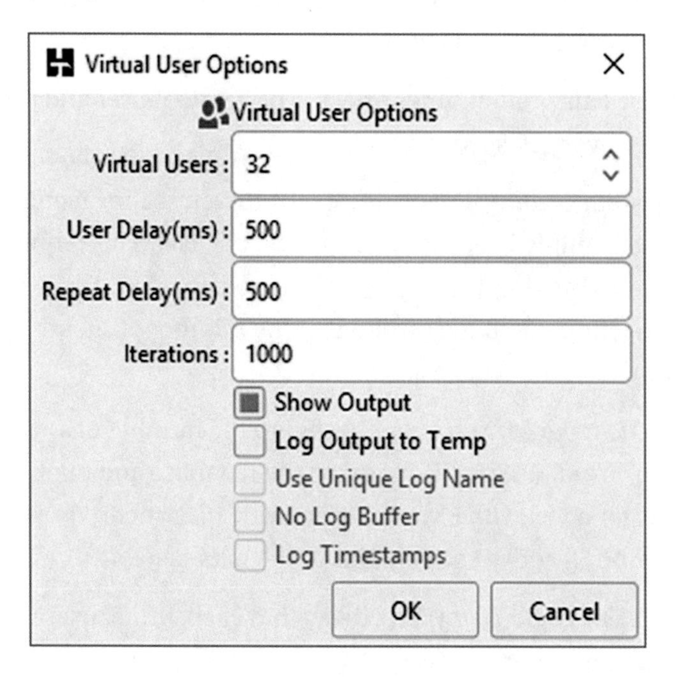

Figure 12-12. *HammerDB Virtual User Options window*

Note While running a timed workload script using HammerDB, one additional virtual user will be automatically created to monitor the workload.

The following are the virtual option parameters that you can configure:

- *Virtual Users*: Specify the number of virtual users that will be created and used for testing.
- *User Delay (ms)*: This is the amount of time in milliseconds to wait before the next user logs in. This is to prevent all virtual users from trying to log in at the same time.
- *Repeat Delay (ms)*: This is the amount of time a virtual user will wait between each run of the driver script.

- *Iterations*: This is the number of times that the driver script will be run by each virtual user.
- *Show Output*: Set this to true to show the output report for each virtual user.

Note If the Show Output option is enabled, it will put additional load on your system and so might affect your benchmark results.

- *Log Output to Temp*: This setting causes the logs from each virtual user to be written to a text file in the temp folder.
- *Use Unique Log Name*: This ensures a unique filename is created for the log file.
- *Log Timestamps*: This sets a timestamp for each log entry in the log file.

Autopilot Options

HammerDB provides an automated stress test function that can be configured using the autopilot options. This feature executes the tests with a specific number of virtual users in an automated fashion. To use the Autopilot feature in HammerDB, follow these steps:

1. Select TPROC-C, then Autopilot, and then Options in the Benchmark box.
2. Select the Autopilot Enabled option.
3. Set the Minutes per Test in Virtual User Sequence field to five minutes or higher, as shown in Figure 12-13.

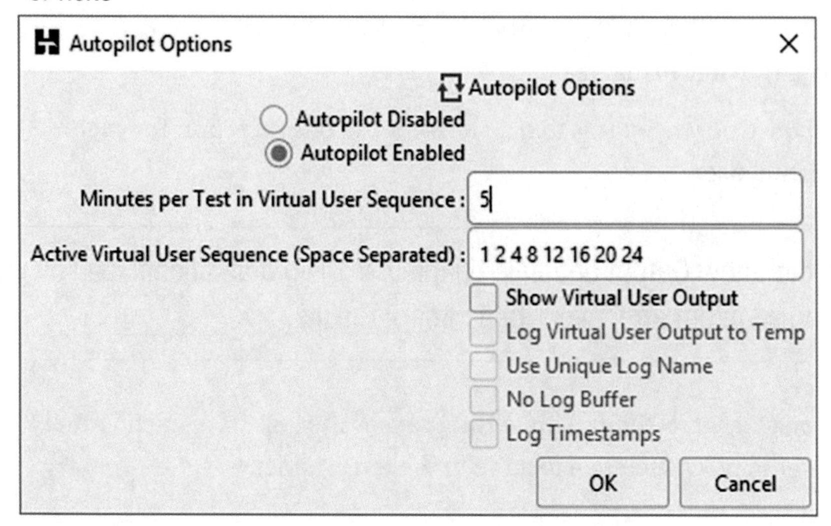

Figure 12-13. *HammerDB Autopilot Options window*

4. Set the Active Virtual User Sequence to control the number of virtual users to execute the sequence of tests separated by the time specified in Minutes per Test in Virtual User Sequence.

Tip It is recommended that Active Virtual User Sequence is configured to suit the workload that is being evaluated.

Run the Virtual Users and Execute the TPROC-C Workload

The final step is to run the virtual users and execute the SQL Server TPROC-C workload against the schema in the database that was created. To run virtual users and execute the TPROC-C workload in HammerDB, follow these steps:

1. Expand the TPROC-C ➤ Virtual User tree menu.
2. Click Run, as shown in Figure 12-14.

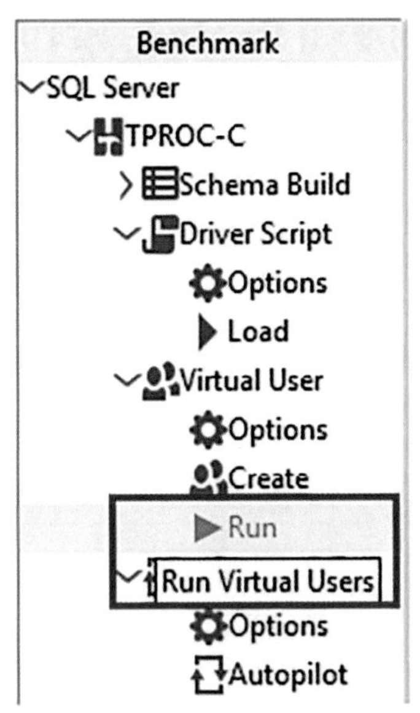

***Figure 12-14.** HammerDB running SQL Server TPROC-C benchmark workload*

Schema Build Performance Metrics

We are going to do the test on the schema build and benchmark against three different Azure SQL deployment options, as shown in Figure 12-15.

Database ↑↓	Status ↑↓	Pricing tier
tpcc_bc	Online	Business Critical: Standard-series (Gen5), 8 vCores
tpcc_gp	Online	General Purpose: Standard-series (Gen5), 8 vCores
tpcc_hs	Online	Hyperscale: Standard-series (Gen5), 8 vCores

***Figure 12-15.** Azure SQL Database deployment options*

All our Azure SQL Database deployments will use the same logical SQL Server instance, and the HammerDB tool will be run from a virtual machine in the same Azure region as the logical server. The virtual machine is connecting to the logical server via the public endpoint.

Azure SQL Database General Purpose Service Tier (GP_Gen5_8)

First, we are going to run a schema build operation against the Azure SQL Database instance with the General Purpose service tier with eight vCores. This is the lowest specification database deployment we are going to use in these tests. Therefore, we expect the performance results to reflect that; see Table 12-1.

Table 12-1. *Azure SQL Database with General Purpose Service Tier (GP_Gen5_8) Deployment Specifications*

Compute Size (Service Objective)	GP_Gen5_8
Hardware	Gen5
vCores	8
Memory (GB)	41.5
Max data size (GB)	2048
Max log size (GB)	614
Storage type	Remote SSD
Read I/O latency (approximate ms)	5 to 10
Write I/O latency (approximate ms)	5 to 7
Max log rate (MBps)	36

1. Before we build the schema, we need to enable the transaction log output so we can see the transaction counts during and after the build is finished, as shown in Figure 12-16 (Transactions ➤ Options).

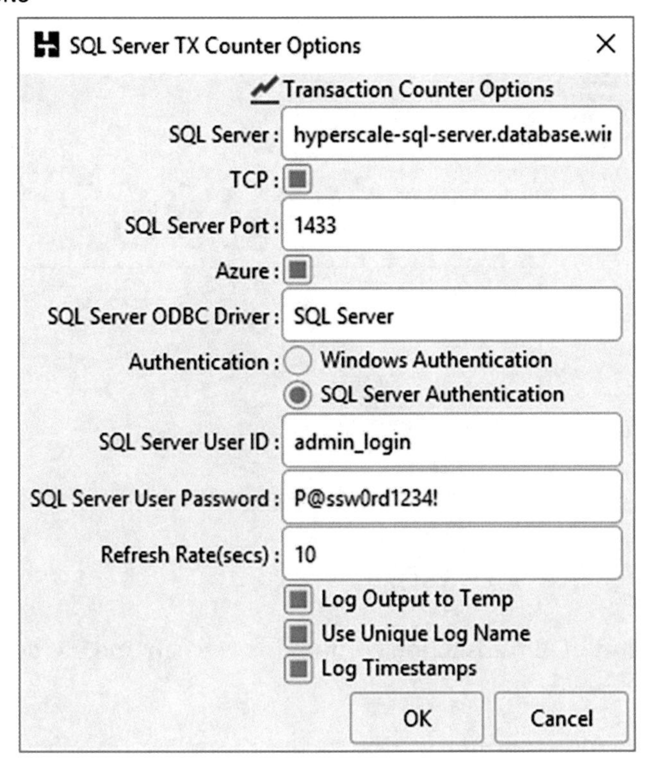

Figure 12-16. *HammerDB transaction counter options*

Once we start the schema build or a benchmark, we can see an ongoing transaction per minute count on the Transaction Counter tab.

2. Run the schema build against the Azure SQL Database instance with the General Purpose service tier deployment. Figure 12-17 shows the schema build progress against the General Purpose service tier with a transaction counter enabled.

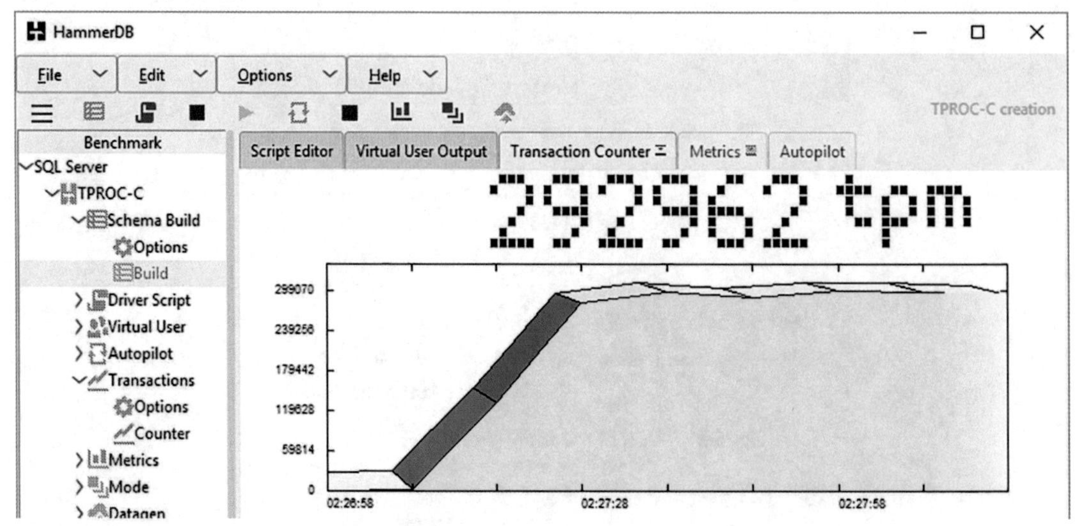

Figure 12-17. *Schema build progress*

Within the HammerDB transaction counter log, we can find all the transaction timestamps.

- The transactions started at 02:26:58.

```
Count 0 MSSQLServer tpm @ Mon Oct 31 02:26:58 +1300 2022
```

- The transactions finished at 02:53:51.

```
Count 15474 MSSQLServer tpm @ Mon Oct 31 02:53:51 +1300 2022
```

- The peak was 305,778 transactions per minute.

```
Count 305778 MSSQLServer tpm @ Mon Oct 31 02:33:19 +1300 2022
```

- The overall execution time was approximately 27 minutes.

We can examine resource usage through the Azure Portal in the logical SQL Server instance on the Metrics blade by using the following metrics:

- CPU percentage (max)
- Data IO percentage (max)
- Log IO percentage (max)
- Workers percentage (max)

Figure 12-18 shows the metrics for the logical server while running this test.

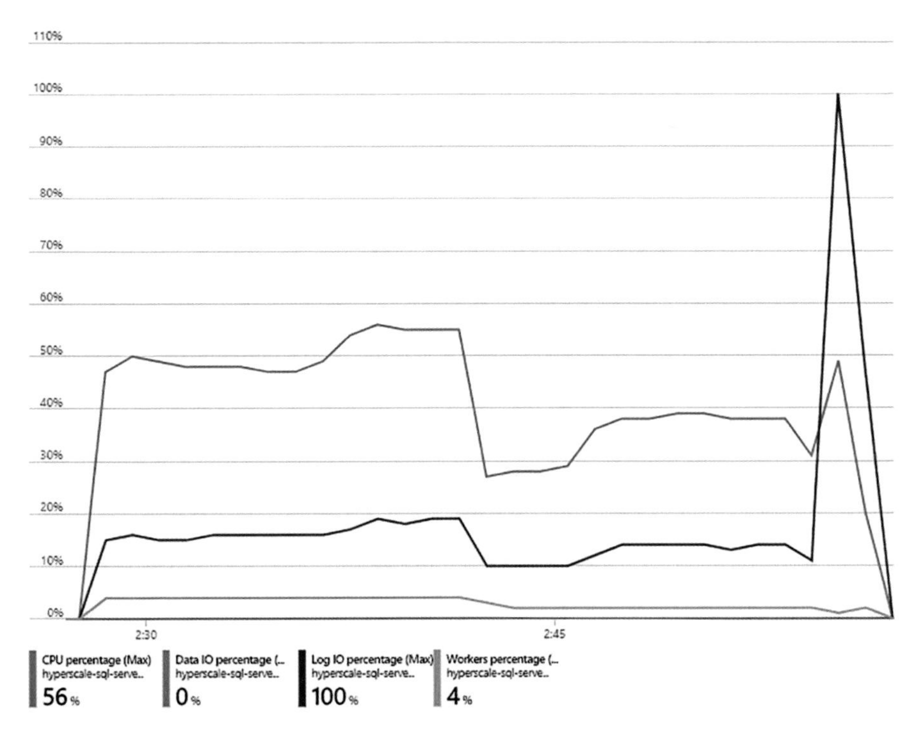

Figure 12-18. *Azure Portal metrics graph for the General Purpose service tier*

Tip For more information on monitoring an Azure SQL Database instance and a logical SQL server, please see Chapter 13.

Azure SQL Database Business Critical General Service Tier (BC_Gen5_8)

Next, we are going to run a schema build against the Azure SQL Database Business Critical service tier; see Table 12-2.

Table 12-2. *Azure SQL Database with Business-Critical Service Tier (BC_Gen5_8) Deployment Specifications*

Compute Size (Service Objective)	**BC_Gen5_8**
Hardware	Gen5
vCores	8
Memory (GB)	41.5
Max data size (GB)	2048
Max log size (GB)	614
Max local storage size (GB)	4096
Storage type	Local SSD
Read I/O latency (approximate ms)	1 to 2
Write I/O latency (approximate ms)	1 to 2
Max log rate (MBps)	96

Note The Business Critical service tier has database data and log files on an SSD drive attached to the node and an approximate storage I/O latency of 1 to 2 milliseconds.

Figure 12-19 shows the schema build progress against the Business Critical service tier with the transaction counter enabled.

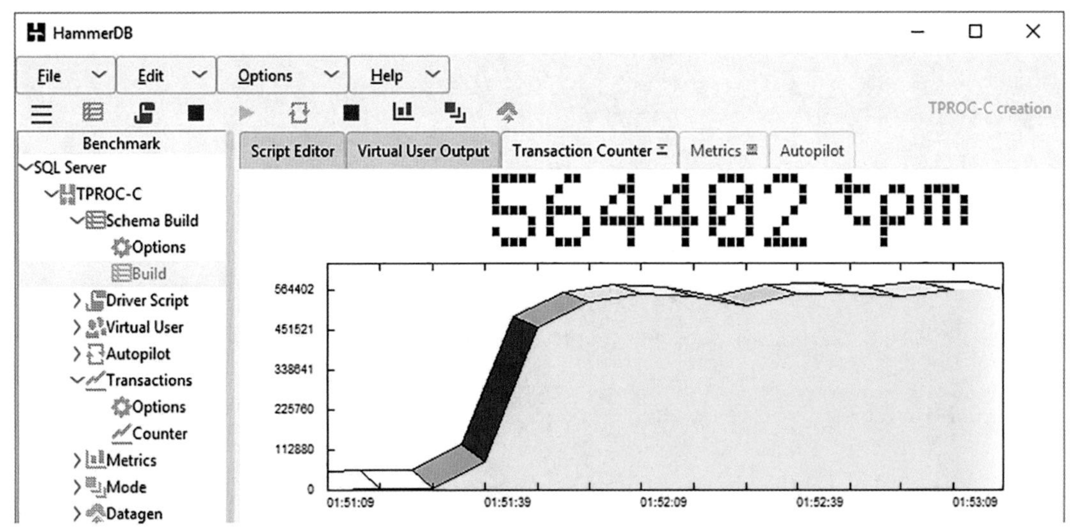

Figure 12-19. *Schema build progress*

Within HammerDB Transaction Counter Log, we can find all transaction timestamps.

- The transactions started at 01:51:0.

```
Count 0 MSSQLServer tpm @ Mon Oct 31 01:51:09 +1300 2022
```

- The transactions finished at 02:06:00.

```
Count 42528 MSSQLServer tpm @ Mon Oct 31 02:06:00 +1300 2022
```

- The peak was 565,254 transactions per minute.

```
Count 565254 MSSQLServer tpm @ Mon Oct 31 01:53:29 +1300 2022
```

- The overall execution time was approximately 15 minutes.

We can examine resource usage through the Azure Portal in the logical SQL server on the Metrics blade by using CPU percentage (max), Data IO percentage (max), Log IO percentage (max), and Workers percentage (max). See Figure 12-20.

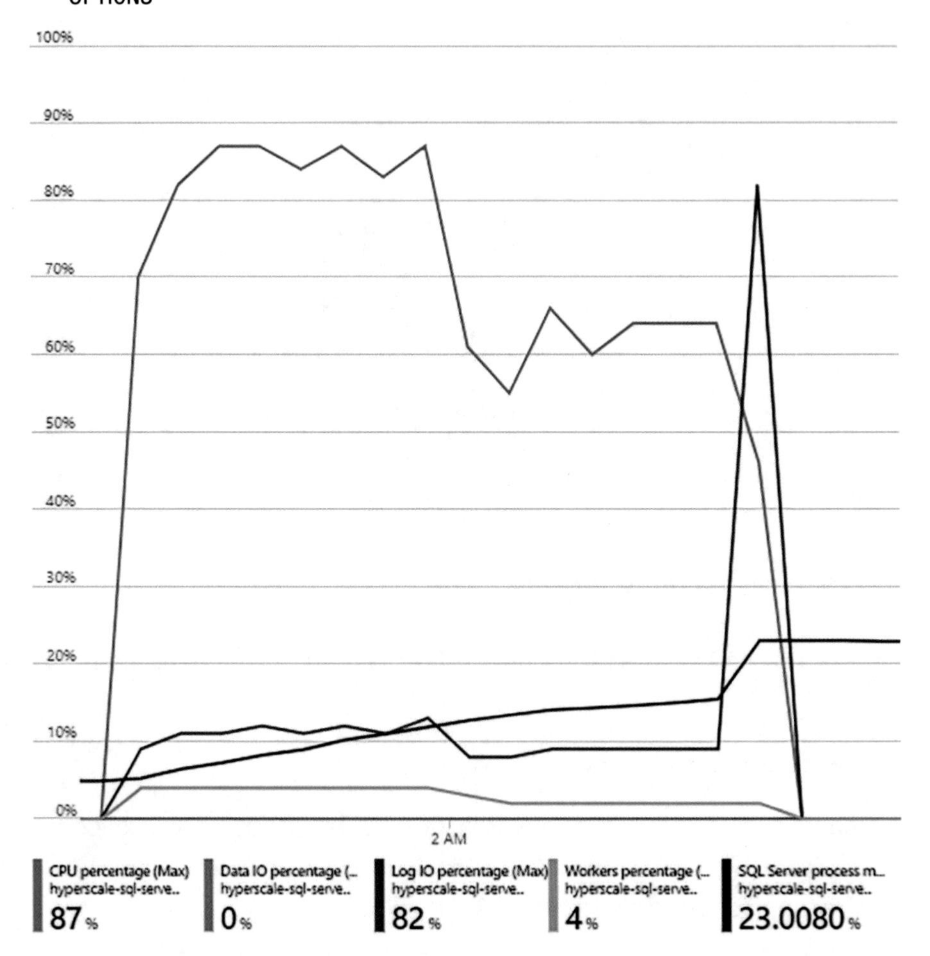

Figure 12-20. *Azure Portal metrics graph for the Business Critical service tier*

Azure SQL Database Hyperscale Service Tier (HS_Gen5_8)

Finally, we are going to run a schema build operation against the Azure SQL Database Hyperscale service tier deployment. See Table 12-3.

Table 12-3. *Azure SQL Database with Hyperscale Service Tier (HS_Gen5_8) Deployment Specifications*

Compute Size (Service Objective)	HS_Gen5_8
Hardware	Gen5
vCores	8
Memory (GB)	41.5
RBPEX Size	3x Memory
Max data size (TB)	100
Max log size (TB)	Unlimited
Max local SSD IOPS	32000
Max log rate (MBps)	105
Local read I/O latency (approximate ms)	1 to 2
Remote read I/O latency (approximate ms)	1 to 5
Write IO latency (approximate ms)	3 to 5
Storage type	Multitiered

Tip For Hyperscale, performance depends heavily on your workload type. Bear in mind that latency numbers are approximate and not guaranteed; those numbers are based on a typical steady workload.

Figure 12-21 shows the schema build progress against the Hyperscale service tier with the transaction counter enabled.

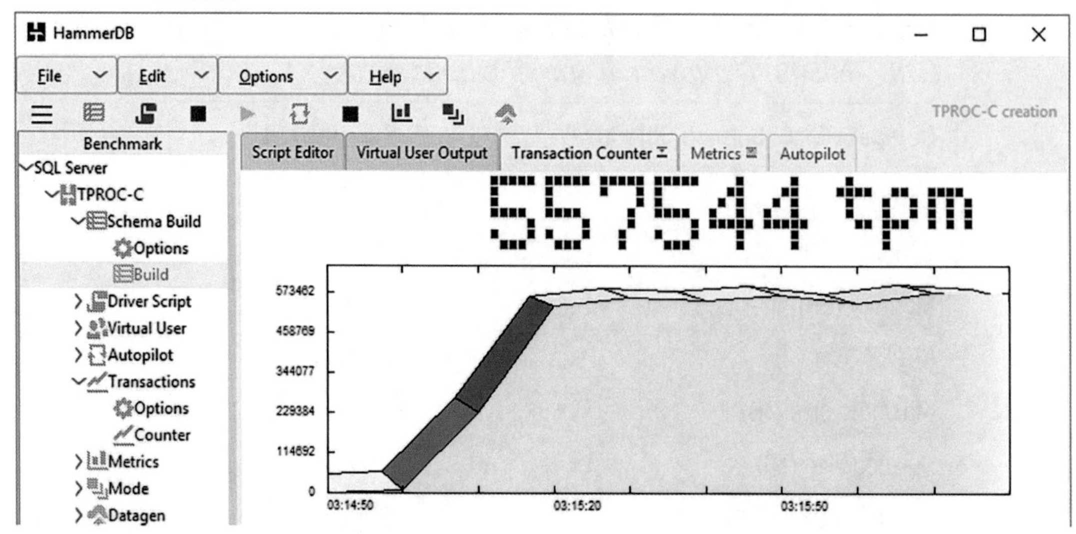

Figure 12-21. *Schema build progress*

Within the HammerDB transaction counter log, we can find all the transaction timestamps.

- The transactions started at 03:14:50.

```
Count 0 MSSQLServer tpm @ Mon Oct 31 03:14:50 +1300 2022
```

- The transactions finished at 03:30:32.

```
Count 53922 MSSQLServer tpm @ Mon Oct 31 03:30:32 +1300 2022
```

- The peak was 573,462 transactions per minute.

```
Count 573462 MSSQLServer tpm @ Mon Oct 31 03:16:10 +1300 2022
```

- The overall execution time was approximately 15 minutes.

We can examine resource usage through the Azure Portal in the logical SQL server on the Metrics blade by using CPU percentage (max), Data IO percentage (max), Log IO percentage (max), and Workers percentage (max). See Figure 12-22.

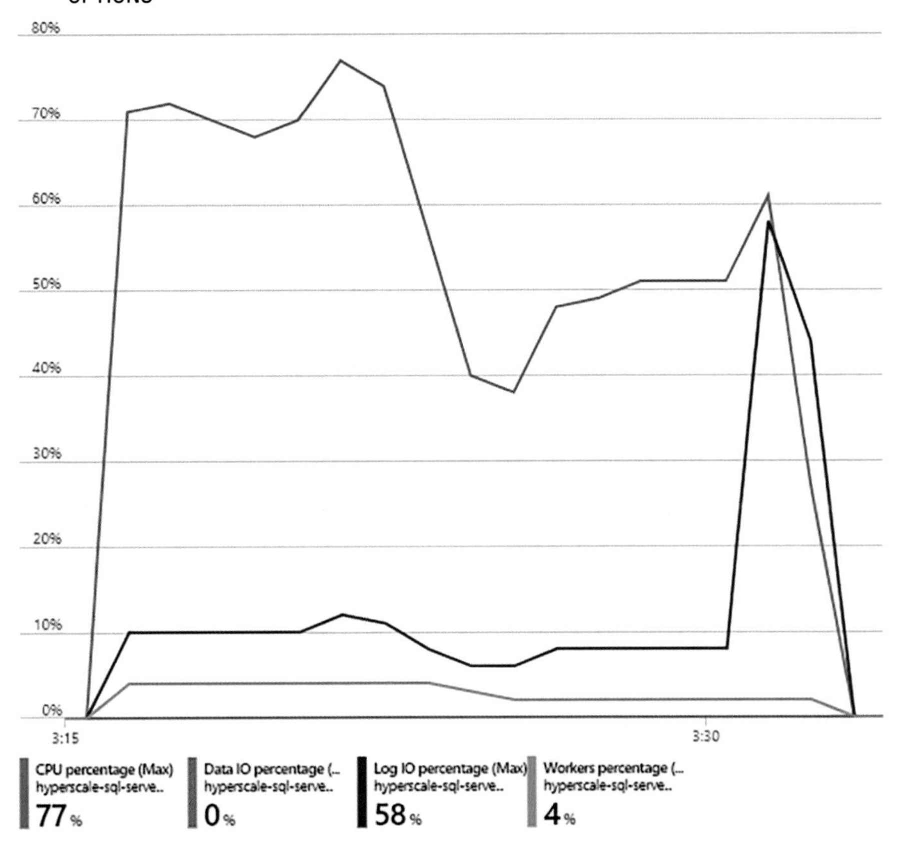

***Figure 12-22.** Azure Portal metrics graph for the Hyperscale service tier*

Summary of Schema Build Performance Results

As expected, we have similar schema build performance results for the Business Critical and Hyperscale service tiers. Although the performance of the Hyperscale service tier may vary, it is heavily dependent on the type of workload we are running. This will be more noticeable during TPROC-C workload execution. See Table 12-4.

***Table 12-4.** HammerDB Schema Build Performance Comparison Using 50 Warehouses and 32 Virtual Users*

Hammer DB Schema Build	GP_Gen5_8	BC_Gen5_8	HS_Gen5_8
Total execution time (minutes)	27	15	15
Transactions per minute	305778	565254	573462

TPROC-C Workload Metrics

In the next stage of this evaluation process, we will run the TPORC-C workload against all three Azure SQL Database service tiers. We will specify the following parameters for TPROC-C Driver options:

1. Set Total Transactions per User to 1000 since we are using the Timed Driver Script option. This should a high enough value so that our virtual users do not complete their transactions before the timed test is complete.
2. Set Minutes of Ramp-up Time to 2 minutes. This should be set to a long enough time for a workload to reach a steady transaction rate.
3. Set Minutes for Test Duration to 8 minutes because we are using the Timed Driver Script option with the specified duration time.
4. Enable the Use All Warehouses option as we want virtual users to select a different warehouse for each transaction to ensure greater I/O activity. See Figure 12-23.

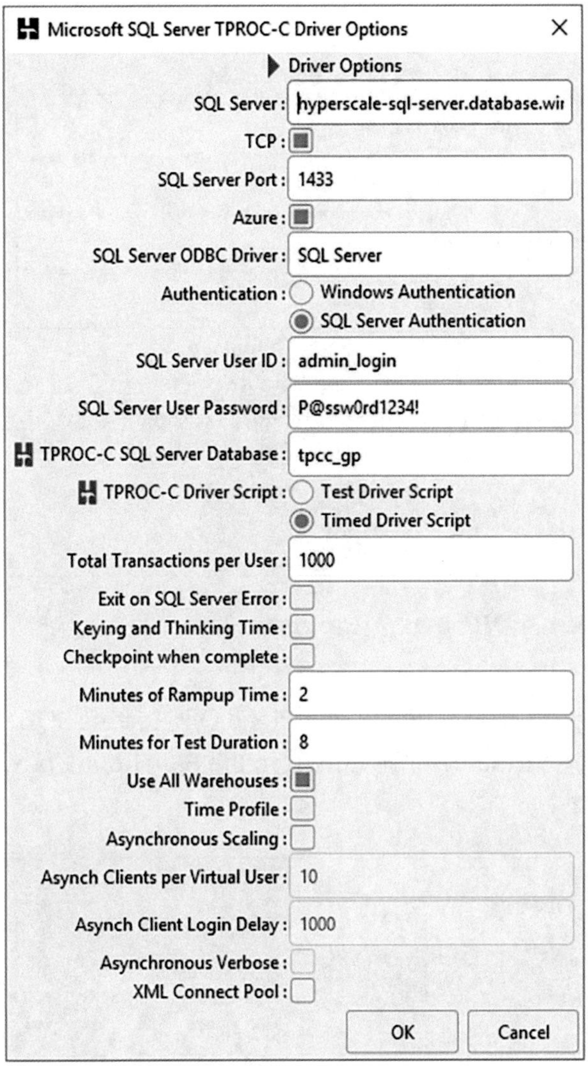

Figure 12-23. *HammerDB TPROC-C driver options*

Next, we will specify the following parameters for the TPROC-C Virtual User options.

5. Set the number of virtual users to 32. This is the number of virtual users that will be created to execute the tests.

6. Set the iterations to 100. This is the number of times that the driver script is going to be run by each virtual user. See Figure 12-24.

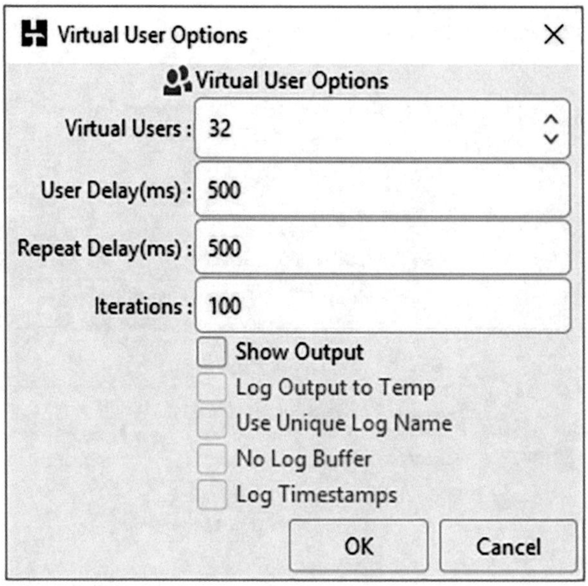

Virtual User Options
Virtual User Options
Virtual Users : 32
User Delay(ms) : 500
Repeat Delay(ms) : 500
Iterations : 100
Show Output
Log Output to Temp
Use Unique Log Name
No Log Buffer
Log Timestamps
OK
Cancel

Figure 12-24. *HammerDB virtual user options*

7. Run the SQL Server TPROC-C workload by selecting SQL Server ➤ TPROC-C ➤ Virtual User ➤ Run from the Benchmark box. See Figure 12-25.

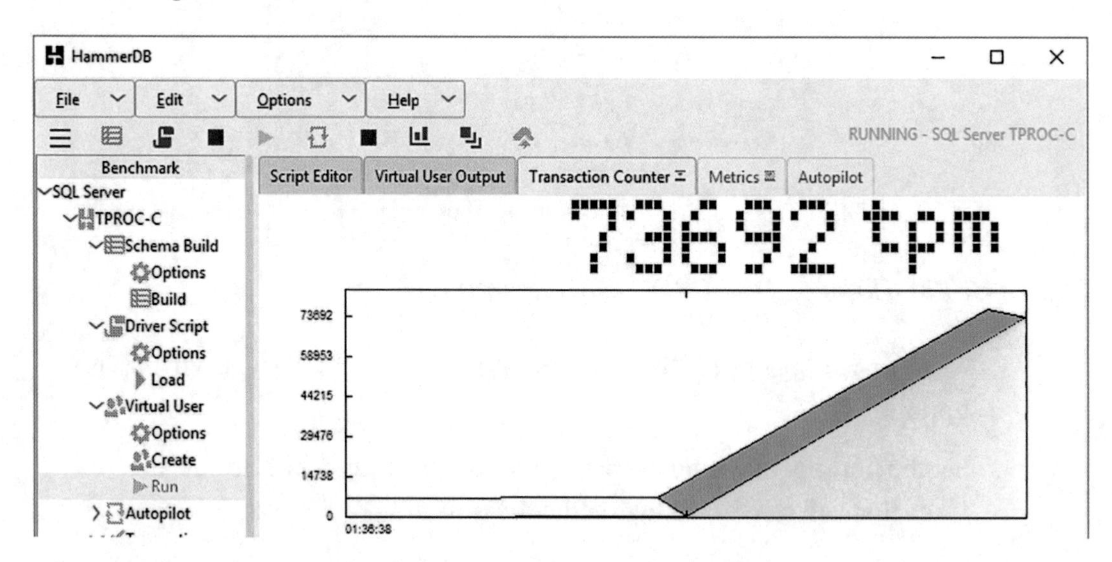

Figure 12-25. *HammerDB running the SQL Server TPROC-C workload*

The virtual users will be created, and the driver script iterations will begin.

Azure SQL Database General Purpose Service Tier (GP_Gen5_8)

You can run the following T-SQL statement to see the resource limitations of your selected Azure SQL deployment:

```
/* SLO, Server Name, CPU Limit, MAX Log Throughput, MAX DB Size*/
SELECT
slo_name as 'Service Level Objective'
, server_name as 'Server Name'
, cpu_limit as 'CPU Limit'
, database_name as 'DB Name'
, primary_max_log_rate/1024/1024 as 'Log Throughput MB/s'
, max_db_max_size_in_mb/1024/1024 as 'MAX DB Size in TB'
FROM sys.dm_user_db_resource_governance
GO
```

Table 12-5 shows the T-SQL query result from the sys.dm_user_db_resource_governance DMV, showing actual configuration and capacity settings for Azure SQL Database deployment.

Table 12-5. *Actual Configuration and Capacity Settings Query Result*

Service Level Objective	Server Name	CPU Limit	DB Name	Log Throughput MB/s	MAX DB Size in TB
SQLDB_GP_GEN5_8_SQLG5	hyperscale-sql-server	8	tpcc_gp	36	4

To see the maximum amount of committed memory for the selected Azure SQL Database deployment, run the following T-SQL statement:

```
/* Memory Size - Depends on compute vCPU number */
SELECT
CAST((memory_limit_mb) /1024. as DECIMAL(7,2)) as 'Max Memory in GB'
FROM sys.dm_os_job_object
GO
```

Table 12-6 shows the query result from sys.dm_os_job_object showing the maximum amount of committed memory for the Azure SQL Database deployment.

Table 12-6. *Maximum Amount of Committed Memory*

Max Memory in GB
41.52

Within the HammerDB transaction counter log, we can find all the transaction timestamps.

- The transactions started at 01:36:38.

```
Count 0 MSSQLServer tpm @ Wed Nov 02 01:36:38 +1300 2022
```

- The transactions finished at 01:46:49.

```
Count 85806 MSSQLServer tpm @ Wed Nov 02 01:46:49 +1300 2022
```

- The peak was 167,670 transactions per minute.

```
Count 167670 MSSQLServer tpm @ Wed Nov 02 01:43:59 +1300 2022
```

- The overall execution time was 10 minutes, which includes the Timed Driver Script option taking 8 minutes plus 2 minutes of ramp-up time.

The tests were completed using 32 virtual users achieving 59,112 new orders per minute (NOPM) from 139,351 transactions per minute (TPM). Figure 12-26 shows the Azure Portal metrics during TPROC-C workload execution against the Azure SQL Database General Purpose service tier (GP_Gen5_8).

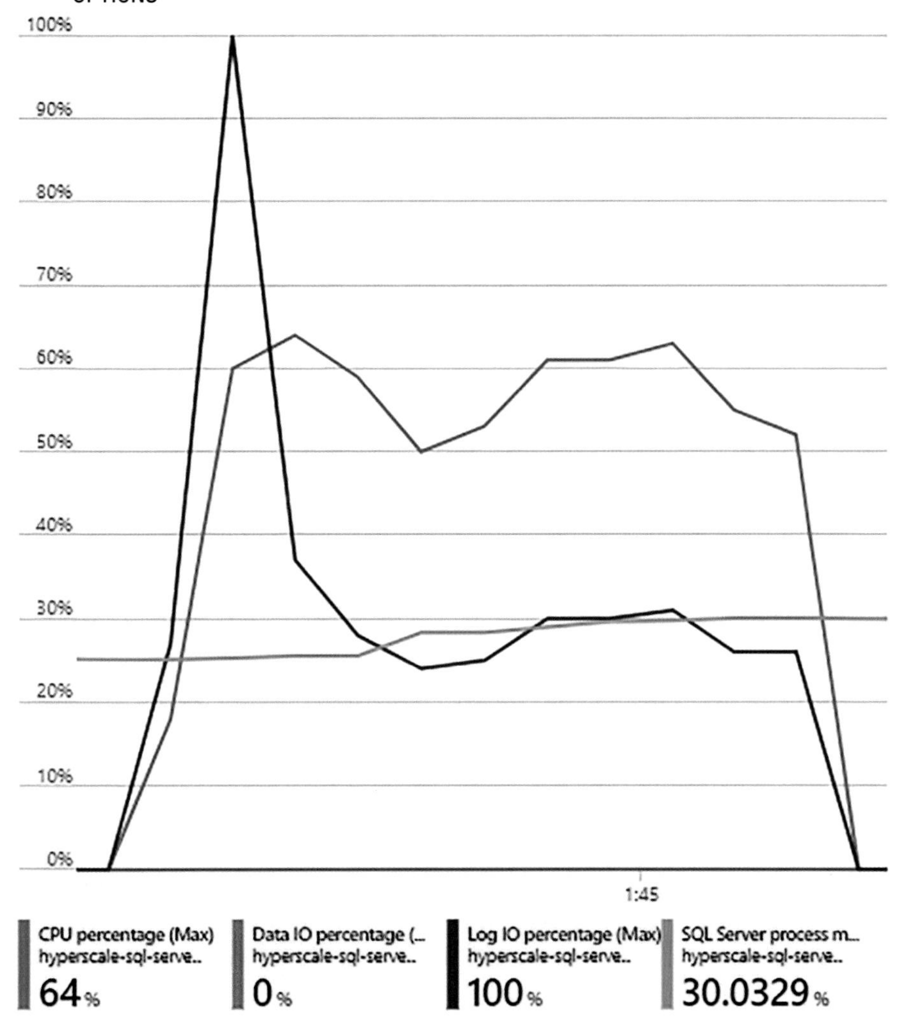

***Figure 12-26.** Azure Portal metrics graph for the General Purpose service tier*

Table 12-7 lists more details about accumulated waits during the TPROC-C workload execution for the General Purpose service tier deployment.

Table 12-7. *Query Store Wait Categories Statistics Accumulated During SQL Server TPROC-C Workload Execution (GP_Gen5_8)*

Wait category id	wait category	avg wait time	min wait time	max wait time	▼ total wait time	execution count
16	Parallelism	3653.79	215	35543	1260558	345
4	Latch	0.06	0	158012	1129443	17576426
5	Buffer Latch	0.05	0	1986	1014261	19840990
3	Lock	0.03	0	1822	375271	10798853
6	Buffer IO	0.02	0	2020	348908	19929954
1	CPU	0.01	0	6653	154165	26155843
14	Tran Log IO	0.01	0	1935	150507	17880810
23	Log Rate Governor	0.01	0	25323	89700	12492362
17	Memory	0	0	3958	15857	23674247

Azure SQL Database Business Critical Service Tier (BC_Gen5_8)

To examine the performance of our Business Critical Azure SQL Database service tier deployment, we will use the same T-SQL as in the previous example using the `sys.dm_user_db_resource_governance` DMV. See Table 12-8.

Table 12-8. *DMV sys.dm_user_db_resource_governance Query Result Showing Actual Configuration and Capacity Settings for Azure SQL Database Deployment*

Service Level Objective	Server Name	CPU Limit	DB Name	Log Throughput MB/s	MAX DB Size in TB
SQLDB_BC_GEN5_ON_GEN8IH_8_INTERNAL_GPGEN8HH_128ID	hyperscale-sql-server	8	tpcc_bc	96	4

To see the maximum amount of committed memory for selected Azure SQL Database deployment, run the following T-SQL statement:

```
/* Memory Size - Depends on compute vCPU number */
SELECT
CAST((mmory_limit_mb) /1024. as DECIMAL(7,2)) as 'Max Memory in GB'
FROM sys.dm_os_job_object
GO
```

Table 12-9 shows the query result from `sys.dm_os_job_object` that shows the maximum amount of committed memory for Azure SQL Database deployment.

Table 12-9. *Maximum Amount of Committed Memory*

Max Memory in GB
41.52

Within the HammerDB transaction counter log, we can find all the transaction timestamps.

- The transactions started at 02:56:40.

  `Count 0 MSSQLServer tpm @ Wed Nov 02 02:56:40 +1300 2022`

- The transactions finished at 03:07:02.

  `Count 41008 MSSQLServer tpm @ Wed Nov 02 03:07:02 +1300 2022`

- The peak was 362,496 transactions per minute.

  `Count 362496 MSSQLServer tpm @ Wed Nov 02 03:05:22 +1300 2022`

- The overall execution time was 10 minutes, which includes the Timed Driver Script option taking 8 minutes plus 2 minutes of ramp-up time.

The tests were completed using 32 virtual users achieving 130,662 new orders per minute (NOPM) from 308,240 transactions per minute (TPM). Figure 12-27 shows the Azure Portal metrics during the TPROC-C workload execution against the Azure SQL Database Business Critical General service tier (BC_Gen5_8).

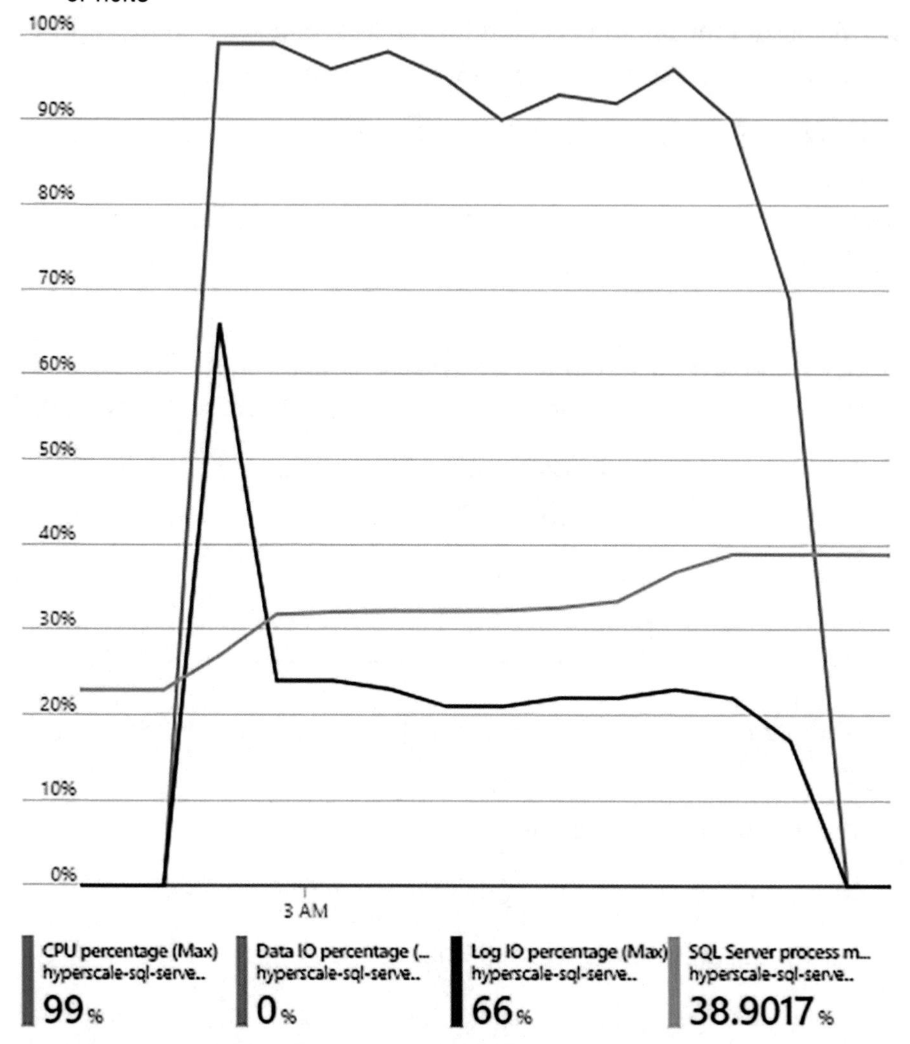

Figure 12-27. *Azure Portal metrics graph for the Business Critical service tier*

You can find more details on the accumulated waits during the TPROC-C workload execution for the Business Critical service tier deployment in Table 12-10.

***Table 12-10.** Query Store Wait Categories Statistics Accumulated During SQL Server TPROC-C Workload Execution (BC_Gen5_8)*

wait category id	wait category	avg wait time	min wait time	max wait time	▼ total wait time	execution count
16	Parallelism	2111.54	187	18097	1433735	679
4	Latch	0.01	0	63083	408065	33050174
5	Buffer Latch	0.01	0	147	213091	37758377
3	Lock	0.01	0	3998	177909	23540270
14	Tran Log IO	0	0	215	133753	33536619
1	CPU	0	0	18808	96438	51186847
23	Log Rate Governor	7119.5	5430	8809	14239	2
17	Memory	0	0	1808	13212	31823559

Azure SQL Database Hyperscale Service Tier (HS_Gen5_8)

We are going to run the same TPROC-C workload test using two different scenarios.

- With data pages cached in the buffer pool (in memory) and RBPEX cache
- With an empty cache

When enough data is cached, running the same TPROC-C workload should give us similar overall results to the Business Critical service tier.

As with the previous two examples, first we will review the resource limitations of our Hyperscale deployment. Table 12-11 shows the query result from the `sys.dm_user_db_resource_governance` DMV, showing actual configuration and capacity settings for the Azure SQL Database deployment.

***Table 12-11.** Actual Configuration and Capacity Settings Query Result*

Service-Level Objective	Server Name	CPU Limit	DB Name	Log Throughput MB/s	Max DB Size in TB
SQLDB_HS_GEN5_ON_GEN8IH_8_INTERNAL_GPGEN8HH_128ID	hyperscale-sql-server	8	tpcc_hs	105	100

To see the maximum amount of committed memory for the selected Azure SQL Database deployment, run the following T-SQL statement:

```
/* Memory Size - Depends on compute vCPU number */
SELECT
CAST((memory_limit_mb) /1024. as DECIMAL(7,2)) as 'Max Memory in GB'
FROM sys.dm_os_job_object
GO
```

Table 12-12 shows the query result from `sys.dm_os_job_object`, showing the maximum amount of committed memory for the Azure SQL Database deployment.

***Table 12-12.** Maximum Amount of Committed Memory*

Max Memory in GB
41.52

For the Azure SQL Hyperscale Database deployment with eight vCores, the memory size is 41.5GB, which results in 125GB resilient buffer pool size. To see the size of compute replica RBPEX cache, we can run the following T-SQL statement:

```
/* Resilient Buffer Pool File Size in GB = 3 * Memory size */
SELECT size_on_disk_bytes / 1024 / 1024 / 1024 AS "Resilient Buffer Pool
File Size in GB" from sys.dm_io_virtual_file_stats(0,NULL);
GO
```

Table 12-13 shows the query result from DMV function `sys.dm_io_virtual_file_stats` and returns the number of bytes used on the disk for this file, which presents the compute node RBPEX size for the Azure SQL Database deployment.

Table 12-13. *Hyperscale Service Tier Compute Node RBPEX Size*

Resilient Buffer Pool File Size in GB
125

Tip The Azure SQL Hyperscale deployment compute node RPB size is equal to three times the maximum memory.

In this scenario, we will run the TPROC-C workload immediately after the schema build is complete. Bear in mind that after the schema build is done, there are pages already cached in RBPEX. You can use the following SQL statement to fetch the RBPEX usage statistics and cache size:

```
SELECT
DB_NAME (database_id) AS 'DB Name'
, file_id AS 'File ID'
, num_of_reads AS 'Total Reads',
CASE WHEN num_of_reads = 0 THEN 0 ELSE (num_of_bytes_read / num_of_reads)
END AS 'Avg Bytes Per Read',
CASE WHEN num_of_reads = 0 THEN 0 ELSE (io_stall_read_ms / num_of_reads)
END AS 'Read Latency',
num_of_writes AS 'Total Writes',
CASE WHEN num_of_writes = 0 THEN 0 ELSE (num_of_bytes_written / num_of_
writes) END AS 'Avg Bytes Per Write',
CASE WHEN num_of_writes = 0 THEN 0 ELSE (io_stall_write_ms / num_of_writes)
END AS 'Write Latency',
CASE WHEN (num_of_reads = 0 AND num_of_writes = 0) THEN 0 ELSE (io_stall /
(num_of_reads + num_of_writes)) END AS 'Overall Latency',
CASE WHEN (num_of_reads = 0 AND num_of_writes = 0) THEN 0 ELSE ((num_of_
bytes_read + num_of_bytes_written) / (num_of_reads + num_of_writes)) END AS
'Avg Bytes Per IO'
,size_on_disk_bytes  / 1024 / 1024 / 1024 AS 'RBPX Cache Size in GB'
FROM sys.dm_io_virtual_file_stats(0,NULL)
```

Table 12-14 shows RBPEX usage statistics example before the schema build was executed against the Azure SQL Hyperscale database.

Table 12-14. *RBPEX Usage Statistics Before Schema Build*

DB Name	File ID	Total Reads	Read Latency	Total Writes	Write Latency	Avg Bytes Per I/O	RBPX Cache Size in GB
tpcc_hs	0	11	0	173	0	11887	125

From these results, we can see that the RBPEX cache is almost empty. After the schema build is complete, the cache is going to look a bit different. Table 12-15 shows that the number of writes in the RBPEX cache is high.

Table 12-15. *RBPEX Usage Statistics After Schema Build Is Finished*

DB Name	File ID	Total Reads	Read Latency	Total Writes	Write Latency	Avg Bytes Per I/O	RBPX Cache Size in GB
tpcc_hs	0	167	0	30958	2	417250	125

Next, we will run the TPROC-C workload against our Azure SQL Hyperscale database. For the TPROC-C workload execution, we will use the same parameters that we used in the previous section. This will take approximately 10 minutes including the Timed Driver Script duration of 8 minutes and 2 minutes of ramp-up time.

After the TPROC-C Workload execution is finished, the RBPEX cache of our Azure SQL Hyperscale database is going to look significantly different, as demonstrated in Table 12-16.

Table 12-16. *RBPEX Usage Statistics After TPROC-C Workload Is Finished*

DB Name	File ID	Total Reads	Read Latency	Total Writes	Write Latency	Avg Bytes Per I/O	RBPX Cache Size in GB
tpcc_hs	0	337	0	19121176	2	27730	125

As we can see, the number of the writes are increasing in RBPEX cache, which is expected due to the HammerDB TPROC-C workload mostly consisting of update SQL statements. The performance of our Azure SQL Database Hyperscale service tier deployment is going to heavily depend on the workload type.

Within the HammerDB transaction counter log, we can find all transaction timestamps.

- The transactions started at 03:48:10.

  ```
  Count 0 MSSQLServer tpm @ Wed Nov 02 03:48:10 +1300 2022
  ```

- The transactions finished at 03:58:41

  ```
  Count 70788 MSSQLServer tpm @ Wed Nov 02 03:58:41 +1300 2022
  ```

- The peak was 316,650 transactions per minute.

  ```
  Count 316650 MSSQLServer tpm @ Wed Nov 02 03:51:20 +1300 2022
  ```

- The overall execution time was 10 minutes, which includes the Timed Driver Script option taking 8 minutes plus 2 minutes of ramp-up time.

The test completed for 32 virtual users, achieving 120,896 new orders per minute (NOPM) from 285,210 transactions per minute (TPM). Figure 12-28 shows the metrics for the logical SQL server from the Azure Portal during the TPROC-C workload execution against the Azure SQL Database Hyperscale service tier (HS_Gen5_8).

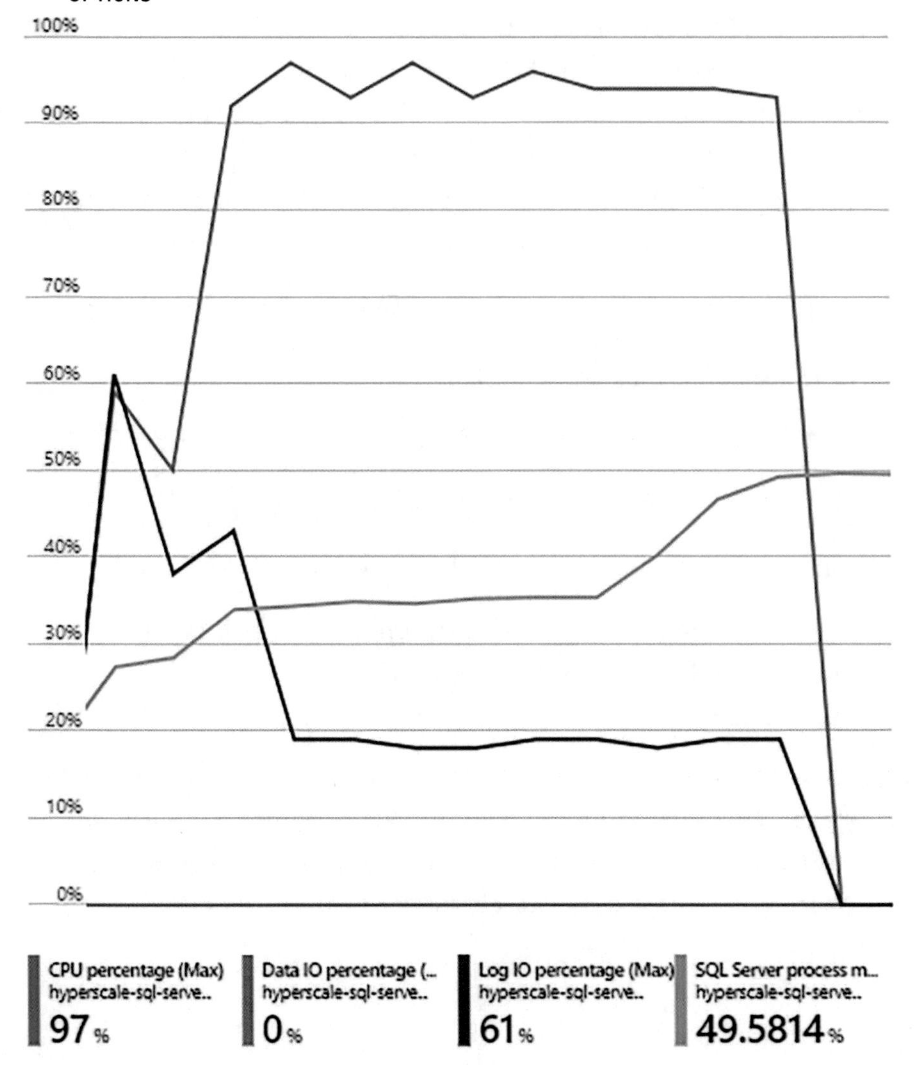

Figure 12-28. *Azure Portal metrics graph for Hyperscale service tier*

Table 12-17 shows the results for the query store wait category statistics accumulated during the SQL Server TPROC-C workload execution (HS_Gen5_8).

Table 12-17. *Query Store Wait Category Statistics (HS_Gen5_8)*

Wait category id	wait category	avg wait time	min wait time	max wait time	▼ total wait time	execution count
16	Parallelism	2340.19	179	17144	1453257	621
6	Buffer IO	0.02	0	1097	518556	33684629
3	Lock	0.01	0	921	226597	20814752
14	Tran Log IO	0.01	0	1444	188372	30944538
1	CPU	0	0	18596	89766	44335823
17	Memory	0	0	1755	14075	39146384
23	Log Rate Governor	6101	4840	7362	12202	2

To better understand which wait types belong to each wait category in the query store, please see Table 12-18.

Table 12-18. *Query Store Wait Categories*

Integer Value	Wait Category	Wait Types Include in the Category
0	Unknown	Unknown
1	CPU	SOS_SCHEDULER_YIELD
2	Worker Thread	THREADPOOL
3	Lock	LCK_M_%
4	Latch	LATCH_%
5	Buffer Latch	PAGELATCH_%
6	Buffer IO	PAGEIOLATCH_%
7	Compilation	RESOURCE_SEMAPHORE_QUERY_COMPILE
8	SQL CLR	CLR%, SQLCLR%

(*continued*)

Table 12-18. *(continued)*

Integer Value	Wait Category	Wait Types Include in the Category
9	Mirroring	DBMIRROR%
10	Transaction	XACT%, DTC%, TRAN_MARKLATCH_%, MSQL_XACT_%, TRANSACTION_MUTEX
11	Idle	SLEEP_%, LAZYWRITER_SLEEP, SQLTRACE_BUFFER_FLUSH, SQLTRACE_INCREMENTAL_FLUSH_SLEEP, SQLTRACE_WAIT_ENTRIES, FT_IFTS_SCHEDULER_IDLE_WAIT, XE_DISPATCHER_WAIT, REQUEST_FOR_DEADLOCK_SEARCH, LOGMGR_QUEUE, ONDEMAND_TASK_QUEUE, CHECKPOINT_QUEUE, XE_TIMER_EVENT
12	Preemptive	PREEMPTIVE_%
13	Service Broker	BROKER_% (but not BROKER_RECEIVE_WAITFOR)
14	Tran Log IO	LOGMGR, LOGBUFFER, LOGMGR_RESERVE_APPEND, LOGMGR_FLUSH, LOGMGR_PMM_LOG, CHKPT, WRITELOG
15	Network IO	ASYNC_NETWORK_IO, NET_WAITFOR_PACKET, PROXY_NETWORK_IO, EXTERNAL_SCRIPT_NETWORK_IOF
16	Parallelism	CXPACKET, EXCHANGE, HT%, BMP%, BP%
17	Memory	RESOURCE_SEMAPHORE, CMEMTHREAD, CMEMPARTITIONED, EE_PMOLOCK, MEMORY_ALLOCATION_EXT, RESERVED_MEMORY_ALLOCATION_EXT, MEMORY_GRANT_UPDATE
18	User Wait	WAITFOR, WAIT_FOR_RESULTS, BROKER_RECEIVE_WAITFOR
19	Tracing	TRACEWRITE, SQLTRACE_LOCK, SQLTRACE_FILE_BUFFER, SQLTRACE_FILE_WRITE_IO_COMPLETION, SQLTRACE_FILE_READ_IO_COMPLETION, SQLTRACE_PENDING_BUFFER_WRITERS, SQLTRACE_SHUTDOWN, QUERY_TRACEOUT, TRACE_EVTNOTIFF

(continued)

Table 12-18. (*continued*)

Integer Value	Wait Category	Wait Types Include in the Category
20	Full Text Search	FT_RESTART_CRAWL, FULLTEXT GATHERER, MSSEARCH, FT_METADATA_MUTEX, FT_IFTSHC_MUTEX, FT_IFTSISM_MUTEX, FT_IFTS_RWLOCK, FT_COMPROWSET_RWLOCK, FT_MASTER_MERGE, FT_PROPERTYLIST_CACHE, FT_MASTER_MERGE_COORDINATOR, PWAIT_RESOURCE_SEMAPHORE_FT_PARALLEL_QUERY_SYNC
21	Other Disk IO	ASYNC_IO_COMPLETION, IO_COMPLETION, BACKUPIO, WRITE_COMPLETION, IO_QUEUE_LIMIT, IO_RETRY
22	Replication	SE_REPL_%, REPL_%, HADR_% (but not HADR_THROTTLE_LOG_RATE_GOVERNOR), PWAIT_HADR_%, REPLICA_WRITES, FCB_REPLICA_WRITE, FCB_REPLICA_READ, PWAIT_HADRSIM
23	Log Rate Governor	LOG_RATE_GOVERNOR, POOL_LOG_RATE_GOVERNOR, HADR_THROTTLE_LOG_RATE_GOVERNOR, INSTANCE_LOG_RATE_GOVERNOR

In our test example, there are waits related to the parallelism (`CXCONSUMER`, `CXPACKET`) and buffer I/O waits (`PAGEIOLATCH_EX`, `PAGEIOLATCH_SH`). If we examine this in further detail, we can see that the Tran Log IO wait category with the `WRITELOG` wait type is the dominant wait, as shown in Figure 12-29. We already mentioned that the HammerDB TPROC-C workload is mostly consisting of update SQL statements, WRITELOG wait type will be expected as a predominant wait type in both of our testing scenarios.

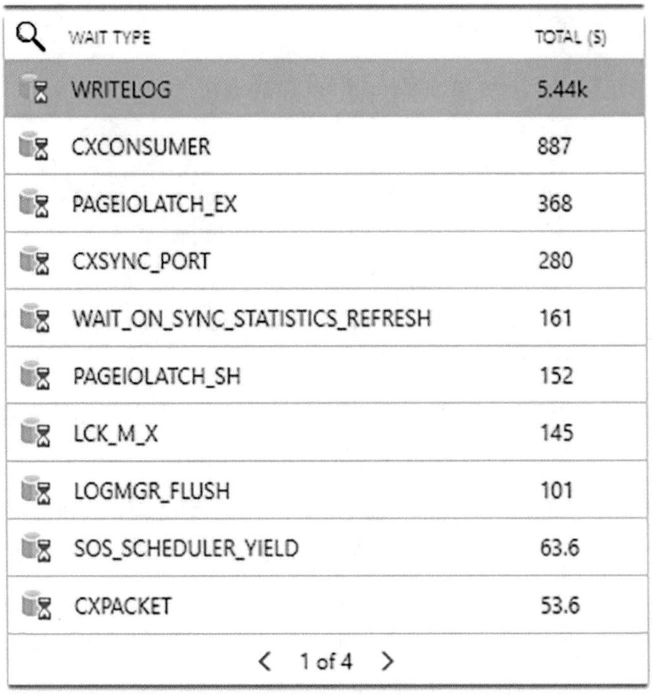

WAIT TYPE	TOTAL (S)
WRITELOG	5.44k
CXCONSUMER	887
PAGEIOLATCH_EX	368
CXSYNC_PORT	280
WAIT_ON_SYNC_STATISTICS_REFRESH	161
PAGEIOLATCH_SH	152
LCK_M_X	145
LOGMGR_FLUSH	101
SOS_SCHEDULER_YIELD	63.6
CXPACKET	53.6

< 1 of 4 >

Figure 12-29. *Azure SQL analytics log preview (HS_Gen5_8)*

The predominant waits are `WRITELOG` and `CXCONSUMER` waits:

- The `WRITELOG` wait type presents a delay in hardening the log. In Hyperscale, this log flush delay happens during the wait time between the Log Writer and writing the log to the Log Landing Zone on Azure Premium Storage.
- The `CXPACKET` and `CXCONSUMER` wait types show us that in our Hyperscale database, when running queries in parallel, the work isn't being equally distributed across the available threads.

In this scenario, we will empty the RBPEX cache of the compute node after the schema build execution is finished. See Table 12-19.

Table 12-19. *Compute Node RBPEX Cache*

DB Name	File ID	Total Reads	Read Latency	Total Writes	Write Latency	Avg Bytes Per I/O	RBPX Cache Size in GB
tpcc_hs	0	0	0	1094	0	1048576	125

From the query result shown in Table 12-19, you can see that the total writes are almost gone on the compute node's RBPEX cache. The next step would be to execute the TPROC-C workload again.

Within the HammerDB transaction counter log, we can find all the transaction timestamps.

- The transactions started at 04:25:47.

  ```
  Count 0 MSSQLServer tpm @ Wed Nov 02 04:25:47 +1300 2022
  ```

- The transactions finished at 04:35:59.

  ```
  Count 196914 MSSQLServer tpm @ Wed Nov 02 04:35:59 +1300 2022
  ```

- The peak was 276,756 transactions per minute.

  ```
  Count 276756 MSSQLServer tpm @ Wed Nov 02 04:32:28 +1300 2022
  ```

- The overall execution time was approximately 10 minutes, which includes the Timed Driver Script option with 8 minutes duration time plus 2 minutes of ramp-up time.

The test completed for 32 virtual users, achieving 100,884 new orders per minute (NOPM) from 238,153 transactions per minute (TPM). Figure 12-30 shows the test results for the Azure SQL Database Hyperscale service tier (HS_Gen5_8).

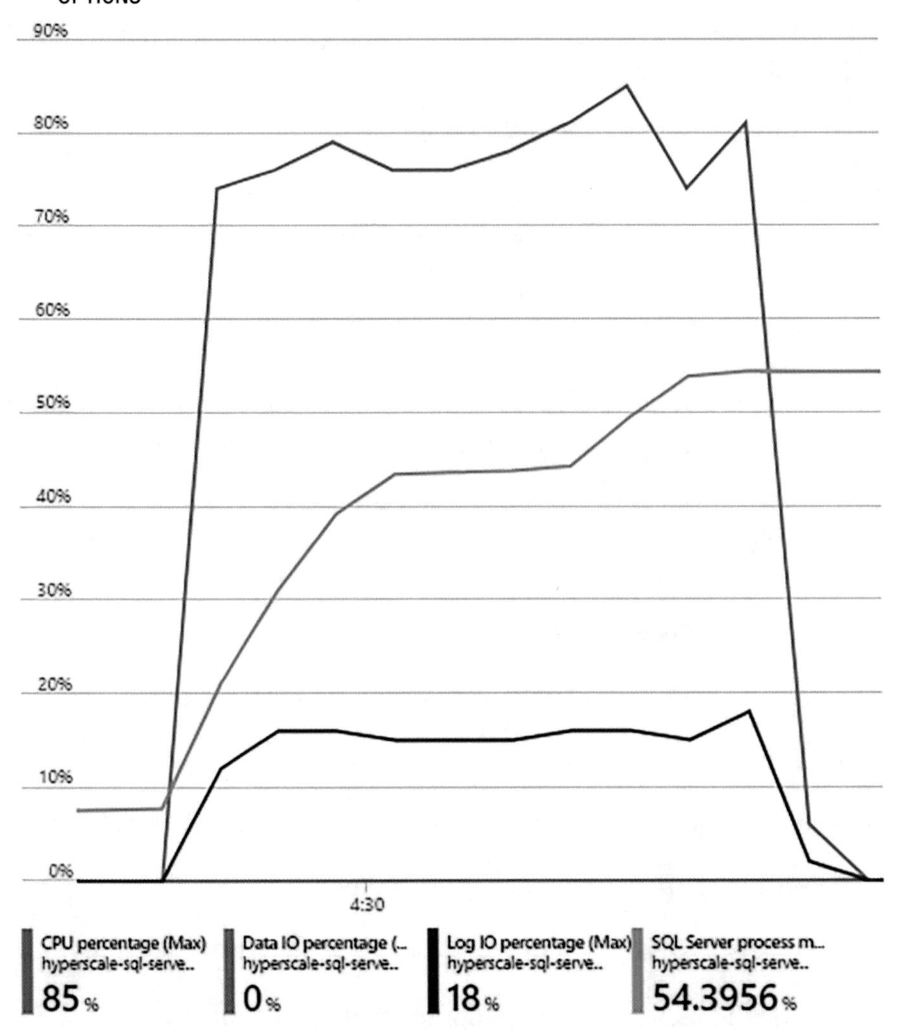

***Figure 12-30.** Azure Portal metrics graph for Hyperscale service tier*

Table 12-20 shows the query wait statistics accumulated during the SQL Server TPROC-C workload execution.

Table 12-20. *Query Store Wait Categories (HS_Gen5_8)*

wait category id	wait category	avg wait time	min wait time	max wait time	▼ total wait time	execution count
6	Buffer IO	0.11	0	9736	3500500	32906931
16	Parallelism	2168.15	372	14166	951819	439
3	Lock	0.02	0	9712	307538	18767349
14	Tran Log IO	0	0	291	95127	21389117
1	CPU	0	0	431	40516	34575645
5	Buffer Latch	0	0	2540	34428	24562761
17	Memory	0	0	69	11928	29523276
4	Latch	0	0	126	1300	14581460
11	Idle	0	0	10	10	786718

We can see that there are waits related to the Tran Log IO wait category (`WRITELOG` wait type). In addition, in the dominant wait category, we can see Buffer IO waits (`PAGEIOLATCH_EX` and `PAGEIOLATCH_SH`). See Figure 12-31.

WAIT TYPE	TOTAL (S)
WRITELOG	4.47k
PAGEIOLATCH_SH	1.82k
PAGEIOLATCH_EX	1.53k
CXCONSUMER	690
LCK_M_X	294
CXSYNC_PORT	201
WAIT_ON_SYNC_STATISTICS_REFRESH	121
LOGMGR_FLUSH	66.7
SOS_SCHEDULER_YIELD	41.7
CXPACKET	35.3

< 1 of 4 >

***Figure 12-31.** Azure SQL analytics log preview (HS_Gen5_8)*

The predominant waits are the Buffer IO category waits:

- The `PAGEIOLATCH` wait type occurs when the page is not available in the buffer pool or RBPEX cache of the compute node and a GetPage@LSN request is sent to the page server. If the page server has the page, the page will be returned to the compute node. However, if the page does not exist on the page server RBPEX, then the GetLogBlock request will be sent to the Log service. Every page request will create a `PAGEIOLATCH` wait type.

We can conclude from these two scenarios that Hyperscale heavily relies on workload type and capability to utilize the compute node buffer pool and RBPEX caching mechanism. If pages are read from the compute node caching layers, performance will be like the Azure SQL Database Business-Critical service tier, but with the many other advantages the Hyperscale service tier has to offer.

Summary of TPROC-C Workload Performance Results

As expected, we have similar schema build performance results for the Business Critical and Hyperscale service tiers. However, the performance of the Hyperscale service tier may vary and is heavily dependent on the type of workload being run. That difference will be most noticeable during the TPROC-C workload execution. Table 12-21 shows resource limits of Azure SQL Database deployments we used in these test examples and their specifications comparison.

Table 12-21. *Azure SQL Database Resource Limits Specifications*

Compute Size (Service Objective)	GP_Gen5_8	BC_Gen5_8	HS_Gen5_8
Hardware	Gen5	Gen5	Gen5
vCores	8	8	8
Memory (GB)	41.5	41.5	41.5
RBPEX size	N/A	N/A	3x memory
Max data size (TB)	2	2	100
Max log size (GB)	614.4	614.4	Unlimited
Max log rate (MBps)	36	96	105
Local read IO latency (approximate ms)	N/A	1 to 2	1 to 2*
Remote read I/O latency (approximate ms)	5 to 10	N/A	1 to 5*
Local write I/O latency (approximate ms)	N/A	1 to 2	3 to 5
Remote Write IO latency (approximate ms)	5-7	N/A	N/A
Storage type	Remote SSD	Local SSD	Shared/multitiered

** Latency can vary and mostly depends on the workload and usage patterns. Hyperscale is a multitiered architecture with caching at multiple levels. If data is being accessed primarily via a cache on the compute node, latency will be similar as for Business Critical or Premium service tiers.*

Table 12-22 shows TPROC-C Workload performance results comparison.

Table 12-22. *Azure SQL Database TPROC-C Workload Performance Comparison with 50 Warehouses and 32 Virtual Users to Build the Schema*

Hammer DB TPROC-C Workload	GP_Gen5_8	BC_Gen5_8	HS_Gen5_8 (Scenario 1)	HS_Gen5_8 (Scenario 2)
Total execution time (minutes)	10 (8 Timed Driver Script option + 2 minutes ramp-up)	10 (8 Timed Driver Script option + 2 minutes ramp-up)	10 (8 Timed Driver Script option + 2 minutes ramp-up)	10 (8 Timed Driver Script option + 2 minutes ramp-up)
New orders per minute (NOPM)	59112	130662	120896	100884
Transactions per minute	139351	308240	285210	238153
Peak transactions per minute	167670	362496	316650	276756

Summary

In this chapter, we talked about an overall performance comparison between the traditional Azure architecture and Hyperscale architecture. We deployed and ran benchmark tests against three different Azure SQL Database service tiers.

- Azure SQL Database General Purpose service tier
- Azure SQL Database Business Critical service tier
- Azure SQL Database Hyperscale service tier

For benchmark tests, we used the TPROC-C OLTP workload implemented in the HammerDB benchmark tool. The HammerDB TPROC-C workload is derived from TPC-C and designed to be reliable and tuned to produce consistently repeatable and accurate results. We compared schema build and PROC-C workload performance results against all three deployments, hosting them all in the same Azure region and on the same logical SQL server.

In the following chapter, we will cover the basics of monitoring the performance and health of a Hyperscale database. Also, we are going to explain how to adjust your Hyperscale deployment tier to your ongoing workload requirements utilizing automatic PaaS scaling ability.

PART III

Operation and Management

CHAPTER 13

Monitoring and Scaling

In the previous chapters, we covered many important topics related to the initial stages of adopting Azure SQL Hyperscale. Up until now, this book has focused on the design and deployment of a typical highly available Hyperscale environment. However, we now need to start looking at how we continue to effectively operate the Hyperscale environment post-deployment. As your usage of a Hyperscale database grows and changes, being able to proactively prevent issues and reactively diagnose problems will become increasingly important.

In this chapter, we will cover the basics of monitoring the performance and health of the Hyperscale database as well as examining how Hyperscale databases can be manually or automatically scaled. Over time, the usage patterns and needs of the workloads using a Hyperscale database are likely to change, so understanding how we can scale to support these workloads will be important.

Many of these tasks may be familiar to you if you've spent time operating Microsoft SQL Server, but the approach to monitoring and scaling a PaaS service often differs because of the reduced access to the underlying infrastructure. Therefore, we need to understand the various approaches available to us to monitor Azure SQL Hyperscale databases.

Tip All the features and settings outlined in this chapter can be implemented using infrastructure as code. It is strongly recommended that for production systems that you configure these via infrastructure as code once you're familiar with their operation. However, the examples in this chapter are limited to the Azure Portal only for brevity.

Z. Barać and D. Scott-Raynsford, *Azure SQL Hyperscale Revealed*,
https://doi.org/10.1007/978-1-4842-9225-9_13

Monitoring Platform Metrics

The first place to start when thinking about monitoring a Hyperscale database is platform metrics. These are metrics that are emitted by the underlying Azure SQL database resource and exposed to you through the Azure Portal via the Metrics Explorer and from the REST APIs. They can be used to monitor the current performance of the Hyperscale database as well as setting up automated notifications to be sent when the metrics exceeds a threshold.

Viewing Metrics with the Metrics Explorer

Platform metrics can be viewed in the Metrics Explorer, which can be found on the Metrics blade when the Hyperscale database resource is selected in the Azure Portal. Figure 13-1 shows an example of the Metrics Explorer displaying several of the metrics that are emitted by the Hyperscale database and the underlying infrastructure.

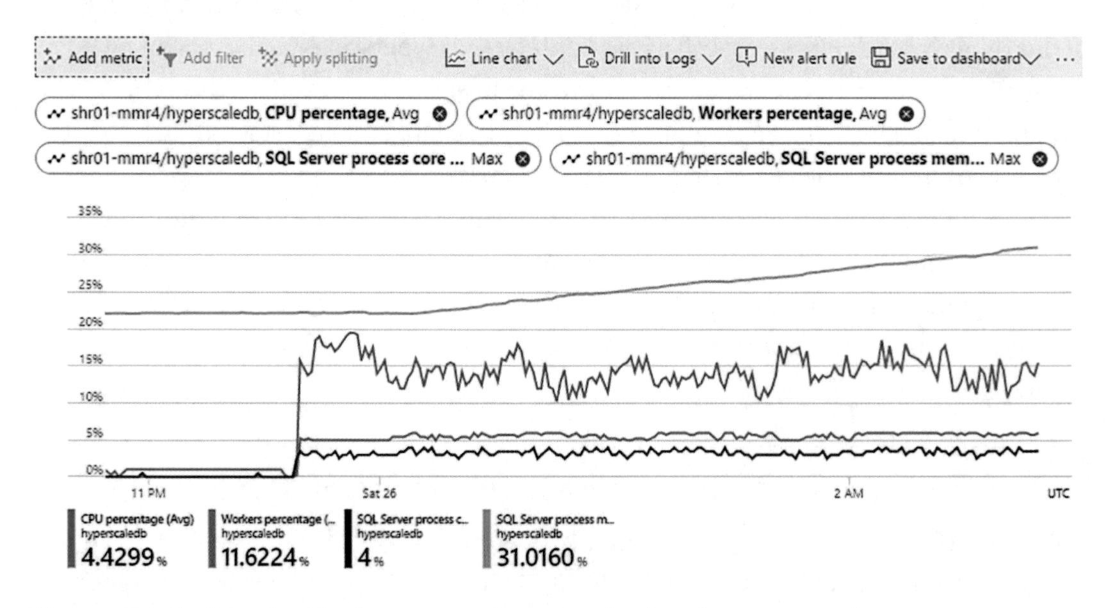

Figure 13-1. *Metrics Explorer on a Hyperscale database example*

The Metrics Explorer allows you to select one or more metrics to be displayed from an available list that apply to the Hyperscale database resource. Some metrics can also be filtered or grouped together to enable greater detail to be shown. For example, it is possible to group by or filter the TLS version of the successful or failed SQL connections metrics.

The following are some of the metrics that are currently emitted for a Hyperscale database:

- *Compute metrics*: These measure the compute consumption of the Hyperscale database and include CPU limit, CPU percentage, CPU used, and SQL Server core process percent.
- *I/O metrics*: The input/output utilization by the Hyperscale database and include data IO percentage and log IO percentage.
- *Memory, deadlocks, and workers metrics*: These report the amount of memory and workers being used in the Hyperscale database and include Deadlocks, In-Memory OLTP Storage, SQL Server core process memory, SQL Server core process memory percentage, and Workers percentage.
- *Network, connection, and session metrics*: These are metrics related to the networking, connections, and sessions to the database and include Blocked by Firewall, Failed connections, Successful connections, Sessions count, and Sessions percentage.
- *Data storage metrics*: These metrics provide information on the storage use by data and log files and include Data storage size, Data space allocated, Data backup storage size, Log backup storage size, Tempdb Data File Size Kilobytes, and Tempdb Log File Size Kilobytes.
- *Log storage metrics*: These are storage metrics that relate to the use of the SQL log and include Log backup storage size, Tempdb Log File Size Kilobytes, and Tempdb Percent Log used.

Tip For more information on Azure Metrics explorer, review `https://learn.microsoft.com/azure/azure-monitor/essentials/metrics-getting-started`.

Once you've configured a Metrics Explorer graph with the information you're interested in, you can pin this to an Azure Dashboard or send it to an Azure Monitor Workbook using the "Save to dashboard" menu. Figure 13-2 shows the Pin to Dashboard button in the "Save to dashboard" menu. This will save a live version of the graph to the dashboard.

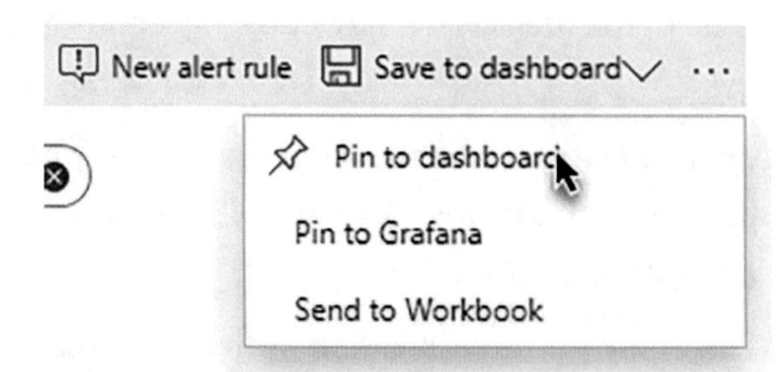

***Figure 13-2.** Saving the Metrics Explorer graph*

The platform metrics that are provided through this method are usually retained for only 93 days. So, if you need to retain them for a longer period for reporting purposes or if you need to combine these metrics with other diagnostic information, then you may need to send these to a Log Analytics workspace.

Streaming Metrics to a Log Analytics Workspace

A Log Analytics workspace can be used to collect and query large amounts of log and metric data. Metrics from the resources can be transformed into log records and can be automatically sent to one or more Log Analytics workspaces. Configuring metrics to be sent to a Log Analytics workspace is done in the same way that logs are sent.

1. In the Azure Portal, select a Hyperscale database resource.
2. Select the "Diagnostic settings" blade.
3. Click "Add diagnostic settings."
4. Enter a name for the diagnostic setting in the "Diagnostic setting" box.
5. Select the "Send to Log Analytics workspace" box and select the Log Analytics workspace to send the metrics to.
6. Select the checkbox for each type of metric you want to send to the Log Analytics workspace.
7. Click Save.

Once the metrics are sent to the Log Analytics workspace, they can be queried and visualized like any other data in the workspace. It is possible to combine the metrics data with other diagnostic data in complex queries that can be used for visualization via Azure Dashboards or Azure Monitor workbooks or alerting.

Important Sending metrics to a Log Analytics workspace incurs a cost, just like sending diagnostic logs. To understand and managed costs of your Log Analytics workspace, see `https://learn.microsoft.com/azure/azure-monitor/logs/analyze-usage`.

The metrics that are sent to the Log Analytics workspace will land in the "AzureMetrics" table. By default, data in this table will be retained for only 30 days. If you want to retain the data for a longer period, you will need to set the retention period on either of the following:

- The "AzureMetrics" table
- The entire Log Analytics workspace

It is also possible to stream the metric logs to an Azure Storage account or Azure Event Hub for either long-term storage and reporting or export into other systems.

Alerting on Platform Metrics

One of the most useful features of Azure Monitoring is the ability to create alert rules that enable a notification to be sent or action to be taken whenever the rule evaluates to be true.

For example, you may want to send an SMS message and execute an Azure Automation Runbook to scale up the Hyperscale database whenever the average CPU percentage exceeds 90 percent for 5 minutes and the Worker percentage exceeds 80 percent for 15 minutes.

When an alert rule is triggered, it will execute one or more action groups. An action group contains a list of notifications to send and actions to execute. The following are the notification types that can be configured:

- *Email Azure Resource Manager role*: This sends an email to all users who have a specific RBAC role (e.g., owner) on the resource.
- *Email*: This sends an email to a single email address.
- *SMS message*: This sends an SMS message to a single mobile phone number.

- *Push*: This sends a notification to the Azure mobile app for a single user identified by their email address.
- *Voice*: This makes a phone call to a phone number and plays a message. Currently only U.S. phone numbers are supported.

When the action group is executed, any defined actions will also be executed. The following are the types of actions that can be executed:

- *Automation Runbook*: This executes either a PowerShell script or Python contained in an Azure Automation Runbook. This can be a good way to implement autoscaling on a Hyperscale database.
- *Azure Function*: This calls an Azure Function. This is another easy way to implement autoscaling on a Hyperscale database.
- *Event Hub*: This sends a message to an Azure Event Hub.
- *ITSM*: This creates a ticket in an IT Service Management system, such as ServiceNow.
- *Logic App*: This runs an Azure logic app workflow.
- *Webhook*: This calls a webhook.
- *Secure Webhook*: This calls a webhook that requires authentication by Azure AD.

To create a simple alert rule that sends an SMS message when the average workers exceed 80 percent for 15 minutes, follow these steps:

1. In the Azure Portal, select a Hyperscale database resource.
2. Select the Metrics blade.
3. Click "Add metric" and select "Workers percentage."

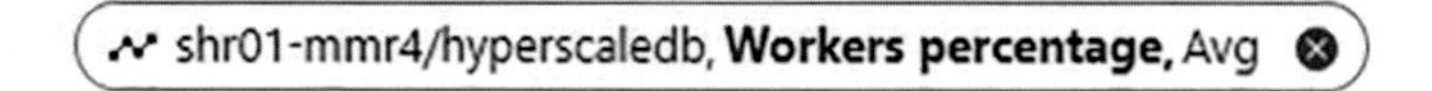

4. Click "New alert rule."
5. The threshold should be set to static as we want to simply evaluate this against a threshold value rather than dynamically determine if the pattern of this metric has changed.

6. Ensure the Aggregation type is set to Average and the Operator is set to "Greater than."

7. Enter **80** in the Threshold value.

   You may notice that the preview window will show you a graph of the metric over the last six hours and the threshold line as well as highlighting anywhere over the period the metric crossed it. Figure 13-3 shows the alert rule metric preview at 15 percent.

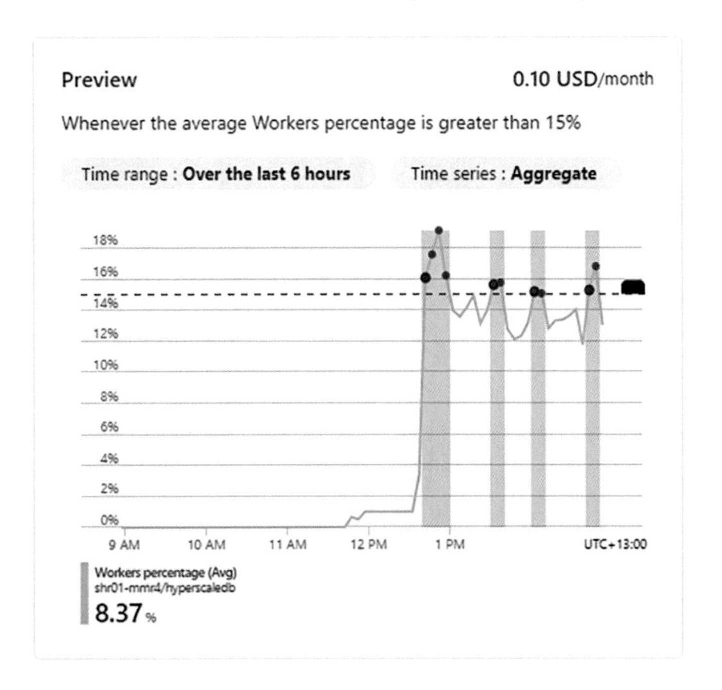

Figure 13-3. *The alert rule metric preview*

8. Set "Check every" to 5 minutes and "Lookback period" to 15 minutes.

9. Select the Actions tab.

10. Click "Create action group."

11. Actions groups can be used by more than one alert rule, so it is common to create general-purpose action groups.

12. Specify the action group name and display name; in this case, we're setting these to Notify DBA.

13. Select the Notifications tab.

14. Select Email/SMS message/Push/Voice in the "Notification type" drop-down.

15. Select the SMS box.

16. Enter the country code and phone number of the mobile phone to send the SMS to and click OK.

17. In the Name box, enter the name for the notification. For example, enter **SMS to DBA**.

18. Because we're not going to execute any actions, we can skip to the "Review + create" tab.

19. Click the Create button to create the action group.

20. Select the Details tab.

21. Set Severity to Warning and enter the alert rule name and alert rule description.

Alert rule details

Severity * ⓘ 2 - Warning

Alert rule name * ⓘ Workers 80 Percent

Alert rule description ⓘ Hyperscale DB Workers at 80% for the last 15-minutes

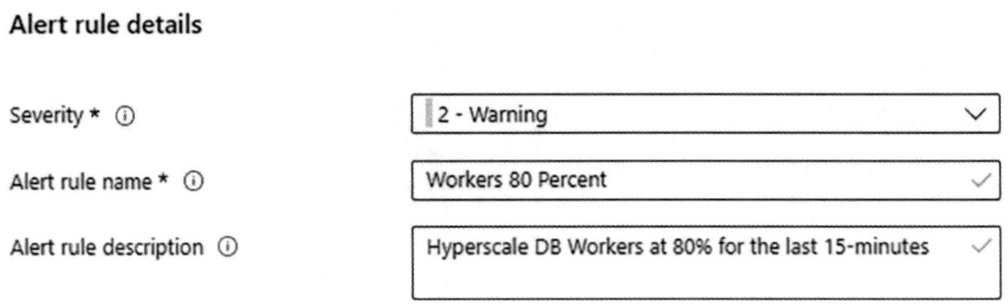

22. Select the "Review + create" tab and click the Create button.

After a few seconds our alert rule will be created. Our DBA will now be sent an SMS every five minutes that the average number of workers on the database exceeds 80 percent for 15 minutes. Figure 13-4 shows the SMS message received by our DBA whenever the alert rule triggers.

Notify DBA:Fired:Sev2 Azure Monitor Alert Workers 80 Percent on hyperscaledb

Figure 13-4. *The SMS message received when the alert rule triggers*

This may get very annoying for our DBA. Thankfully, our DBA will receive an SMS notifying them they've been joined to the Notify DBA group and that they can stop notifications from this group. Figure 13-5 shows the SMS message that is received by any mobile device in an action group.

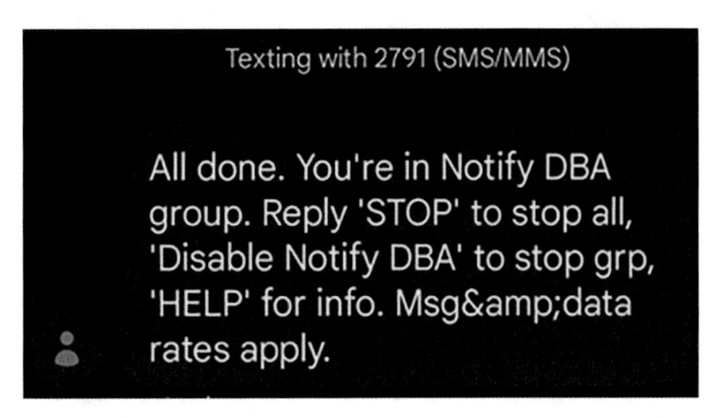

Figure 13-5. *The SMS message received by the action group members*

Important Overuse of alert notifications or false-positive notifications can reduce the value of them. This will often result in users ignoring the notification if they are too frequent or occur for normal conditions or conditions that will automatically resolve.

If you find that you're sending too many false positive notifications, consider using a dynamic threshold for your alert rule. This will use machine learning algorithms to continuously learn the metric behavior pattern and calculate the thresholds automatically.

When the alert rule is triggered, a record in the Alerts tab displays the details of the alert. Figure 13-6 shows the Alerts summary page for the Hyperscale database.

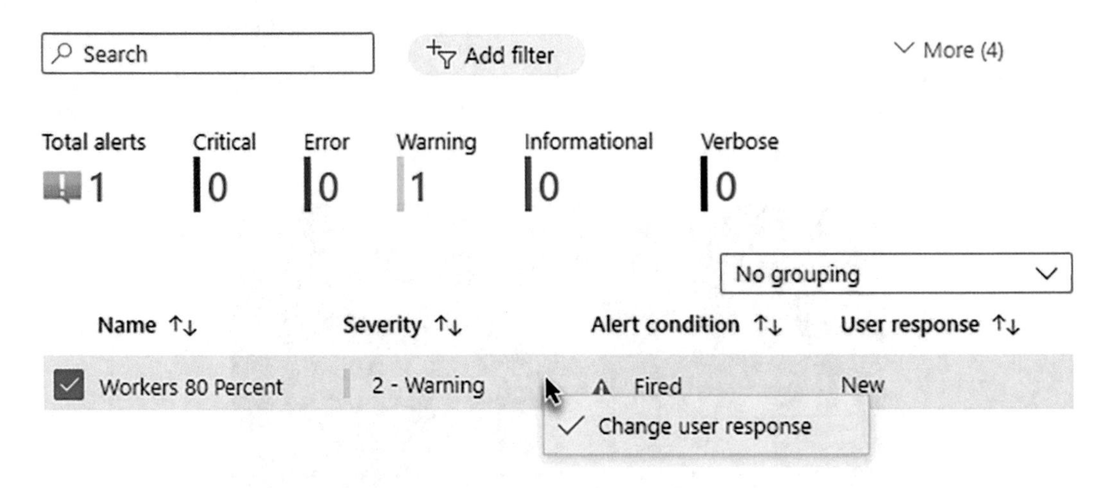

***Figure 13-6.** Acknowledging an alert in the Alerts Summary page*

Clicking an individual alert shows the Alert Details page, which will show the details of the metric conditions that caused the alert rule to trigger. Figure 13-7 shows part of the Alert Details page indicating why the alert fired. In this example, the threshold was set to 15 percent.

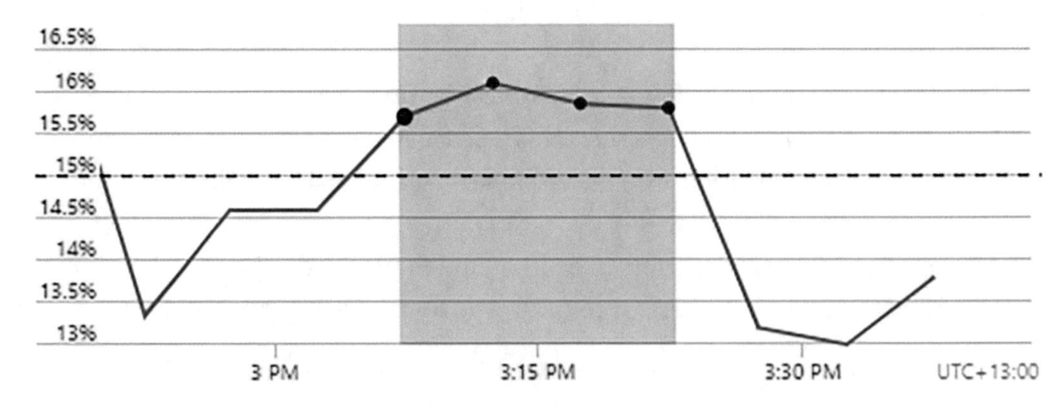

***Figure 13-7.** Why did the alert fire on the Alert Details page?*

Clicking the "Change user response" button for the alert can be used to indicate that the alert has been acknowledged and responded to.

Monitoring and Tuning Database Performance

It is important to regularly review the performance of workloads using your Hyperscale database to ensure both the database schema and queries are optimal. This will help you manage the cost of your Hyperscale environment by reducing the need to scale up or scale out resources.

Monitoring Query Performance

Monitoring query performance on a Hyperscale database is an important management task that should be performed regularly. The Query Performance Insight tool in the Azure Portal helps you do the following:

- Identify top long-running queries
- Identify top resource-consuming queries
- Make recommendations to improve the performance of any long-running or resource-consuming queries

To review the Query Performance Insight tool in the Azure Portal, follow these steps:

1. Select the Query Performance Insights blade under Intelligent Performance in a Hyperscale database resource.
2. Select the "Resource consuming queries" or "Long running queries" tabs to review the top queries.
3. Click the individual query rows to show the details of the query.

Query Performance Insight requires that Query Store is active on your database. It's automatically enabled for all Hyperscale databases by default, but if it is not, you'll be prompted to enable it when you open the tool.

It is recommended that you use this tool regularly to identify the top problematic queries and review the recommendations that could be implemented to improve performance.

Tip For more information on the Query Performance Insights tool, see `https://learn.microsoft.com/azure/azure-sql/database/query-performance-insight-use`.

Performance Recommendations

Built in to the Azure SQL Database services is the "Performance recommendations" tool. This tool continuously analyzes and makes intelligent recommendations based on recent query performance from your workload. The recommendations will identify the following:

- Optimize the layout of nonclustered indexes by creating missing indexes to improve query performance and drop duplicate indexes to save on disk space.
- Fix database schema issues where database schema and query definitions do not match, causing queries to fail.
- Fix database parameterization issues where lack of proper query parameterization leads to excessive resource usage for query compilation.

To review the performance recommendations in the Azure Portal, follow these steps:

1. Select the "Performance recommendations" blade under Intelligent Performance in a Hyperscale database resource.
2. Select a recommendation to review it.
3. Click Apply to apply the recommendation to your database or click Discard to flag it to be ignored.

Performing these tasks manually can become arduous, which is where the automatic tuning can be very useful.

Automatically Tuning Database Performance

Azure SQL Database instances include an automatic tuning feature that will constantly monitor the performance recommendations on your database and automatically apply them.

Enabling this feature on a database can do the following:

- *Force plan*: This identifies any queries using an execution plan that is slower than the previous good plan. It then forces queries to use the last known good plan instead of the regressed plan.

- *Create index*: This identifies indexes that can improve the performance of your workload and create them. Once created, it will automatically verify that the query performance has improved.
- *Drop index*: This drops indexes that have been used over the last 90 days and duplicate indexes. Unique indexes, including indexes supporting primary key and unique constraints, are never dropped.

To enable automatic tuning on an individual Hyperscale database in the Azure Portal, follow these steps:

1. Select the Automatic tuning blade under Intelligent Performance in a Hyperscale database resource.
2. Select the desired state of each automatic tuning option to On.

Configure the automatic tuning options

Option	Desired state	Current state
FORCE PLAN	ON OFF INHERIT	**ON** Forced by user
CREATE INDEX	ON OFF INHERIT	**ON** Forced by user
DROP INDEX	ON OFF INHERIT	**ON** Forced by user

3. Click the Apply button.

You can enable automatic tuning at the logical server level if you want and then set the desired state to `inherited` to centralize the configuration of these settings.

Your Hyperscale database will now be automatically tuned with the options you've selected. For more information on automatic tuning, see `https://learn.microsoft.com/azure/azure-sql/database/automatic-tuning-enable`.

Gathering Insights from the Database

As we saw earlier in this chapters, it is quite easy to review and alert on platform metrics emitted from the Hyperscale resource. However, there are many other rich sources of monitoring information that are accessible only from within the SQL database itself. These are usually surfaced up via dynamic managed views (DMVs).

For example, to obtain wait statistics from the database, the `sys.dm_db_wait_stats` view must be queried.

Since these DMVs are accessible only by connecting to the Hyperscale database and running the query, it requires a lot of additional work to collect and report on them. To simplify and automate the process of pulling the data from DMVs, a new preview feature called SQL Insights is available.

This tool requires the deployment of an Azure virtual machine that runs a Collection Agent. The Collection Agent collects one or more Hyperscale databases and pulls the DMV data down and pushes it into a Log Analytics workspace. A single collector machine can pull data from multiple Hyperscale databases.

Note SQL Insights supports not just Hyperscale databases. It can connect to and query DMVs from any SQL deployment model in Azure.

Once SQL Insights has been set up to extract the DMV information into the Log Analytics workspace, you can do the following:

- Run complex queries on it for reporting or analytics.
- Create alert rules on it to trigger action groups when conditions are met.

At the time of writing, Azure SQL Insights is in preview. For more information about SQL Insights, see `https://learn.microsoft.com/azure/azure-sql/database/sql-insights-overview`.

SQL Analytics

The tools discussed so far in this chapter have been mostly focused on monitoring individual Hyperscale databases. However, many larger environments have tens, hundreds, or even thousands of SQL databases. They often have multiple different deployment models for SQL, including both IaaS and PaaS. Therefore, it is useful to have a single view of all the SQL resources across an entire environment. Azure SQL Analytics provides this by collecting and visualizing key performance metrics with built-in intelligence for performance troubleshooting.

Azure SQL Analytics is deployed as a solution into an Azure Log Analytics workspace. To install the solution in the Azure Portal, follow these steps:

1. Enter **Azure SQL Analytics** into the search box at the top of the Azure Portal.
2. Select the Azure SQL Analytics (Preview) entry under Marketplace.
3. Select a resource group to contain the metadata for the solution.
4. Set the Azure Log Analytics Workspace option to a workspace that will have the solution installed.

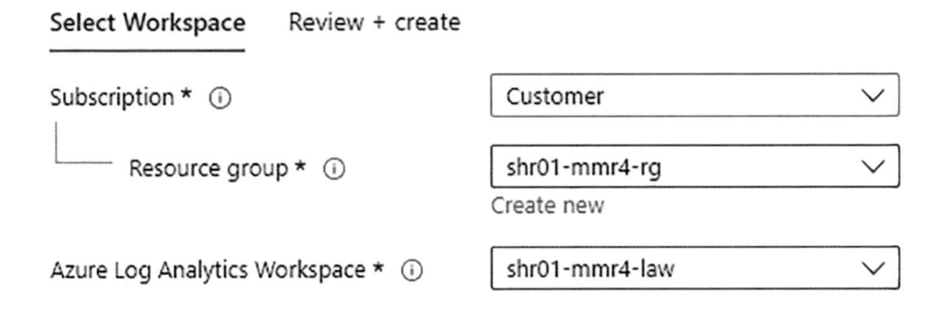

5. Click "Review + create" and then click Create.
6. After a few seconds, the deployment will start. It should take only a few seconds to deploy the solution. Once the deployment has been completed, the solution will be available in the Log Analytics workspace.

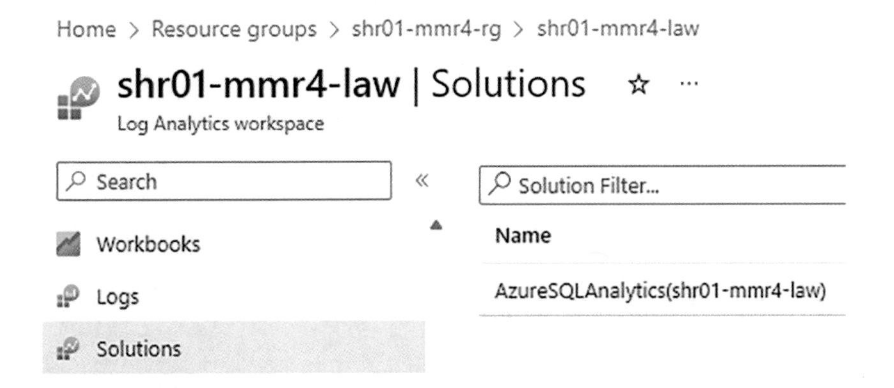

7. Selecting the solution will display the solution summary. Figure 13-8 shows the solution summary of a SQL Analytics solution.

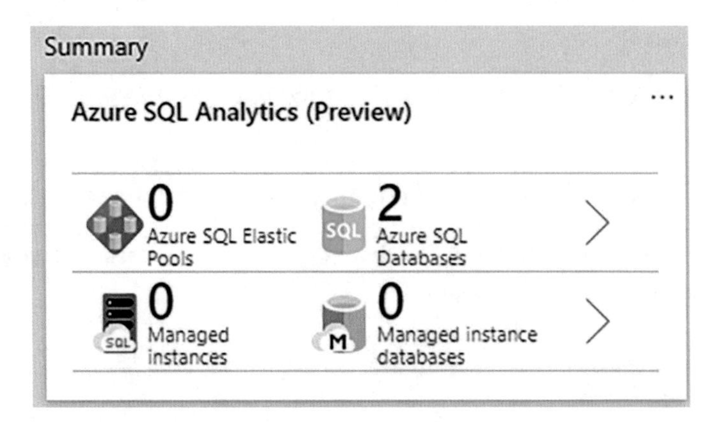

***Figure 13-8.** The SQL Analytics solution summary*

8. Clicking the Azure SQL Databases will show the SQL Analytics dashboard for all Azure SQL Databases, including our Hyperscale databases.

You can navigate through the SQL Analytics dashboard to review individual databases. Figure 13-9 shows part of the SQL Analytics dashboard view of a single Hyperscale database.

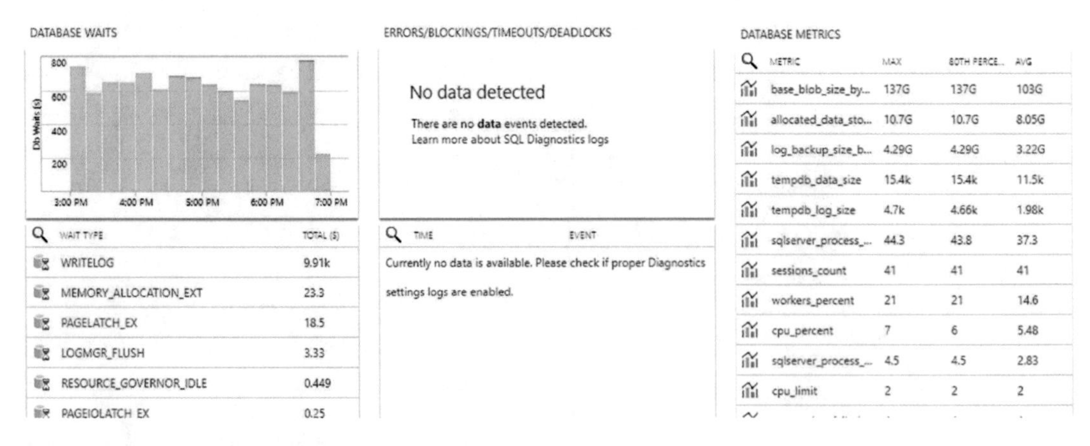

***Figure 13-9.** The SQL Analytics dashboard*

At the time of writing, Azure SQL Analytics is in preview, but you can find out more information about it at `https://learn.microsoft.com/azure/azure-monitor/insights/azure-sql`.

Scaling a Hyperscale Database

Over time, the use of your Hyperscale database may change. Typically, you'll not be running into the storage limits of a Hyperscale database, but you may find that your database requires additional CPU or memory to continue to meet your performance requirements. In this case, you may find you need to scale your Hyperscale database.

Note Before increasing the number of vCores on your SQL Hyperscale database, you should first investigate query performance and investigate automatic tuning. Once you've investigated these, you might decide that the only way to support the workload demands is to increase the number of vCores assigned to the Hyperscale database.

You may decide that you need to scale up the database by increasing the number of vCores assigned to it. Scaling up (or down) the number of vCores is an online operation. This means your Hyperscale database will remain operational throughout the process. A new Hyperscale database engine will be provisioned with the desired number of vCores, and once ready, a cutover will occur. Any open connections will be terminated, and any uncommitted transactions will be rolled back with the cutover to the new database engine.

Important You can expect a short connection break when the scale-up or scale-down process is complete. If you have implemented retry logic for standard transient errors, your workload won't be impacted by the failover. However, you should wait for any long-running transactions or operations to complete first.

It is also possible to scale out a Hyperscale database to add read replicas. These can be either named replicas or high-availability secondary replicas. These replicas can be used only to execute read operations.

Tip Because Hyperscale replicas in the same region share the underlying storage infrastructure, the storage replication lag is near zero, which makes this an attractive read scale-out option and one of the most important features of Hyperscale. It is not completely zero because of the way Hyperscale leverages Resilient Buffer Pool Extension caches (described in Chapter 2).

Leveraging read-scale out requires your workload to be designed to leverage it. This usually means the workload is configured to send read operations to either of the following:

- A named replica with the SQL connection string specifying the logical server and named replica database name.
- A high-availability secondary replica with the SQL connection string containing `ApplicationIntent=ReadOnly`. These types of replicas are primarily intended to enable high-availability but can be used for read scale-out. If more than one high-availability replica is present, the read-intent workload is distributed arbitrarily across all available high-availability replicas.

If your workload can't be changed to separate read operations in this way, then you won't be able to leverage scale-out, so increasing the number of vCores will be your only option.

Manually Scaling Up a Hyperscale Database

Manually scaling up or down the number of vCores can be done via the Azure Portal or using code with Azure PowerShell or the Azure CLI. If you are managing your Hyperscale environment using declarative infrastructure as code such as Azure Bicep or Terraform, then you should be changing these files and applying them.

To manually scale up or down the number of vCores in the Azure Portal, do the following:

1. Select the "Compute + storage" blade in the Hyperscale database resource.
2. Change the number of vCores to the desired value.
3. Click Apply.

Your new compute resources will be provisioned, and then a cutover will occur.

Manually Scaling Out a Hyperscale Database

There are two ways to scale out a Hyperscale database.

- Adding a named replica on the same logical server or another logical server
- Increasing the number of high-availability secondary replicas

Both of these tasks can be performed in the Azure Portal or using code. Again, leveraging declarative infrastructure as code is recommended.

Note that adding replicas does not force your workloads to reconnect to the database, so it can be done easily.

To add a new named replica through the Azure Portal, follow these steps:

1. Select the Replicas blade in the Hyperscale database resource.
2. Click "Create replica."
3. Change the replica type to "Named replica."

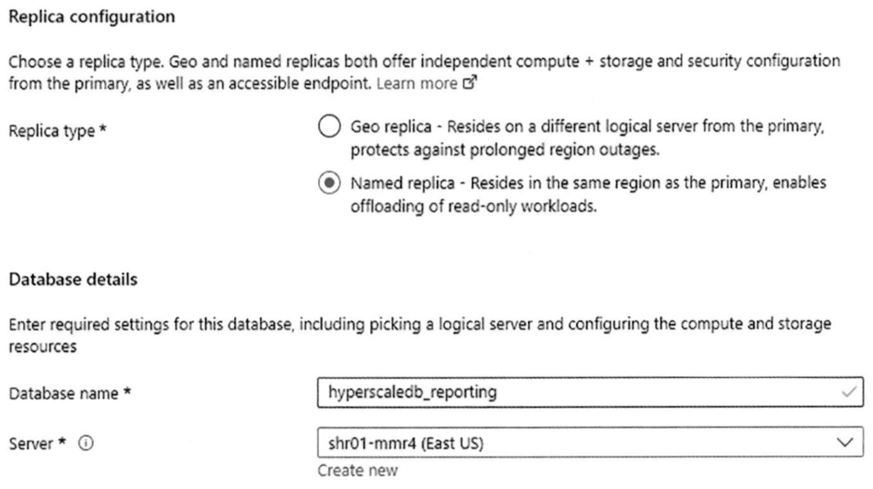

Replica configuration

Choose a replica type. Geo and named replicas both offer independent compute + storage and security configuration from the primary, as well as an accessible endpoint. Learn more

Replica type *

Geo replica - Resides on a different logical server from the primary, protects against prolonged region outages.

Named replica - Resides in the same region as the primary, enables offloading of read-only workloads.

Database details

Enter required settings for this database, including picking a logical server and configuring the compute and storage resources

Database name * hyperscaledb_reporting

Server * shr01-mmr4 (East US)

Create new

4. Select the Server (or create a new one) and configure "Compute + storage" for the new replica.
5. Click Review + Create.
6. Click Create.

After a few seconds, the new named replica will be created. It may take a few minutes for the deployment to complete as new resources will be provisioned.

To increase the high-availability secondary replicas in the Azure Portal, follow these steps:

1. Select the "Compute + storage" blade in the Hyperscale database resource.

2. Change High-Availability Secondary Replicas to the desired value (up to four).

3. Click Apply.

The new high-availability replica will be deployed.

Autoscaling a Hyperscale Database

Currently there isn't a built-in method to autoscale a Hyperscale database. However, it is possible to implement your own autoscaling feature using a combination of either `sys.dm_db_resource_stats` to monitor resources or alerts on Azure Metrics. Autoscaling is something you might want to implement once you're familiar with the usage of your Hyperscale database but is not necessarily something you should implement out of the box.

Although this is beyond the scope of this book, you can see a possible implementation of an autoscaling feature on the Azure SQL Database blog at `https://devblogs.microsoft.com/azure-sql/autoscaling-with-azure-sql-hyperscale/`.

Summary

In this chapter, we looked at the many tools and methods that should be used to monitor a Hyperscale database. We reviewed the tools for monitoring and reporting on the platform metrics emitted from the Hyperscale database. The options for sending the platform metrics to a Log Analytics workspace for enhanced reporting scenarios and increased retention were also examined. We reviewed the built-in tooling for monitoring the query performance and automatically turning a Hyperscale database. The SQL Insights feature is a way to provide a deeper level of monitoring on one or more Hyperscale and other Azure SQL Database instances. SQL Analytics was discussed as a way to monitor SQL at scale in Azure. Finally, the different approaches and methods of scaling a Hyperscale database were highlighted.

In the next chapter, we will look at disaster recovery for a Hyperscale database as well as backing it up and restoring it.

CHAPTER 14

Backup, Restore, and Disaster Recovery

In the previous chapter, we covered the basics of monitoring the performance and health of a Hyperscale database. We explained Hyperscale database's manual and automatically scaling features and showed how to scale to support different usage patterns and workloads.

In this chapter, we will explain the advantages of Azure SQL Hyperscale deployments in detail in the examples. We will cover fast database backup and restore operations and explain why these operations provide some of the greatest advantages of Hyperscale compared to other Azure SQL Database service tiers.

Hyperscale Database Backups

As discussed earlier, with the Hyperscale architecture, backups are based on file snapshots and kept on Azure Blob Storage. This enables backups to be performed very quickly regardless of the size of the database files. Restore operations are almost as fast as backups and are not dependent on database size.

Tip Backups and restores do not have any I/O impact on the compute layer of the Azure SQL Database Hyperscale service tier.

Every Azure SQL Database deployment supports automated backups. Traditional Azure SQL architecture requires full, differential, and log backups. However, the backup policy for the Azure SQL Database Hyperscale service tier differs from other service tiers in that the backups are snapshot based so have a different backup frequency and schedule.

Z. Barać and D. Scott-Raynsford, *Azure SQL Hyperscale Revealed*,
https://doi.org/10.1007/978-1-4842-9225-9_14

Tip Differential backups are not supported for Azure SQL Hyperscale deployment.

Backup Retention Policy

In the Azure Portal, under the "Data management" section of the logical SQL Server instance hosting your Hyperscale database, you can configure the backup retention policy. Figure 14-1 shows Azure SQL backup retention policies configuration on the SQL Server ➤ Data management ➤ Backups ➤ Retention policies blade.

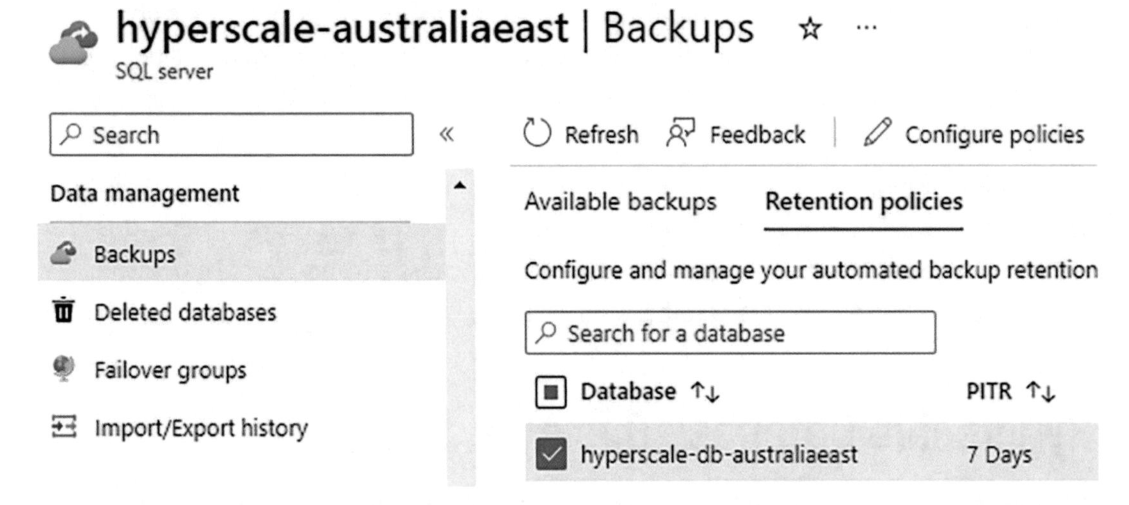

Figure 14-1. Backup retention policy

By default, every Azure SQL Hyperscale deployment has a short-term retention period of seven days.

Tip Short-term retention means the backups are available for 1 to 35 days.

At the time of writing, options for point-in-time restore (PITR) backup retention periods of longer than seven days and long-term retention are available in preview. Figure 14-2 shows the logical SQL Server instance backup retention policy configuration pane.

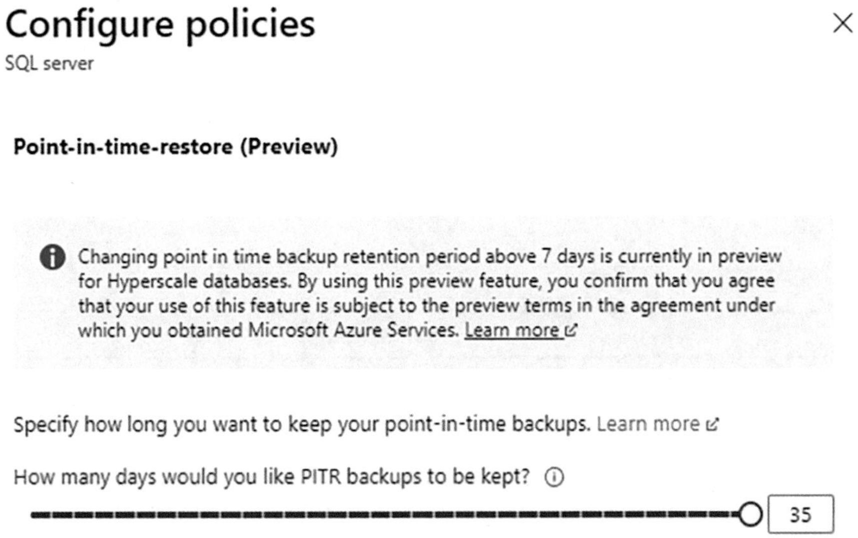

Configure policies
SQL server

Point-in-time-restore (Preview)

Changing point in time backup retention period above 7 days is currently in preview for Hyperscale databases. By using this preview feature, you confirm that you agree that your use of this feature is subject to the preview terms in the agreement under which you obtained Microsoft Azure Services. Learn more

Specify how long you want to keep your point-in-time backups. Learn more

How many days would you like PITR backups to be kept?

35

Figure 14-2. *SQL Server backup retention policy configuration*

For the long-term backup retention policy configuration, you can specify the period of time that weekly, monthly, and yearly long-term retention (LTR) backups will be retained for. When the long-term backup retention policy is enabled, LTR backups will be copied to a different Azure storage account for long-term retention. If you're using geo-replication, your backup storage costs for LTR won't increase because your backups are not produced by secondary replicas.

Tip For more information on the long-term retention of backups, see `https://learn.microsoft.com/azure/azure-sql/database/long-term-retention-overview`,

Figure 14-3 shows the long-term backup retention policy configuration pane.

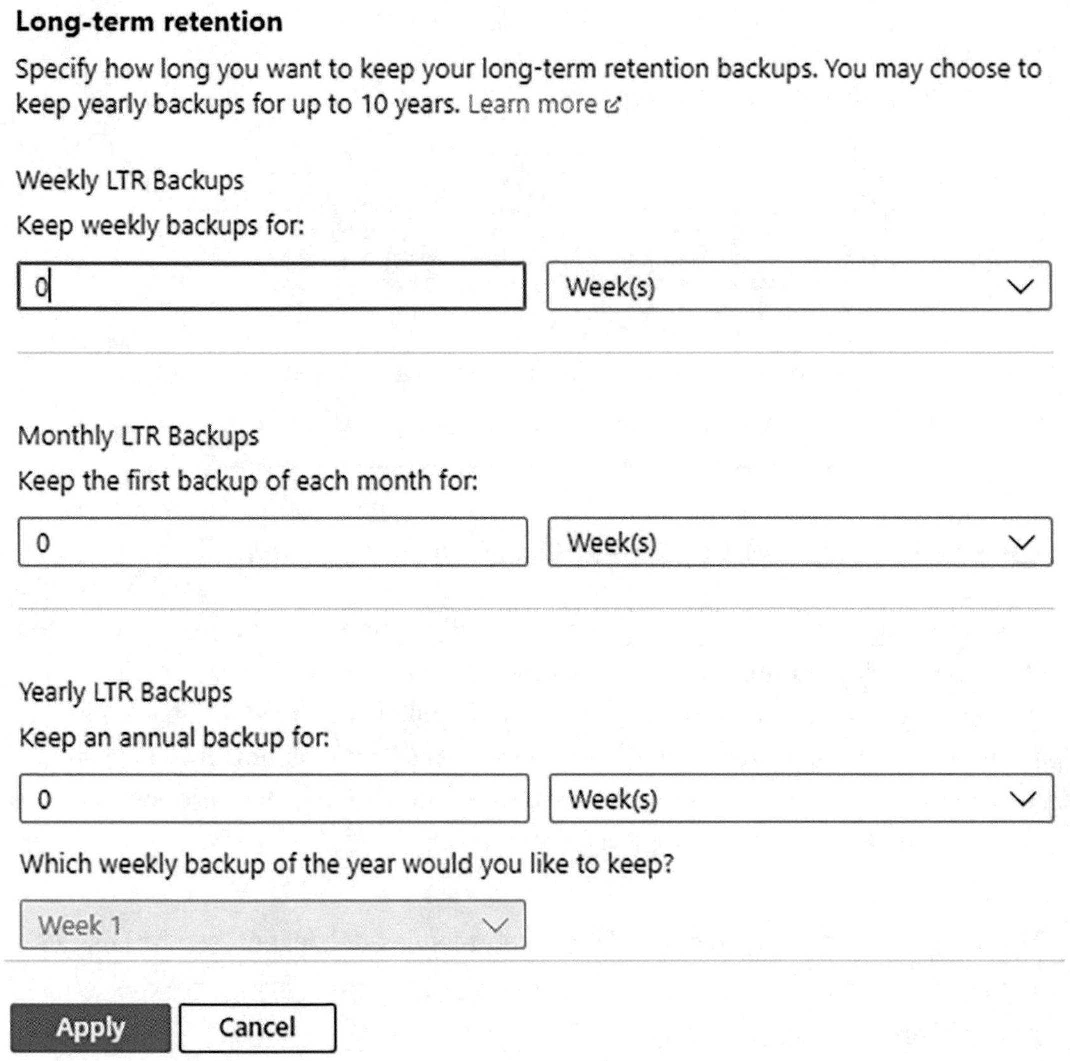

Long-term retention

Specify how long you want to keep your long-term retention backups. You may choose to keep yearly backups for up to 10 years. Learn more

Weekly LTR Backups

Keep weekly backups for:

0 | Week(s)

Monthly LTR Backups

Keep the first backup of each month for:

0 | Week(s)

Yearly LTR Backups

Keep an annual backup for:

0 | Week(s)

Which weekly backup of the year would you like to keep?

Week 1

Apply | Cancel

Figure 14-3. *Long-term backup retention policy*

Tip The LTR capability for Hyperscale databases allows yearly backups to be retained for up to 10 years. This option is currently in preview for Hyperscale databases.

Backup Storage Redundancy

Backup storage redundancy should fit your SLA requirements. The LTR capability offers you automatically created full database backups to enable point-in-time restore (PITR). During the deployment of a Hyperscale database you need to choose the redundancy of the storage that will be used to contain the PITR and LTR backups.

The following backup storage redundancy options are available:

- *Locally redundant backup storage*: Backups will not be available in the case of a zonal or regional outage.
- *Zone-redundant backup storage*: Backups will be available in the case of a zonal outage, but not a regional outage.
- *Geo-redundant backup storage (RA-GRS)*: Backups will be available in the case of a regional outage, but only from the paired region.
- *Geo-zone-redundant backup storage (RA-GZRS)*: Backups will be available in both a zonal outage and a regional outage.

Note Geo-redundant storage and geo-zone-redundant storage actually use Azure Standard Storage using RA-GRS and RA-GZRS, respectively. This ensures the read-only copy of the backups are available on a regional endpoint. For the remainder of this chapter, we'll refer to RA-GRS and RA-GZRS as *GRS* for simplicity.

The zone-redundant and geo-zone-redundant backup storage options are available only in regions with availability zone (AZ) support. Also, RA-GZRS storage is not available in all regions, so you may not be able to select this if it is not available in your chosen region.

Not all Azure regions provide availability zone support. If you deploy your Hyperscale database to a logical SQL Server instance located in a region without support for availability zones, you will see only the options shown in Figure 14-4.

Backup storage redundancy

Choose how your PITR and LTR backups are replicated. Geo restore or ability to recover from regional outage is only available when geo-redundant storage is selected.

Backup storage redundancy ⓘ

- (●) Locally-redundant backup storage
- ( ) Geo-redundant backup storage

Figure 14-4. *Backup storage redundancy options in a region without AZ support*

However, if your logical SQL Server instance is deployed to an Azure region that supports availability zones, then you will see the options shown in Figure 14-5.

Backup storage redundancy

Choose how your PITR and LTR backups are replicated. Geo restore or ability to recover from regional outage is only available when geo-redundant storage is selected.

Backup storage redundancy ⓘ

- (●) Locally-redundant backup storage
- ( ) Zone-redundant backup storage
- ( ) Geo-redundant backup storage
- ( ) Geo-Zone-redundant backup storage

Figure 14-5. *Backup storage redundancy options for availability zones*

Important For a Hyperscale service tier deployment, the backup storage redundancy can be configured at the time of database creation. To change the backup storage redundancy of an existing database, you have to create a new Hyperscale database with a different storage redundancy and migrate to it using a database copy or point-in-time restore.

Monitoring Backup Storage Consumption with Azure Monitor Metrics

To gather and monitor backup storage consumption information, you should use an Azure Monitor metrics report. In the Azure Portal, in the Monitoring section you will see the Metrics blade. Figure 14-6 shows how to access the SQL Database ➤ Monitoring ➤ Metrics blade.

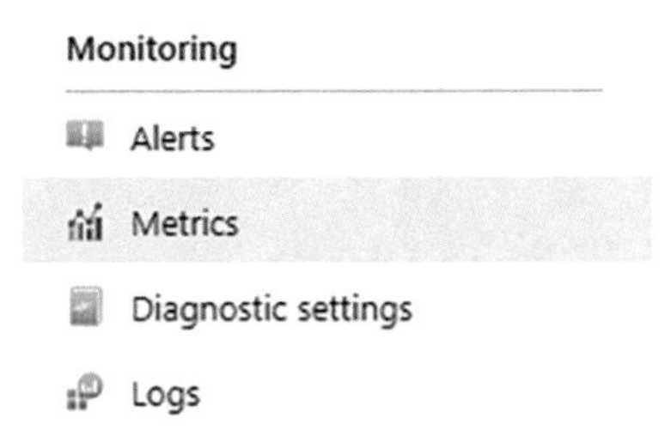

Figure 14-6. *Seeing metrics in the Azure Portal*

Use one of the following metrics to review the backup storage consumption information:

- Data backup storage size (cumulative data backup storage size)
- Log backup storage size (cumulative transaction log backup storage size)

Tip For more information on monitoring a Hyperscale database, including setting up automated alerts, see Chapter 13.

Figure 14-7 shows Azure Monitor data and log backup storage size metrics in the Azure Portal.

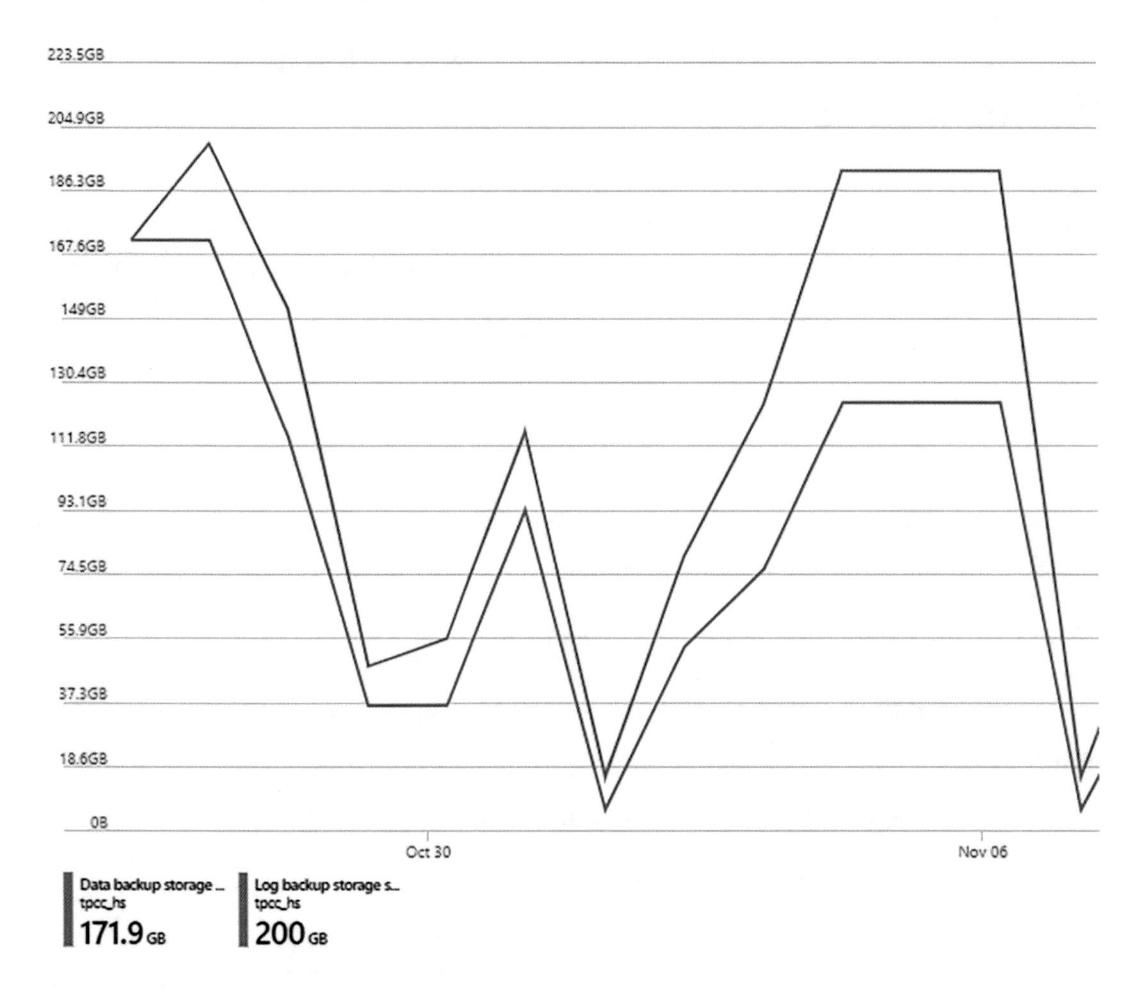

Figure 14-7. *Data and log backup storage size metrics*

Hyperscale Database Restores

There are two common methods of restoring an Azure SQL Database instance depending on the location of the logical SQL Server that the database is restored into.

- *The same region*: The database restores to a logical SQL Server instance within the same region.
- *Geo-restore*: The database restores to a logical SQL Server instance in a different region.

In addition, for Hyperscale databases there two other factors that may affect the performance of the restore operation.

- If the backup storage account redundancy needs to be changed, then the time taken for the storage account with the desired redundancy to be created and the backup files to be copied needs to be factored in.
- The amount of data that is restored to the database has a limited impact on the geo-restore performance and is affected only by the size of the largest file in the database. This is because data files need to be copied in parallel across different Azure regions. If your geo-restore destination Azure region is paired with the source database Azure region, then the geo-restore operation will be much faster.

In the next five examples, we will review the Hyperscale database restore performance to provide more insight into how each of these factors affects restore time. We will use a combination of different database sizes, Azure regions, and backup storage redundancy types.

Tip Even though these tests are real examples, the results may vary slightly across different Azure regions at various times of the day. Your own tests may differ slightly.

Example 1: Restore to Same Region, LRS to LRS

In this example, we are going to measure the restore performance of restoring from a database with locally redundant backup storage to another one with locally redundant backup storage in the same logical SQL Server instance.

- *Source*: hyperscale-sql-server; Australia East; LRS
- *Destination*: hyperscale-sql-server; Australia East; LRS

Figure 14-8 shows restore time for the Hyperscale database tpcc_hs with a used space of 9.21GB and an allocated space of 20GB.

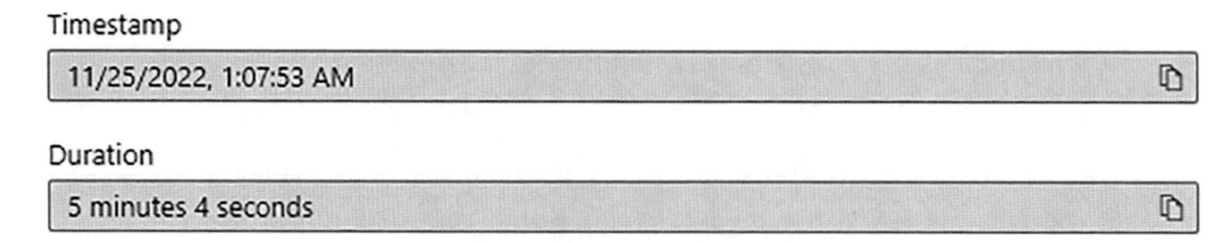

***Figure 14-8.** Restore time for a Hyperscale database (10GB size) from LRS to LRS*

If we double the size of the database, the restore still shows similar performance results. Figure 14-9 shows the restore time for the Hyperscale database tpcc_hs with a used space of 20.4GB and an allocated space of 30GB.

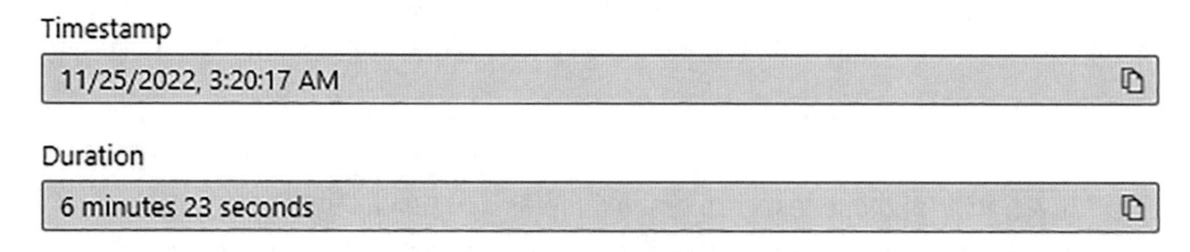

***Figure 14-9.** Restore time for a Hyperscale database (20GB size) from LRS to LRS*

If we triple the size of the database, we are still getting similar results. Figure 14-10 shows the restore time for the Hyperscale database tpcc_hs with a used space of 71.99GB and an allocated space of 120GB.

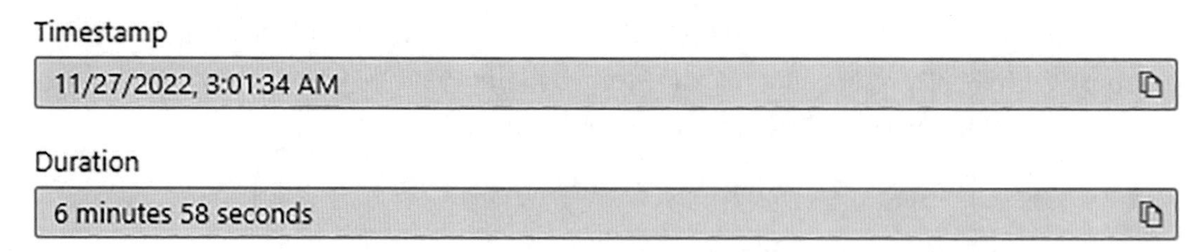

***Figure 14-10.** Restore time for a Hyperscale database (70GB size) from LRS to LRS*

From the first example, we can see that the restore operations for Hyperscale databases are almost as fast regardless of the data size if we are using the same backup storage redundancy within the same region.

Example 2: Restore to Same Region, LRS to GRS

In this example, we are going to restore a Hyperscale database to the same logical SQL Server instance but to a different backup storage redundancy type and measure the restore performance. The backup storage redundancy of the source database will be LRS, and the destination will be GRS.

- *Source*: hyperscale-sql-server; Australia East; LRS
- *Destination*: hyperscale-sql-server; Australia East; GRS

Figure 14-11 shows the Hyperscale database `tpcc_hs` with two page servers managing the data files, with the biggest data file size as 80GB (growth in 10GB increments) and with an allocated space of 120GB.

	file_id	file_name	type_desc	file_size_GB	max_file_size_GB	growth_size_GB
1	1	data_0	ROWS	80	128	10
2	2	log	LOG	1022	1024	2
3	3	data_1	ROWS	40	128	10

Figure 14-11. *Hyperscale database with multiple page servers*

If we restore the same Hyperscale database as before, using the same region and the same logical SQL Server instance for the source and destination but restoring to a destination database that has GRS backup storage redundancy, the restore will take longer than in the previous example. This is because we are changing the redundancy type for our destination backup storage from locally redundant to geo-redundant. Figure 14-12 shows the restore-time performance for the Hyperscale database `tpcc_hs` with two page servers managing data files, with the largest data file size of 80GB, a used space of 71.97GB, and an allocated space of 120GB.

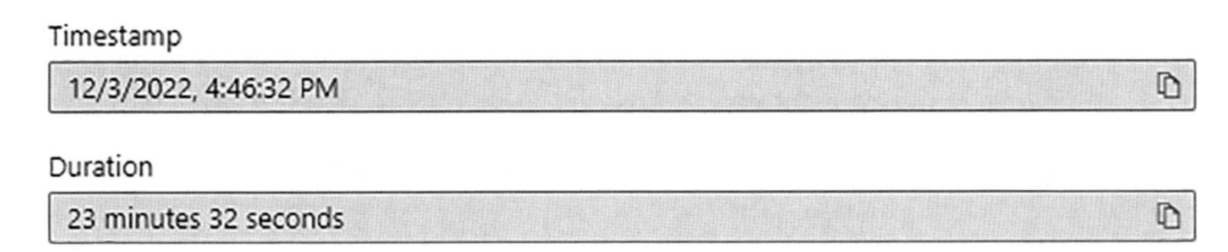

Figure 14-12. *Hyperscale database restore time with the same logical SQL Server instance, from LRS to GRS*

In the case of reallocating your database to geo-redundant storage, all the data files will be copied in parallel. This causes the restore operation to take longer.

Example 3: Restore to the Same Region, GRS to GRS

In this example, we are going to restore a Hyperscale database to the same logical SQL Server instance and measure the restore performance. The backup storage redundancy of both the source and destination databases will be geo-redundant.

- *Source*: hyperscale-sql-server; Australia East; GRS
- *Destination*: hyperscale-sql-server; Australia East; GRS

Figure 14-13 shows the restore-time performance for the Hyperscale database `tpcc_hs` with two page servers managing data files, with the largest data file size of 80GB, a used space of 71.97GB, and an allocated space of 120GB.

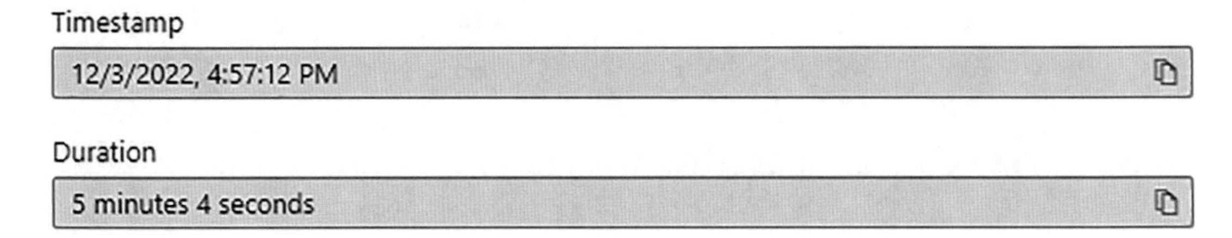

Figure 14-13. *Hyperscale database restore time to the same logical SQL Server instance, from GRS to GRS*

Example 4: Geo Restore to Different Region, GRS to LRS

In this example we are going to restore a Hyperscale database to a different logical SQL Server instance in a different Azure region. The backup storage redundancy of the source database is GRS, and the backup storage redundancy of the destination database will be set to LRS.

- *Source*: hyperscale-sql-server; Australia East; GRS
- *Destination*: hyperscale-geo-sql-server; Australia Central; LRS

Figure 14-14 shows geo-restore time performance for the Hyperscale database `tpcc_hs` with two page servers managing data files, with the largest data file size of 80GB, a used space of 71.97GB, and an allocated space 120GB.

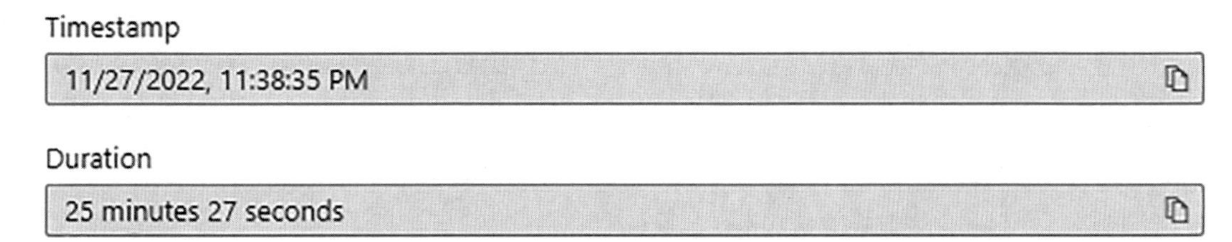

Figure 14-14. *Hyperscale database geo restore time to a different Azure region, from GRS to LRS*

Tip In the case of geo-restore operation, your source database always must be on geo-redundant backup storage.

Example 5: Geo-restore to Different Region, GRS to GRS

In this example, we are going to restore a Hyperscale database to a different logical SQL Server instance in a different Azure region, keeping the same backup storage redundancy type of GRS for both the source and the destination.

- *Source*: hyperscale-sql-server; Australia East; GRS
- *Destination*: hyperscale-geo-sql-server; Australia Central; GRS

Figure 14-15 shows a geo-restore-time performance for the Hyperscale database `tpcc_hs` with two page servers managing data files, with the biggest data file size of 80GB, a used space of 71.97GB, and an allocated space of 120GB.

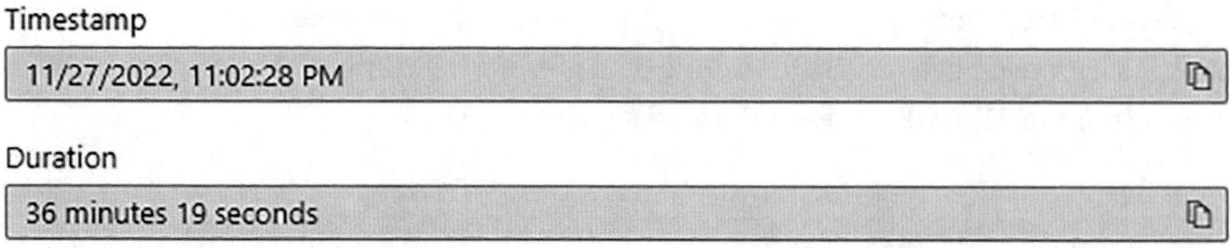

Timestamp

11/27/2022, 11:02:28 PM

Duration

36 minutes 19 seconds

Figure 14-15. *Hyperscale database geo-restore time to a different Azure region, from GRS to GRS*

Note In the case of geo-redundant storage and geo-restore, all data files will be copied in parallel, and the restore operation duration will depend on the size of the largest database file.

Table 14-1 and Table 14-2 compare the restore times for the restore and geo-restore operations for the same database. For these examples, a Hyperscale database with two page servers managing data files, with the largest data file size of 80GB, a used space of 71.97GB, and an allocated space of 120GB.

Table 14-1. *Hyperscale Database Restore Within the Same Azure Region Performance Table*

Restore Time Within Same Region, Same Logical SQL Server		
Backup Storage Redundancy Types	**LRS**	**GRS**
LRS	6 minutes 58 seconds	23 minutes 32 seconds
GRS	26 minutes 34 seconds	5 minutes 4 seconds

Table 14-2. *Hyperscale Database Restore Within the Same Azure Region Performance Table*

Geo-restore Time to a Different Region		
Backup Storage Redundancy Types	**LRS**	**GRS**
GRS	25 minutes 27 seconds	36 minutes 19 seconds

As we can see, changing the backup storage redundancy has a significant impact on restore time, as does performing a geo-restore operation.

Comparing Restore Performance of Hyperscale to Traditional Azure SQL Databases

Finally, we will compare the restore performance of a traditional Azure SQL Database using the premium availability model with the Business Critical service tier (BC_Gen5_8) and a Hyperscale database with the Hyperscale service tier (HS_Gen5_8).

For the comparison, we used a Hyperscale database containing three page servers with data file sizes of 100GB, 97GB, and 20GB. The used space was 174.42GB, and the allocated space was 217.45GB. Figure 14-16 shows the data files' size distribution across the page servers.

Results | Messages

	file_id	file_name	type_desc	file_size_GB	max_file_size_GB	growth_size_GB
1	1	data_0	ROWS	100	128	10
2	3	data_1	ROWS	97	128	10
3	4	data_2	ROWS	20	128	10

Figure 14-16. *Page server file sizes in the Hyperscale database deployment*

The total time of the restore process should be the time necessary to restore the largest data file, in this case 100GB. Figure 14-17 shows the restore performance of a Hyperscale database with data file sizes of 100GB, 97GB, and 20GB with the used space of 174.42GB and an allocated space of 217.45GB.

Timestamp

12/1/2022, 2:03:48 AM

Duration

13 minutes 13 seconds

Figure 14-17. *Hyperscale database restore time*

The next step is a reverse migration from Hyperscale Standard-series (Gen5), with 12 vCores and zone redundancy disabled, to a General Purpose Standard-series (Gen5), with 12 vCores and zone redundancy disabled. Once the reverse migration was complete, the database was restored to the same logical SQL Server instance.

Figure 14-18 shows the restore performance of a General Purpose service tier database with data file sizes of 100GB, 97GB, and 20GB with a used space of 174.42GB and an allocated space of 217.45GB.

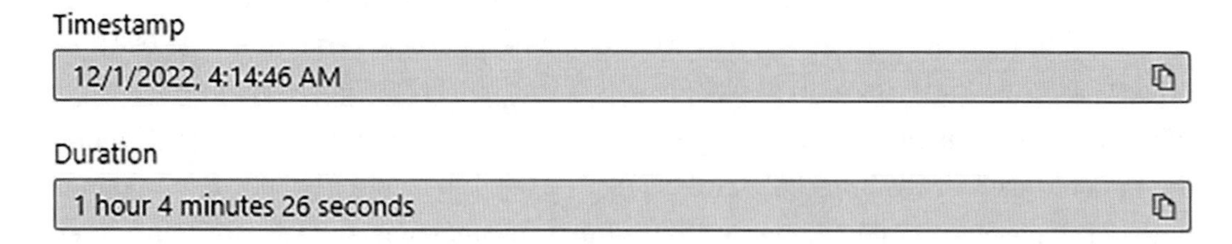

Figure 14-18. *General Purpose database restore-time performance*

The last step was to scale the same database to the Business Critical service tier and then evaluate the restore process again. Once scaling was completed, a database restore was performed to the same logical SQL instance. Figure 14-19 shows the restore performance of the Business Critical service tier database with the data file sizes of 100GB, 97GB, and 20GB with a used space of 174.42GB and an allocated space of 217.45GB.

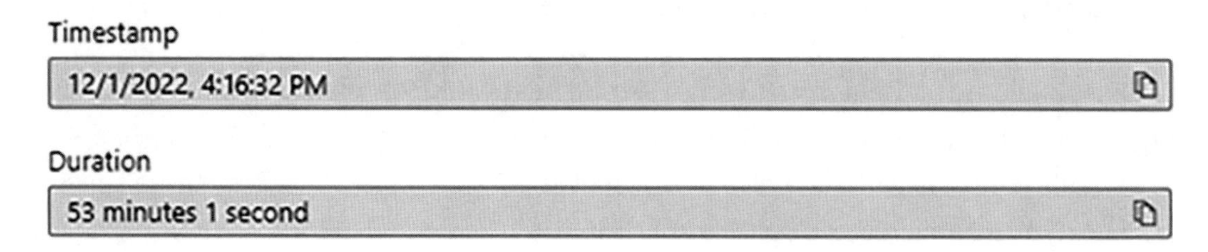

Figure 14-19. *Business Critical database restore-time performance*

If we compare the performance for the same size database between different Azure SQL service tiers, we can clearly see that the Hyperscale restore performance surpasses both the General Purpose and Business Critical restore operations.

Table 14-3. *Restore Time Performance of Same Database on Different Service Tiers from Backups Stored in Geo-Redundant Backup Storage*

Service Tier	Restore Time of a 217GB Database
Hyperscale	13 minutes 13 seconds
Business Critical	53 minutes 1 seconds
General Purpose	1 hour 4 minutes 26 seconds

Disaster Recovery

If you use geo-redundant storage to store your backups, you may use a geo-restore option as a disaster recovery solution for your Hyperscale database. In this case, you can restore your Hyperscale database to a different region as part of your disaster recovery strategy.

Note The geo-restore option is available only when you are using geo-redundant storage as your backup storage redundancy.

Figure 14-20 shows an example of a geo-restore using the Azure Portal to restore the most recent geo-replicated backup.

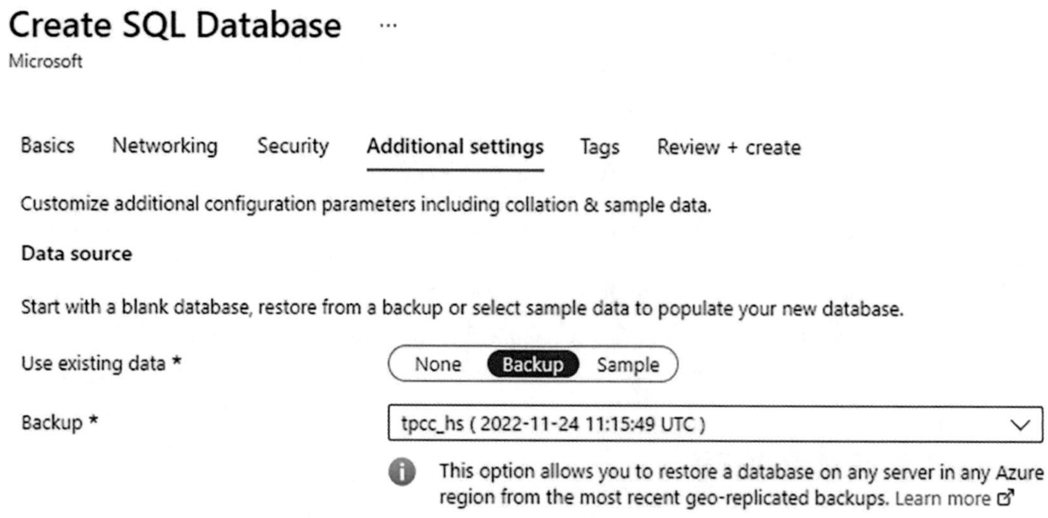

Create SQL Database ...
Microsoft

Basics Networking Security **Additional settings** Tags Review + create

Customize additional configuration parameters including collation & sample data.

Data source

Start with a blank database, restore from a backup or select sample data to populate your new database.

Use existing data * None **Backup** Sample

Backup * tpcc_hs (2022-11-24 11:15:49 UTC)

This option allows you to restore a database on any server in any Azure region from the most recent geo-replicated backups. Learn more

Figure 14-20. *Azure Portal create database from most recent geo-replicated backup*

Another disaster recovery option for the Azure SQL Database Hyperscale service tier is active geo-replication. With the availability of the geo-replica option in the Hyperscale architecture, we can have a readable secondary replica in the same region or a different region. If a geo-replica is in a different region, it can be used for disaster recovery as a failover in the case of a regional outage of the primary region. Figure 14-21 shows an example of the forced failover to a secondary geo-replica.

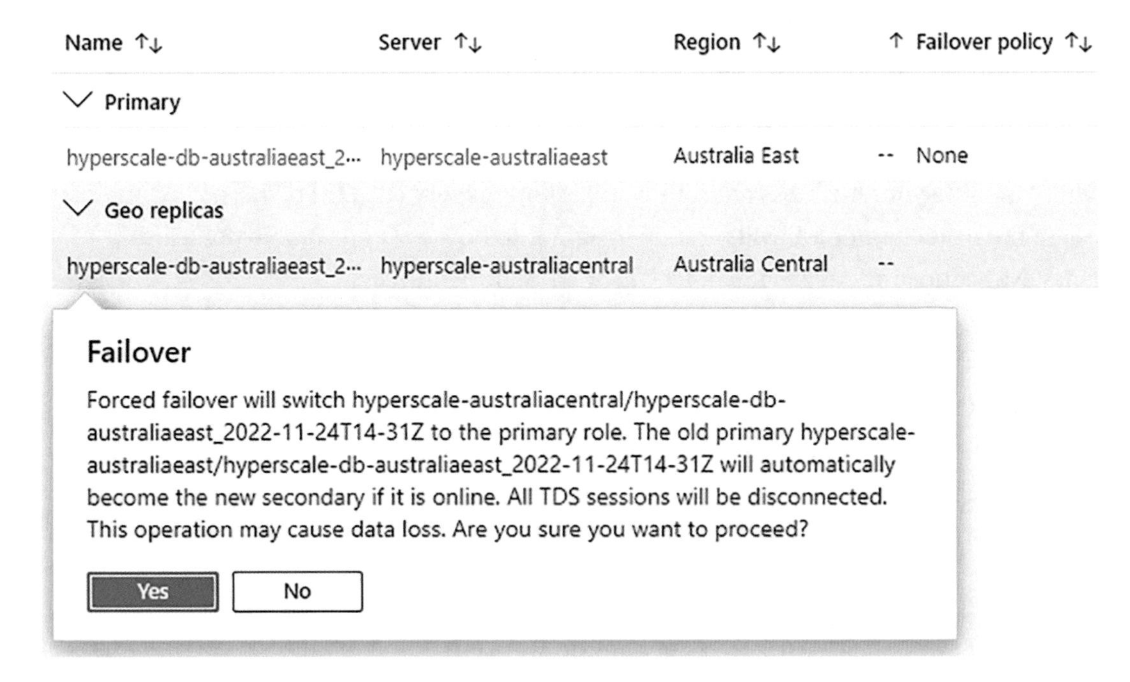

Figure 14-21. *Geo-replica restore example with geo-replica forced failover option*

Tip The geo-replica feature will protect your Hyperscale database against regional or zonal outages.

Summary

In this chapter, we covered the Hyperscale database file snapshot backup process and talked about advantages of this when compared with a traditional database backup process. We also discussed how Hyperscale achieves a fast restore. We talked about these different backup storage redundancy options:

- Locally redundant backup storage
- Zone-redundant backup storage
- Geo-redundant backup storage
- Geo-zone-redundant backup storage

Finally, we demonstrated the performance of database restore operations in different database scenarios and how additional factors affect them. These factors include the source and destination backup storage redundancy and whether you are performing a geo-restore. Understanding how these factors impact restore performance is important when defining your business continuity/disaster recovery (BCDR) plan.

In the next chapter, we will provide insight into how to keep your Hyperscale database secure and review the patching and updating process.

CHAPTER 15

Security and Updating

In the previous chapter, we talked about why backup and restore processes are so efficient and fast in the Hyperscale architecture. We covered all the available backup storage redundancy options. We explained how to use metrics to monitor backup storage consumption. In this chapter, we will move on to discussing the security, updating, and pathing of our SQL cloud services. These are very different but equally important parts of every Azure SQL deployment.

We are going to cover only the basics of securing the data tier within Azure SQL deployments. Keep in mind that Azure security is a much bigger and wider topic overall and includes the built-in capabilities, security features, and other solutions available in the Azure platform.

Azure SQL Database Security Overview

We are going to cover some of the best practices of how to secure the data tier in Azure SQL deployments. We will explain the importance of creating server-level and database-level firewall rules, Azure AD roles, and how to manage user access with SQL authentication and Azure Active Directory (AAD) authentication. In addition, we will talk about some of the most common security features used, such as dynamic data masking, encryption, Microsoft Defender for Cloud, and more.

There are multiple security layers for the network traffic with specifically defined roles to prevent any access to the server and to the customer's sensitive data until that access is explicitly granted.

We can distinguish those layers as follows:

- *Network security*: Public and private access, firewall, virtual network rules, and connection policy

Z. Barać and D. Scott-Raynsford, *Azure SQL Hyperscale Revealed*,
https://doi.org/10.1007/978-1-4842-9225-9_15

- *Access management*: SQL authentication, Azure Active Directory authentication, role-based access control, database engine permissions and row-level security
- *Auditing and threat detection*: SQL auditing, Microsoft Defender for Cloud, and Microsoft Defender for SQL
- *Information protection*: Encryption in transit, encryption at rest with transparent database encryption, the Always Encrypted feature, and data masking

Figure 15-1 shows the Azure SQL database security strategy of securing our sensitive data tier.

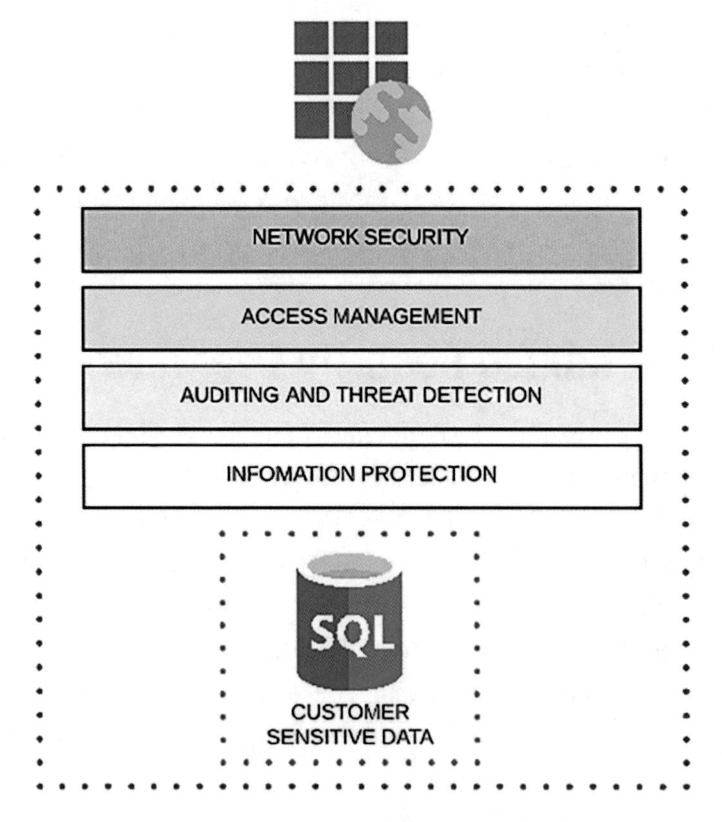

Figure 15-1. *Azure SQL Database security layers overview*

Some of the security features are easily accessible upon deploying a logical SQL Server instance through the Azure Portal. Figure 15-2 shows some of the server-level security features easily available through the Azure Portal security pane.

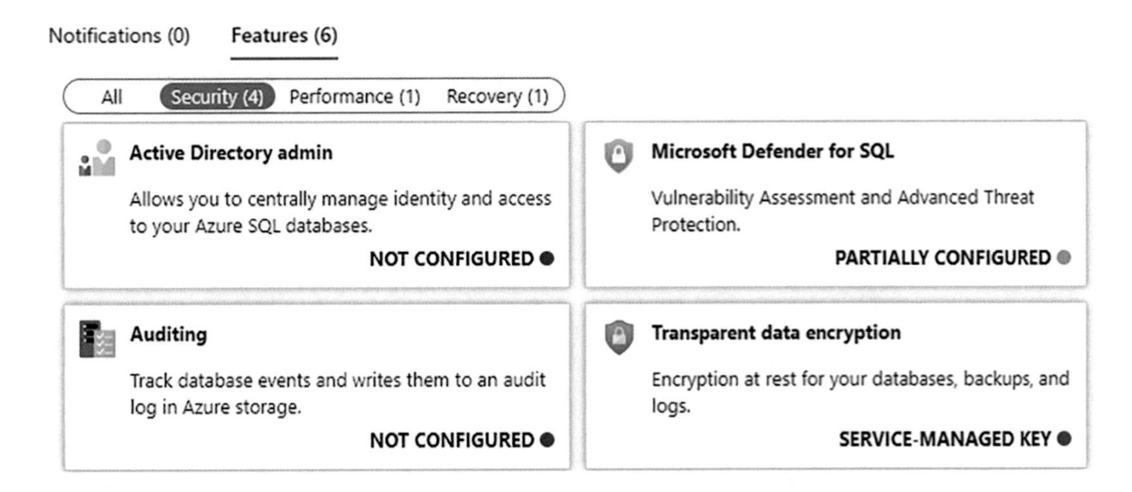

Figure 15-2. *Azure SQL logical server security features in the Azure Portal*

Database-level security features available through the Azure Portal might look a bit different. Figure 15-3 shows some additional database-level features such as dynamic data masking and the Ledger. The Ledger is a powerful feature that keeps an immutable and tamper-proof history of all transactions executed against the tables. This provides the ability to compare data with the historical values kept by the Ledger and digitally verify that it is correct.

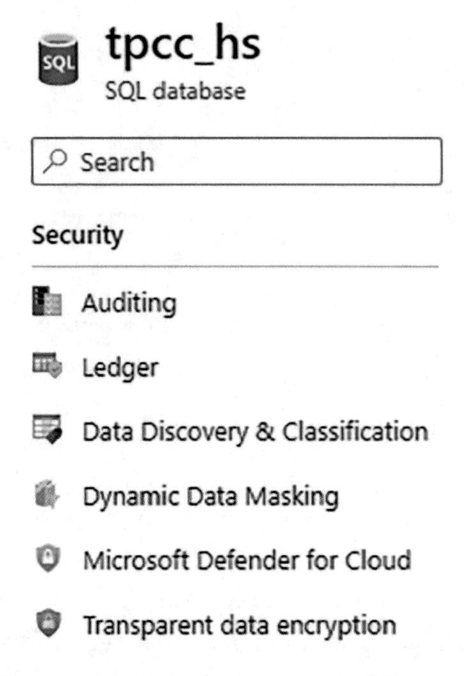

Figure 15-3. *Azure Portal Azure SQL Database-level security features*

Network Security

After a logical SQL Server instance is deployed, it is recommended to restrict network access and allow only limited access to a database. This can be done using the Azure Portal. It is recommended to configure firewall and virtual network rules and explicitly allow connections only from selected virtual networks and specific IP addresses.

Public Network Access

You can choose to disable public network access through the Internet. If this choice is selected, only connections via private endpoints will be accepted. If you choose to enable public access only for selected networks, only IP addresses specified in the firewall rules will be allowed to access your resource. Figure 15-4 shows public network access settings.

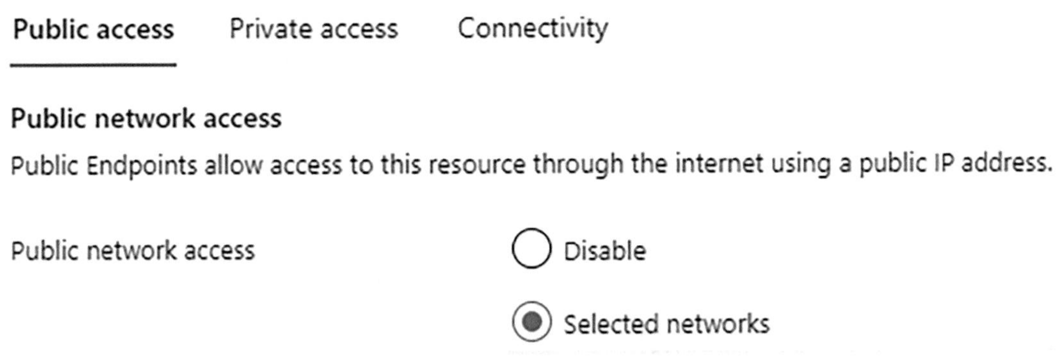

Public access Private access Connectivity

Public network access

Public Endpoints allow access to this resource through the internet using a public IP address.

Public network access ◯ Disable

◉ Selected networks

Figure 15-4. *Azure Portal public network access settings*

Firewall and Virtual Network Rules

For every Azure SQL database deployment option, you can restrict access for the specific IP addresses and subnets using firewall and virtual network rules. In the Azure Portal's Security section for the logical SQL Server hosting your Hyperscale database, you can configure your firewall and virtual network rules. Figure 15-5 shows the logical server networking settings in the SQL Server ➤ Security ➤ Networking ➤ Public access blade.

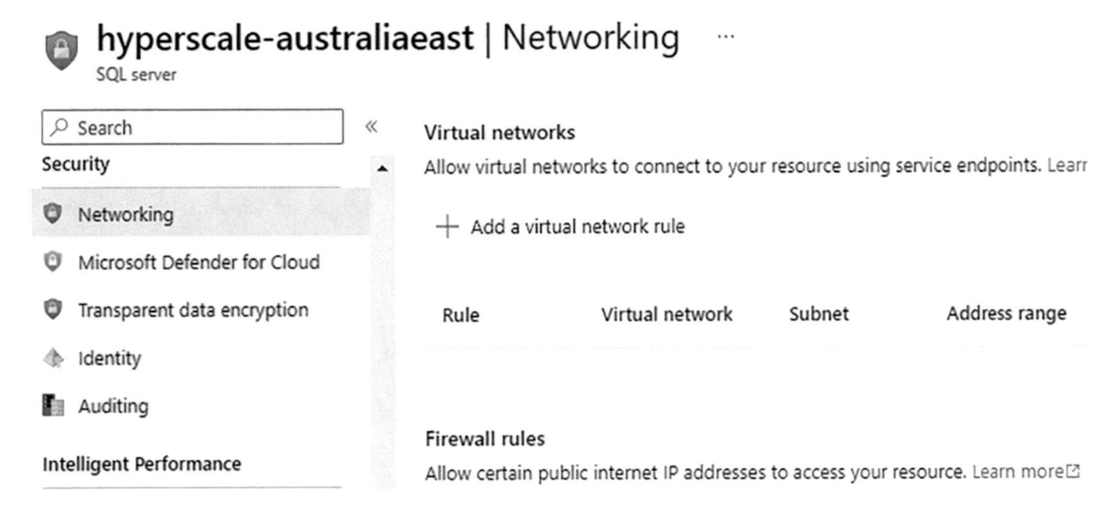

Figure 15-5. *Logical Azure SQL Server Networking pane*

Some of the usual firewall and virtual network rules configuration use case scenarios may include but not be limited to the following:

- Public network access is enabled for the IP addresses configured in the “Firewall rules” section. By default, no public IP addresses are allowed.
- Firewall rules allow public access to your resources only for the specified Internet IP addresses.
- Virtual network rules allow virtual networks to connect through service endpoints.

You can specify some networking rules during Azure SQL database deployments as well, including creating private endpoints for your Azure SQL deployment. Chapter 4 covered a demonstration of connecting a Hyperscale database to a virtual network. Figure 15-6 shows a firewall rules configuration option during the Azure SQL Database deployment process.

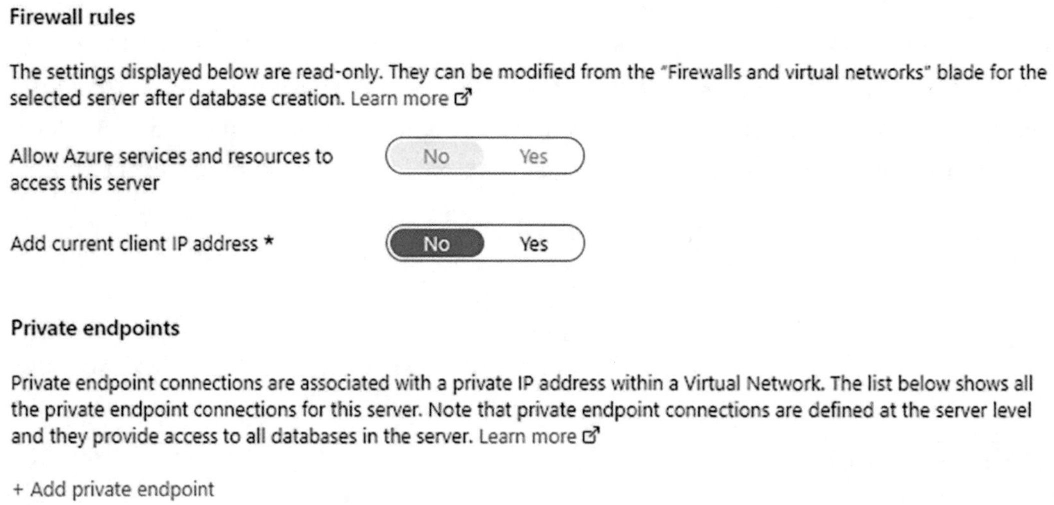

Firewall rules

The settings displayed below are read-only. They can be modified from the "Firewalls and virtual networks" blade for the selected server after database creation. Learn more

Allow Azure services and resources to access this server No Yes

Add current client IP address * No Yes

Private endpoints

Private endpoint connections are associated with a private IP address within a Virtual Network. The list below shows all the private endpoint connections for this server. Note that private endpoint connections are defined at the server level and they provide access to all databases in the server. Learn more

+ Add private endpoint

Figure 15-6. *Configuring firewall rules for an Azure SQL logical server during deployment*

Connection Policy

Define how your clients connect to your logical SQL server and access your Azure SQL database.

- *Default Connection Policy*: By default, all clients coming from inside Azure networks will establish connections directly to the node hosting the database. All other connections coming from outside of Azure will be proxied via the Azure SQL Database gateways.
- *Proxy Connection Policy*: This is usually used only for connections coming from outside of Azure. Nevertheless, you can force all connections to be proxied via a gateway. This is also currently used when connecting to your database from a virtual network via a private endpoint.
- *Redirect Connection Policy*: This is usually used only for connections coming from inside Azure. Nevertheless, you may force all clients to connect directly to the database host.

Figure 15-7 shows the logical SQL Server connection policy configuration shown on the SQL Server ➤ Security ➤ Networking ➤ Connectivity blade.

Connection Policy
Configure how clients communicate with your SQL database server. Learn more
Connection policy
Default - Uses Redirect policy for all client connections originating inside of Azure and Proxy for all client connections originating outside Azure
Proxy - All connections are proxied via the Azure SQL Database gateways
Redirect - Clients establish connections directly to the node hosting the database

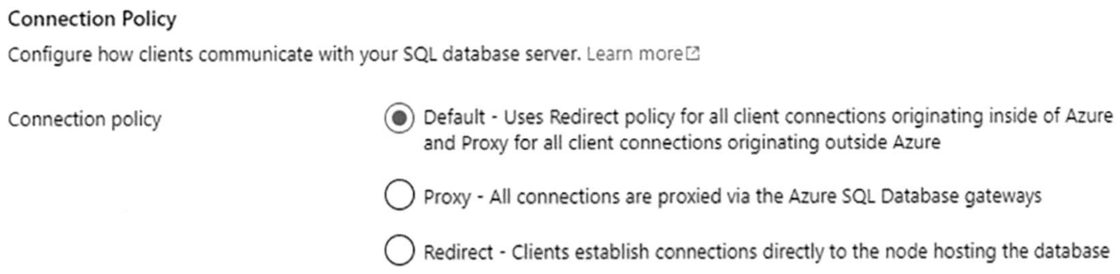

Figure 15-7. *Connection policy for an Azure SQL logical server*

Access Management

Through access management, we manage users and their permissions, as well as verifying the user's identity. When we or one of our workloads connects to a logical server, we can choose to use SQL authentication or Azure Active Directory authentication. If we are using Azure Active Directory authentication, then we can authenticate as a user or a service principal.

Tip Using Azure AD authentication is *strongly recommended*. With many tools, such as SQL Studio Management Studio, this supports multifactor authentication and conditional access to provide improved security. It also enables your workloads to connect using managed identities and service principles. This method of authentication provides the best security features.

Logical Server Authentication

For every Azure SQL Logical Server deployment, you can select the preferred authentication for accessing that logical server.

You may choose from the following methods of authentication:

- *Only Azure AD authentication*: This is the default and is recommended if your workloads support it.
- *Both SQL and Azure AD authentication*: This is useful if your users can use Azure AD to authenticate but you have workloads that still require SQL authentication. It can be useful to enable this as a fallback if for some reason Azure AD authentication is not available.
- *Only SQL authentication*: Use this only if your workloads and tooling don't support Azure AD authentication.

Figure 15-8 shows the available logical Azure SQL Server authentication methods.

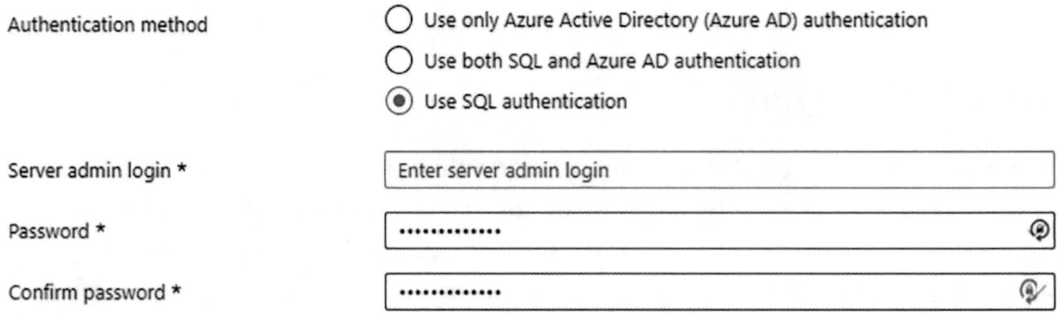

Figure 15-8. *Azure SQL logical server authentication methods*

Role-Based Access Control

Control over the Azure SQL logical server and the Hyperscale database resources in Azure is provided by granting role-based access control (RBAC) roles on the logical server to users, groups, or service principals. Azure provides several built-in roles for Azure SQL logical servers and databases that provide permissions to the Azure resources.

- *SQL DB Contributor*: You can manage SQL databases but cannot connect to them. You can't manage security-related policies or their parent logical servers.
- *SQL Security Manager*: You can manage the security-related policies of SQL servers and databases but not connect to them.
- *SQL Server Contributor*: You can manage SQL servers and databases but not connect to them or to their security-related policies.

Note It is possible to create custom RBAC roles that allow or deny specific permissions on the built-in roles, but it is recommended to follow best practices by ensuring the principle of least privilege is being used.

Configuring whether a user can connect to the logical server is done by setting permissions in the database engine. For more information on this, see `https://learn.microsoft.com/sql/relational-databases/security/authentication-access/getting-started-with-database-engine-permissions`.

Tip In Azure, RBAC can be used to govern the permissions both on the control plane to the resource and on the data plane. The data plane consists of the permissions you require to access the data a resource contains. In the case of Azure SQL Database and logical servers, all data plane access is configured by defining the permissions in the database engine, not via RBAC roles.

Permissions on the Database Engine

Because Hyperscale is fundamentally an Azure SQL Database instance, you can configure it how you would any SQL Server, by setting permissions in the database engine. However, you can't configure server-level permissions for a Hyperscale database. This is considered configuring the permissions within the data plane.

The recommended approach to configuring permissions for users and workloads is to adopt the principle of least privilege. In addition, it is recommended to use object-level permissions where applicable. It is recommended to limit permissions for most users and grant admin permissions only if necessary.

Row-Level Security

Row-Level Security (RLS) is a security mechanism that limits access to rows in a database table based on the logged-in user. This ensures that the logged-in user can only read and write to specific records in each table, based on their permission.

Tip This is useful in multitenant architectures where data for multiple tenants is stored in the same tables and you need to ensure each tenant is only reading and writing data for their tenant.

For more information on row-level security, see `https://learn.microsoft.com/sql/relational-databases/security/row-level-security`.

Auditing and Azure Threat Detection

Auditing and Azure threat detection gives you everything you need to timely detect anomalous activities and potential threats as they occur. Moreover, if it's configured properly, you may receive alerts about SQL injection attacks or any other suspicious database activities immediately when they occur.

Azure SQL Database Auditing

If enabled, the Azure SQL database audit will trigger tracking for database events and write all those events to an audit log stored in the Azure Storage account, Log Analytics workspace, or Event Hub.

Figure 15-9 shows the Azure SQL logical server-level auditing settings in the SQL Server ➤ Security ➤ Auditing blade.

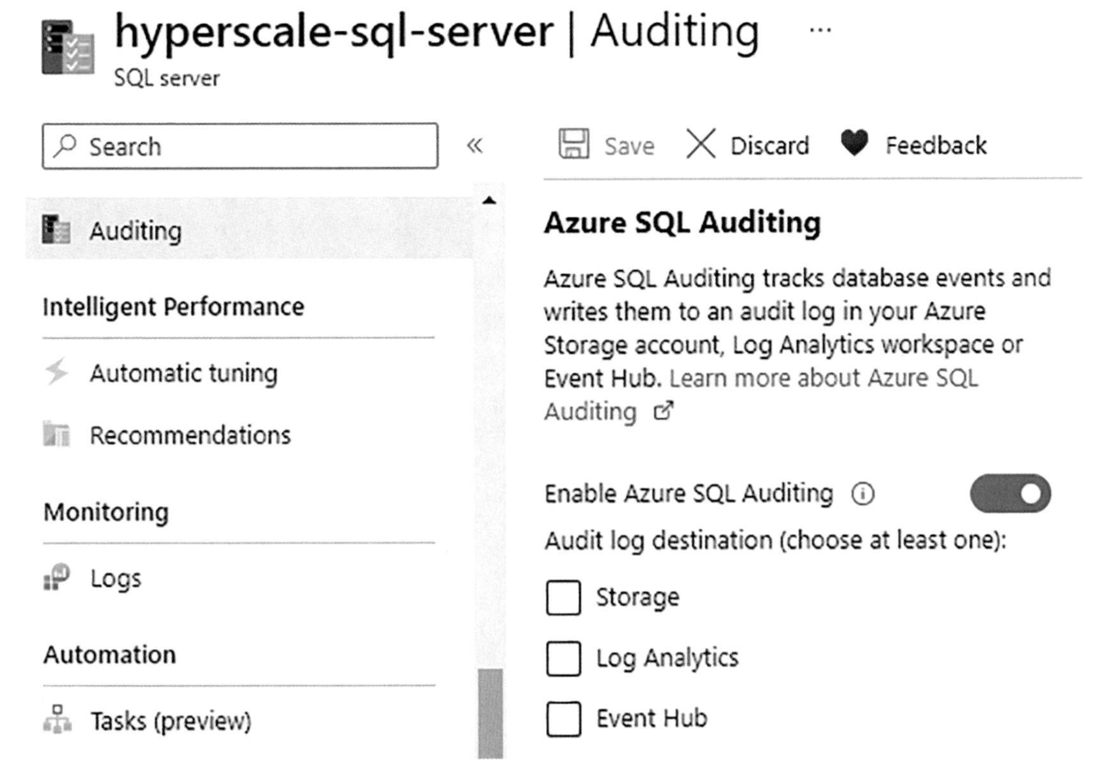

Figure 15-9. *Configuring auditing on the logical server in the Azure Portal*

Tip You can enable auditing on the server level or the database level. If Azure SQL Server auditing is enabled, it will overwrite the database-level settings, and the database will be audited, ignoring the database auditing settings.

Microsoft Defender for Cloud

Microsoft Defender for Cloud is used to manage security across multiple Azure offerings including Azure data services. Microsoft Defender for SQL is part of the Microsoft Defender for Cloud suite of products and incurs a monthly charge. However, a 30-day trial of this service is provided free of charge, enabling you to safely evaluate it.

If enabled, this includes vulnerability assessment recommendations and advanced threat protection. Figure 15-10 shows an example of the potential security vulnerabilities provided by Microsoft Defender for Cloud.

Recommendations	Security alerts	Findings	Enablement Status: **Enabled at the subscription-level** (Configure)
5	0	--	

Recommendations

Defender for Cloud continuously monitors the configuration of your SQL Servers to identify potential security vulnerabilities and recommends actions to mitigate them.

Description	Severity
SQL servers should have an Azure Active Directory administrator provisioned	High
Auditing on SQL server should be enabled	Low
Private endpoint connections on Azure SQL Database should be enabled	Medium
Public network access on Azure SQL Database should be disabled	Medium
SQL servers should have vulnerability assessment configured	High

View additional recommendations in Defender for Cloud >

Figure 15-10. *Microsoft Defender for Cloud recommendations pane example*

Information Protection

In the Information Protection category, we can review all the security features involved with encrypting data both at rest and in use, as well as encrypting connections and masking sensitive data in our Azure SQL deployment.

TLS Encryption in Transit

Logical Azure SQL Server supports encrypted connections using Transport Layer Security (TLS). Within the logical Azure SQL Server networking settings, on the Connectivity tab you may find the Encryption in Transit option, where you can choose the suitable TLS version for your environment. Figure 15-11 shows the logical Azure SQL Server–supported encrypted connections using TLS on the Security ➤ Networking ➤ Connectivity ➤ Encryption in transit blade.

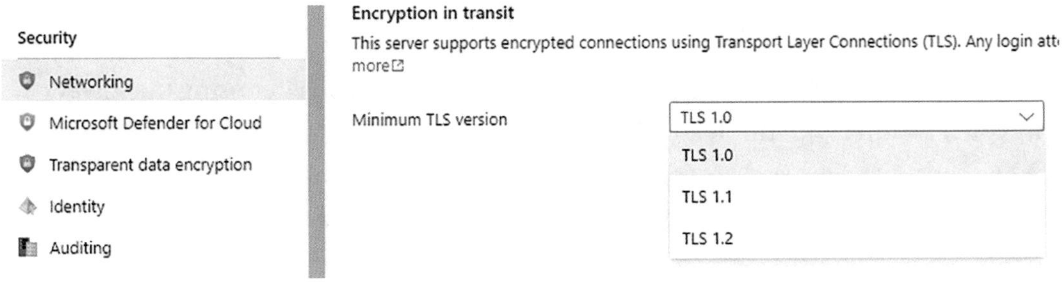

Figure 15-11. *Azure Portal, TLS connections*

Encryption at Rest with Transparent Data Encryption

While using TDE for your Azure SQL Database deployment, you would add additional security for your database files and backups. This feature is enabled by default in any Azure SQL Database service tier. The purpose of transparent data encryption is to encrypt your database files. Figure 15-12 shows the TDE status (Security ➤ Transparent data encryption) and gives you the option to turn it on and off.

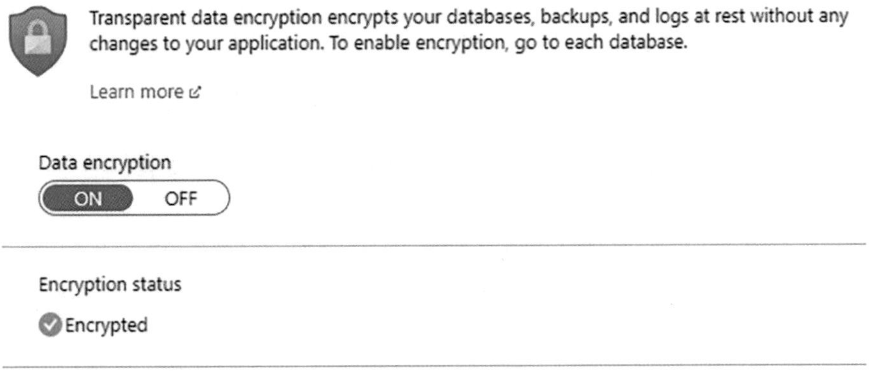

Figure 15-12. *TDE*

Bear in mind that there is no user action necessary to maintain the certificate behind the encryption key, as this is managed by the service, including certificate rotation. Nevertheless, if customers insist on having full control of their encryption keys, they have an option to do so. With bring your own key (BYOK) for TDE, customers can take ownership of key management, including key certificate maintenance and rotation. This was also discussed in Chapter 6.

Always Encrypted: Encryption in Use

If you want to protect the integrity of your data, you should consider using the Always Encrypted feature. With this feature, you can protect specific tables or table columns holding sensitive data, such as credit card details. Data will always be encrypted and is decrypted only by your application using an encryption key you control.

Tip Even users with elevated permissions and access to your data such as database administrators will not be able to see sensitive data without the correct encryption key. For more information on enabling Always Encrypted, see `https://learn.microsoft.com/sql/relational-databases/security/encryption/always-encrypted-database-engine`.

Data Masking

If you want to prevent users from seeing sensitive data, you might consider using dynamic data masking. This feature enables you to create masking rules and avoid exposure of sensitive data to particular users. Figure 15-13 shows the database-level blade found at SQL Database ➤ Security ➤ Dynamic Data Masking.

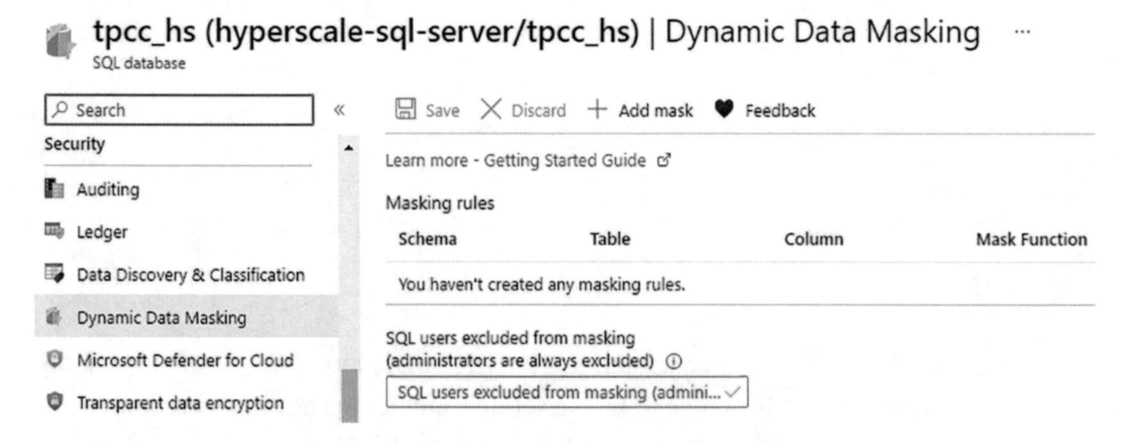

***Figure 15-13.** Azure Portal, dynamic data masking feature*

Within the masking rule you can specify schema, table, and column names with a masking field format. Figure 15-14 shows adding a masking rule page, with the schema, table, and column as requested fields.

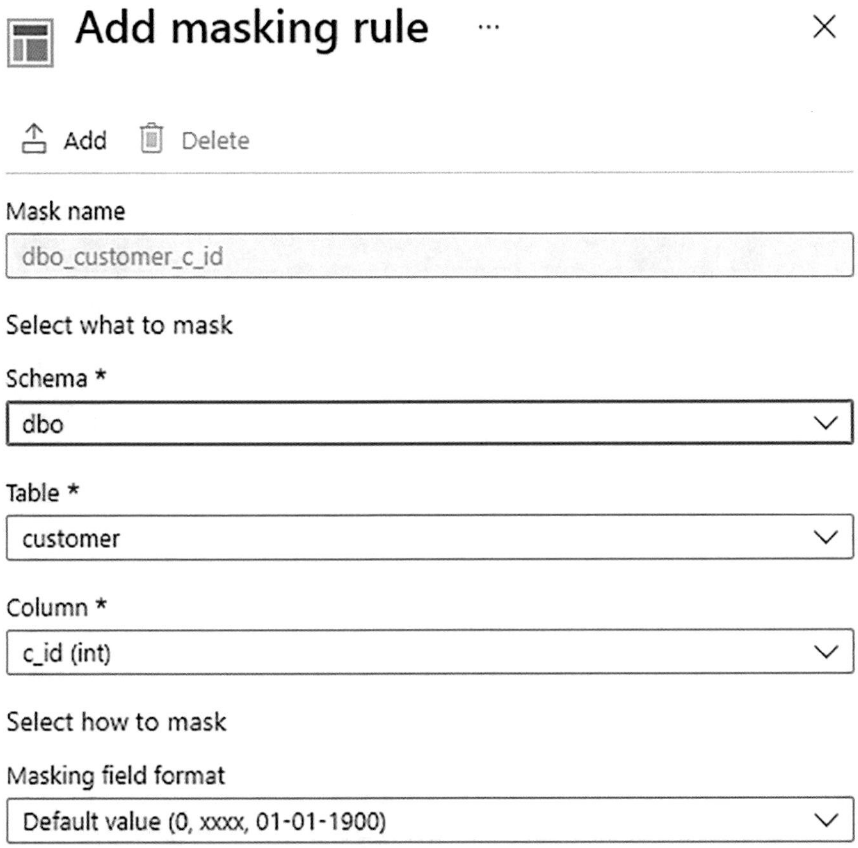

Add masking rule

Add Delete

Mask name

dbo_customer_c_id

Select what to mask

Schema *

dbo

Table *

customer

Column *

c_id (int)

Select how to mask

Masking field format

Default value (0, xxxx, 01-01-1900)

Figure 15-14. *Azure Portal, adding a masking rule*

Additionally, the dynamic data masking feature can recognize fields storing potentially sensitive data and automatically recommend those fields to be masked. Figure 15-15 shows the recommended fields to mask.

Recommended fields to mask

Schema	Table	Column	
dbo	customer	c_credit_lim	Add mask
dbo	customer	c_ytd_payment	Add mask
dbo	customer	c_payment_cnt	Add mask
dbo	customer	c_street_1	Add mask
dbo	customer	c_street_2	Add mask
dbo	customer	c_city	Add mask

Figure 15-15. *Azure Portal, masking fields recommendations*

If masking is enabled for a specific column and a SQL query is executed against that column, you will be able to see only the masked sensitive data in the query results set, while at the same time the data in the database is unchanged.

Updating and Maintenance Events

The maintenance window feature gives you the option to control the time when upgrades or scheduled maintenance tasks are performed on your database. This allows you to reduce the impact of this disruptive patching process on your workload. By default Azure schedules maintenance events to occur outside typical working hours based on the local time of the Azure region. This is usually between 5 p.m. to 8 a.m. Setting the maintenance window is available for only some regions. Figure 15-16 shows how to set the maintenance window in a supported region.

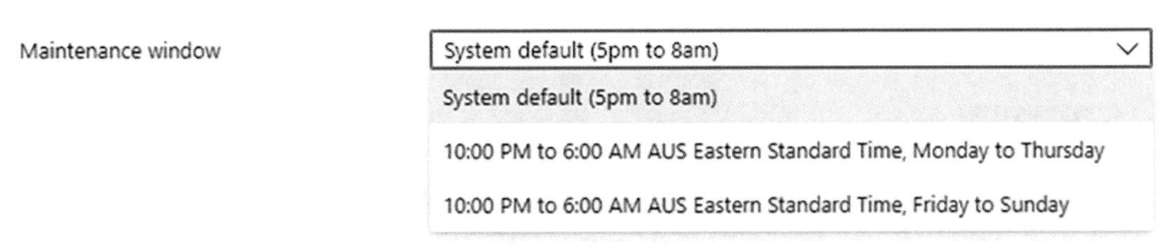

Figure 15-16. *Specifying the maintenance window time*

If you choose a region that does not support configuring the maintenance window, it will look like Figure 15-17.

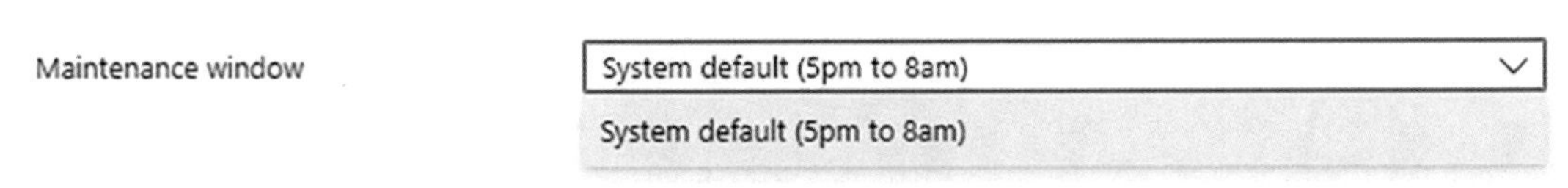

Figure 15-17. *Setting the maintenance window not supported in this region*

During a scheduled maintenance event, your Hyperscale database is completely available, but it can be affected by small reconfigurations that will cause connections to drop and any long-running transactions to not be completed. You should ensure your workload is able to gracefully handle these events.

Tip Bear in mind that maintenance windows are there to ensure that all planned upgrades or maintenance tasks happen within a particular time range. This will not protect your database resource from any unplanned regional or zonal outages.

It is also possible to get advance notifications of planned maintenance events using Azure alert rules. This will cause an alert rule to send a notification via SMS, email, phone, or push event 24 hours before a scheduled maintenance event will occur. This feature is currently in preview.

Summary

In this chapter, we looked at some of the best practices of securing the data tier in Azure SQL deployments. We covered different Azure SQL database security strategies for how to prevent any access to the server and the customer's sensitive data until that access is explicitly granted. We explained the roles of multiple security layers between the network traffic and customer data. We covered network security, access management, auditing, and threat protection and information protection.

In the next chapter, we will talk about how to optimize the total cost of operating the Hyperscale environment.

CHAPTER 16

Managing Costs

Over the previous few chapters, we've looked at many of the management and operational tasks required to ensure resilient, performant, and secure operation of a Hyperscale environment. Another important management task is the need to review and optimize costs.

Most organizations want to know that they're getting the best value for their investment and require regular review of ongoing costs to ensure this. In this chapter, we will look at how we can optimize the total cost of ownership (TCO) of the Hyperscale environment. We will look at the different options available to tune the cost of operating the Azure SQL Database Hyperscale.

Azure Hybrid Benefit

Many organizations have a hybrid approach to using Azure, maintaining resources on-premises as well as running them in Azure. These organizations often purchase Microsoft SQL Server licenses with Software Assurance. Azure Hybrid Benefit (AHB) allows organizations to exchange existing licenses for discounted rates on Hyperscale databases (and other Azure SQL Database tiers and Azure SQL Managed Instances). You can save 30 percent or more on your Hyperscale database by using Software Assurance–enabled SQL Server licenses on Azure.

To see the cost savings for leveraging AHB, use the Azure Pricing Calculator. Figure 16-1 shows the Azure Pricing Calculator without an included SQL license on a Hyperscale database.

Z. Barać and D. Scott-Raynsford, *Azure SQL Hyperscale Revealed*,
https://doi.org/10.1007/978-1-4842-9225-9_16

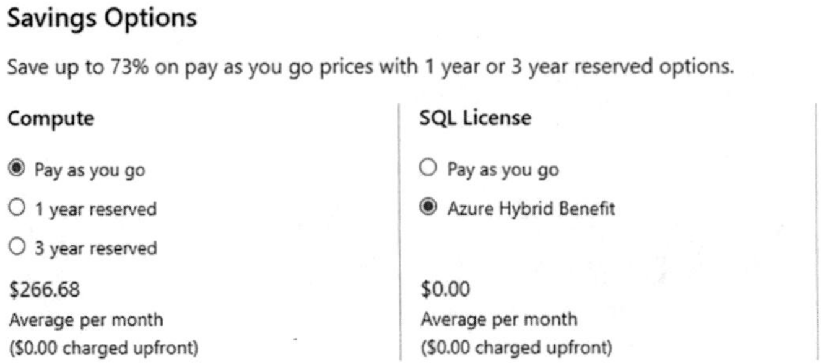

Savings Options

Save up to 73% on pay as you go prices with 1 year or 3 year reserved options.

Compute	SQL License
◉ Pay as you go	○ Pay as you go
○ 1 year reserved	◉ Azure Hybrid Benefit
○ 3 year reserved	
$266.68 Average per month ($0.00 charged upfront)	$0.00 Average per month ($0.00 charged upfront)

Figure 16-1. *Azure Pricing Calculator with AHB selected*

If you qualify for AHB, you can enable this discount when deploying your Hyperscale database through the Azure Portal.

- For new databases, during creation, select "Configure database" on the Basics tab and select the option to save money.

Save money

Already have a SQL Server License? Save with a license you already own with Azure Hybrid Benefit. Actual savings may vary based on region and performance tier. Learn more

◉ Yes ○ No

License type
SQL server

☑ I confirm that I have a SQL server license with Software Assurance to apply this Azure Hybrid Benefit for SQL Server

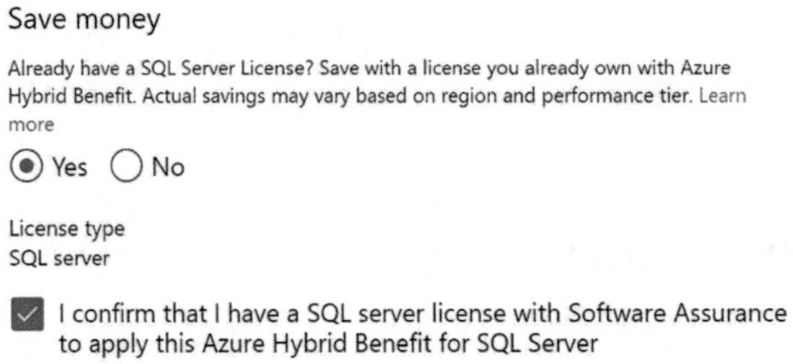

- For existing databases, select the "Compute + storage" blade, under the Settings menu on a Hyperscale database resource, and select the option to save money.

You can also configure this using Azure PowerShell, the Azure CLI, or Azure Bicep, but for the sake of brevity we will not cover those methods here. For more information on Azure Hybrid Benefit, see `https://learn.microsoft.com/azure/azure-sql/azure-hybrid-benefit`.

Reserving Capacity

One of the simplest methods of reducing costs is to leverage reserved capacity. This allows you to commit to using a specified number of vCores on a Hyperscale database in a specified Azure region for either a 1- or 3-year period. This can provide a significant discount on the cost of running a Hyperscale database.

Note Purchasing reserved capacity requires an Azure Enterprise subscription or an Azure *p*ay-as-you-go subscriptio*n*.

Purchasing Reserved Capacity

When you purchase reserved capacity, you will specify the scope of the reservation. This defines where it can be used. For example, a reservation can be scoped to be utilized only by Hyperscale databases in a specific resource group or subscription.

Purchasing reserved capacity is done through the Azure Portal.

1. Select All services ➤ Reservations.
2. Click Add, and then in the Purchase Reservations pane, select SQL Database to purchase a new reservation for Hyperscale.
3. Complete the required fields to specify the type of reservation to create, including the scope, region, quantity, and term.
4. Review the cost of the reservation in the Costs section.
5. Click Purchase.

Your reservation will appear shortly in the Reservations pane.

Important Reservations can be easily canceled, exchanged, or refunded using a self-service process with some limitations. See this page for more details: `https://learn.microsoft.com/azure/cost-management-billing/reservations/exchange-and-refund-azure-reservations`.

For more information on reserving capacity on Azure SQL databases (including Hyperscale), see `https://learn.microsoft.com/azure/azure-sql/database/reserved-capacity-overview`.

Estimating Reserved Capacity Benefits

When you create a reservation in the Azure Portal, you are presented with the cost of the reservation. However, you might not have the appropriate permissions to create a reservation, so an alternative way to evaluate the cost reduction of using reserved capacity is to use the Azure Pricing Calculator. Figure 16-2 shows an example of the Azure Pricing Calculator with a two-vCore Hyperscale database and a 3-year reservation selected.

Savings Options

Save up to 73% on pay as you go prices with 1 year or 3 year reserved options.

Compute	SQL License
○ Pay as you go	◉ Pay as you go
○ 1 year reserved	○ Azure Hybrid Benefit
◉ 3 year reserved	
Compute Payment Options: Monthly	
$120.00	$145.95
Average per month	Average per month
($0.00 charged upfront)	($0.00 charged upfront)

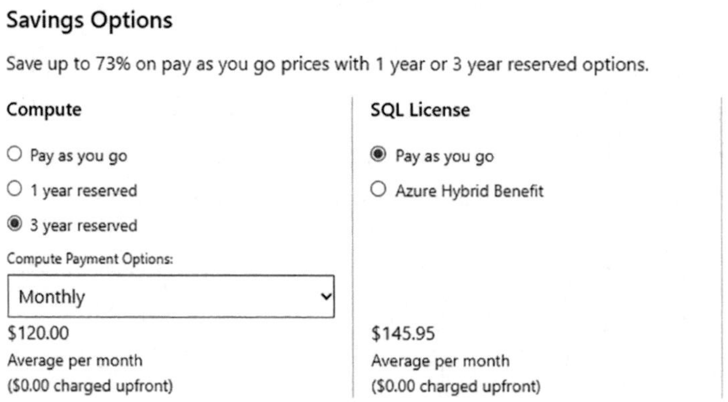

Figure 16-2. *Azure Pricing Calculator with three-year reservation selected*

Utilizing Reservations

Once capacity has been reserved for an Azure region and Hyperscale tier, any Hyperscale database created that matches the scope of any existing reservations will automatically be assigned to use the reservation if there are unutilized vCores in the reservation. Deleting or scaling down a Hyperscale database in the region will return the vCores back to the reservation to be utilized by another Hyperscale database.

It is a good idea to regularly review the utilization of reserved capacity to ensure you're utilizing the capacity you have reserved. To review reserved capacity utilization in the Azure Portal, follow these steps:

1. Enter **Reservations** in the search box at the top of the Azure Portal and select Reservations.

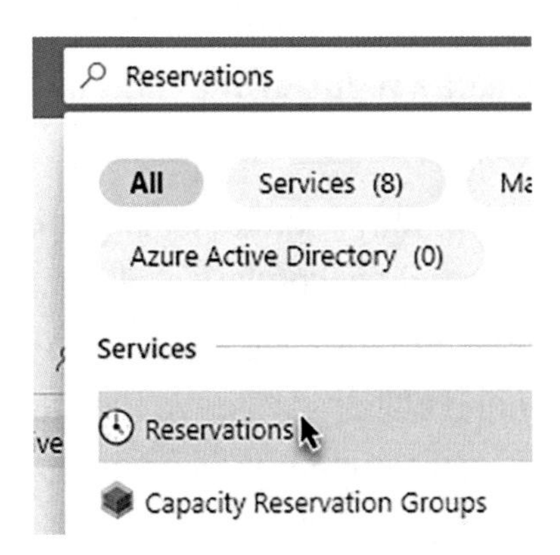

2. The list shows all the reservations where you have the Owner or Reader role. Each reservation displays the last known utilization percentage.

3. Click the utilization percentage to see the details of the reservation and its history.

For more information on reviewing the utilization of your reservations, see `https://learn.microsoft.com/azure/cost-management-billing/reservations/reservation-utilization`.

Scaling In and Scaling Down Replicas

One of the major benefits of Hyperscale is the ability to scale out compute by adding read-only named replicas. This provides us with the ability to direct reporting or other read-only workloads to the named replicas, reducing the load on the primary read/write database. It is common for the read-only workloads to have different compute requirements and utilization patterns from the primary database.

It is also possible to adjust the functional and nonfunctional requirements of a workload to enable it to use a ready-only named replica that can be scaled down or scaled in, removing the replica completely when not needed. For example, changing reporting from an on-demand process to a scheduled process can allow you to leverage a scheduled or automated scaling methodology.

For example, in a retail services environment, the primary database may be used 24 hours a day, but report generation is scheduled as a batch process that runs daily, starting at 1 a.m., and takes roughly four hours. This means the replica could be scaled down or scaled in for 20 hours a day, reducing the compute costs.

Important You should perform only regular scaling operations to tune costs on named replicas. High-availability secondary replicas or geo replicas are not typically used for this purpose. High-availability secondary replicas are intended to be used for high availability and will always have the same number of vCores as the primary database. A geo replica used for failover should always have the same number of vCores as the primary replica.

The following are some common scaling methods of leveraging replica scaling to reduce cost:

- *Scale-down*: Reduce the number of vCores assigned to a replica outside of peak hours on a schedule or autoscaling based on a platform metric.
- *Scale-up*: Temporarily increase the number of vCores assigned to a replica to reduce the completion time of a batch process. Once the process is completed, scale down or scale in the replica. The larger cost of the increased number of vCores will usually be offset by the cost reduction achieved by the scale-down or scale-in.
- *Scale-in*: Remove the replica when not in use. For example, if batch reporting or data export processes are required to run only between the hours of 1 a.m. and 5 a.m., then the named replica can be deleted entirely. Removing a replica will delete the compute associated with it, but not the underlying storage, which is part of the primary database.

Once you've implemented scaling methodology for reducing costs, it is necessary to monitor your workloads to ensure they're performing to expectations.

Important You should ensure that any workloads that utilize the replica will be able to handle the scale operations that you're performing. For example, a workload that doesn't implement retry logic for standard transient errors will experience errors with scale-up and scale-down operations. Any long-running transactions will also be terminated, so your workload should ensure these are completed.

It is also worth considering the TCO for implementing the scaling methodology. When factoring in time to build, monitor, operate, and maintain the infrastructure to enable a replica scaling methodology, this may not result in a net benefit.

Summary

In this chapter, we looked at some of the most common ways to reduce the cost of a Hyperscale environment.

- Leveraging Azure Hybrid Benefits where applicable
- Utilizing reserved capacity
- Implementing scaling patterns on read-only named replicas

In the next chapter, we will look at some common patterns, use cases, and workloads that can benefit from Hyperscale, beyond the increased storage capacity.

PART IV

Migration

CHAPTER 17

Determining Whether Hyperscale Is Appropriate

In the previous chapter, we focused on how to better understand and optimize the costs of the Hyperscale environment. We covered a few of the most effective ways to reduce and tune the operating costs of an Azure SQL Hyperscale database.

The goal of this chapter is to ease the business decision of whether migrating to an Azure SQL Hyperscale database will be the best solution for your production workload. We will cover the various advantages of the Azure SQL Hyperscale architecture compared to other Azure SQL platform-as-a-service (PaaS) deployments.

Key Considerations for Hyperscale

While it is difficult to summarize all the features of a Hyperscale environment, we will identify some of the key differences between the traditional Azure SQL Database service tiers and the new Hyperscale service tier that you should consider when deciding on relational database technologies.

There are several key areas you should consider.

- *Scalability*: How does the solution scale with regard to storage and compute to support your workload growth? Hyperscale offers highly scalable storage and compute options, including support for databases up to 100TB in size. Hyperscale provides the ability to easily and quickly horizontally scale in/out and vertically scale up/down, with minimal disruption to your workload.

Z. Barać and D. Scott-Raynsford, *Azure SQL Hyperscale Revealed*,
https://doi.org/10.1007/978-1-4842-9225-9_17

- *Reliability*: How resilient is the solution to planned and unplanned outages? Azure SQL Database Hyperscale provides several reliability enhancements, such as high-availability secondary replicas and optional zone and geo-redundancy out of the box.
- *Business continuity*: How does the solution cope with regional or zonal outages or unintentional damage to data? Hyperscale leverages significantly faster backups and restores, higher log throughput, and faster transaction commit times, reducing RTO and RPO. Hyperscale also supports regional failover by leveraging geo replicas. Backup file storage can use local, zone, geo redundancy, or geo-zone redundancy, which enables your backups to be available, even in the case of zonal or regional outages.
- *Cost*: What is the cost of running Hyperscale, and how will this change as my solution grows? The cost of Hyperscale is primarily affected by the number of vCores consumed by the primary and secondary replicas and the amount of storage used. Several other factors also affect this.

In the following sections, we'll review these considerations in more detail.

Scalability

Every successful business can thrive only when there are either no scalability limitations or when limits are flexible enough to support future growth. The scalability of your workload can be significantly increased when using fully managed Hyperscale database solutions.

The ability to assign more resources dynamically when necessary is one of the biggest benefits of the fully managed cloud database services. All Azure SQL PaaS deployment options can be easily and dynamically scaled to meet your needs. Azure SQL enables you to adjust performance characteristics by allocating more resources to your databases, such as CPU power, memory, I/O throughput, and storage.

By introducing the Hyperscale architecture, your workload can achieve a level of scalability that is not met by any other Azure SQL deployment. One of the main characteristics of the Hyperscale architecture, the decoupling of remote storage and compute tier, makes vertically scaling resources such as CPU and memory a fast and seamless process and is independent of the size of the database.

When looking at your scalability needs, you should consider the following:

- How will your compute and storage requirements change over time, and will you reach the limits of the service tiers?
- Does your workload gain performance benefits by having more memory per vCore?
- Can your workload scale up efficiently?
- Can your workload be designed to scale out by leveraging named or HA replicas?
- Will your workload achieve diminishing returns from scaling up as you grow? For example, as your storage increases, will adding more vCores not provide a linear increase in performance?

Table 17-1 compares an Azure SQL Hyperscale deployment leveraging the Hyperscale architecture to the General Purpose and Business Critical service tiers leveraging a traditional Azure SQL architecture. It is useful to evaluate the information in this table against your workload scalability requirements.

Table 17-1. *Hyperscale and Traditional Architecture Deployments: Scalability Features Comparison*

	General Purpose (Provisioned)	Business Critical	Hyperscale
Compute size	2 to 128 vCores.	2 to 128 vCores.	2 to 128 vCores.
Memory	Standard-series (Gen5) up to 625 GB memory.	Standard-series (Gen5) up to 625 GB memory.	Standard-series (Gen5) up to 415.23 GB memory. Premium-series (Gen5) up to 625 GB memory. Premium-series (memory-optimized) up to 830.47GB memory.
Log size	Up to 1TB log size.	Up to 1TB log size.	Unlimited log size.
Log throughput	From 9 to 50MBps. 4.5MBps per vCore.	From 24 to 96MBps. 12MBps per vCore.	105MBps, regardless of database size and the number of vCores.
Maximum tempdb size	20GB per vCore, up to 2560GB.	20GB per vCore, up to 2560GB.	32GB per vCore, up to 4096GB.
Storage type	Azure premium remote storage.	Fast locally attached SSD storage to each node.	Multitiered architecture with separate storage components.
Maximum storage size	4TB.	4TB.	100TB.
Read scale-out	No read scale-out.	One read scale-out replica.	Up to four HA/read scale-out replicas.* Up to 30 named replicas.

** You may consider using high-availability replicas for read scale-out scenarios. Nevertheless, you can create up to 30 named replicas suitable for more complex read scale-out scenarios.*

Reliability

All Azure SQL database deployments, including the Hyperscale service tier, are fully managed PaaS database engines. This means that maintenance activities such as patching and upgrading happen automatically and are managed by Microsoft Azure. They are always updated with the latest stable SQL Server version and provide at least 99.99 percent availability guarantee. The inclusion of built-in redundancy, regardless of the Azure SQL deployment option chosen, makes an Azure SQL PaaS database solution reliable, highly available, and resilient to outages. Table 17-2 shows the reliability features of the different service tiers of Azure SQL Database.

Table 17-2. *Hyperscale and Traditional Architecture Deployments: Reliability Features Comparison*

	General Purpose (Provisioned)	Business Critical	Hyperscale
Availability	One replica. No read scale-out. Zone-redundant HA. No local cache.	Three replicas. One read scale-out. Zone-redundant HA. Full local storage.	Up to four high availability replicas. Up to 30 named replicas. One geo-replica. Zone-redundant HA. Partial local cache (RBPEX).
Read-only replicas	Zero read-only replicas built-in.	One read-only replica.	Up to four read-only replicas.
High-availability replicas	One node with spare capacity.	One HA replica.	Four HA replicas.

Business Continuity

All Azure SQL Database deployment options have built-in redundancy. There are several factors to be aware of when considering business continuity.

- *Service redundancy*: This refers to the redundancy of the database and includes high-availability replicas and geo replicas.

- *Backup storage redundancy*: This refers to the availability of the storage storing your backups.

The redundancy type and availability guarantee (service-level agreement) differ depending on the deployment you choose. There are four types of service redundancy.

- *Local redundancy*: Your service won't be protected from a zonal or regional outage.
- *Zone redundancy*: Your service will be protected from a zonal outage, but not a regional outage. This requires the use of zone-redundant availability.
- *Geo redundancy*: Your service will be protected from a regional outage. This requires the use of a geo replica. Geo replicas do protect against zonal outages but will require a complete regional failover.
- *Geo-zone redundancy*: Your service will be protected from zonal and regional outages. This requires the use of zone-redundant availability and a geo replica.

If you choose Azure SQL deployments such as Hyperscale, Business Critical, Premium, or General Purpose, with the Zone Redundant Availability option enabled, you will get at least 99.995 percent availability guaranteed. If not configured with zone redundancy, the availability will be at least 99.99 percent. This availability SLA will be the same for the Basic and Standard service tiers leveraging local redundancy.

When determining the business continuity requirements of your solution, you should consider the following:

- What is the availability your solution requires?
- What is the recovery time objective (RTO) and recovery point objective (RPO) of your solution?
- How tolerant to failure is your workload? For example, will your workload recover seamlessly when moving from one high-availability replica to another?

Table 17-3 compares the business continuity features of the different service tiers of Azure SQL Database.

***Table 17-3.** Hyperscale and Traditional Architecture Deployments: Business Continuity Features Comparison*

	General Purpose (Provisioned)	Business Critical	Hyperscale
Backup frequency	Full backups every week. Differential backups every 12 or 24 hours. Transaction log backups approximately every 10 minutes.**	Full backups every week. Differential backups every 12 or 24 hours. Transaction log backups approximately every 10 minutes.**	Backups are based on storage snapshots of data files.
Backup retention policy	1–35 days (7 days by default).	1–35 days (7 days by default).	1–35 days (7 days by default).
Backup storage redundancy	Locally redundant backup storage. Zone-redundant backup storage.* Geo-redundant backup storage.	Locally redundant backup storage. Zone-redundant backup storage.* Geo-redundant backup storage.	Locally redundant backup storage. Zone-redundant backup storage.* Geo-redundant backup storage. Geo-zone redundant backup.
Restore ability	Restore time depends on database size.	Restore time depends on database size.	Fast restore time regardless of database size.***

** Zone-redundant backup storage is available only in specific Azure regions that have multiple availability zones. Each availability zone represents a separate data center with independent hardware infrastructure.*
*** The frequency of transaction log backups may vary depending on the amount of database activity.*
**** The restore process of a Hyperscale database within the same Azure region is not a size-dependent operation. Nevertheless, geo restores will take more time to finish.*

It is possible to achieve very high levels of availability and almost instant recovery from zonal or regional failures with the right design, but these will usually increase the cost of the solution.

Cost

When determining the total cost of ownership of using a Hyperscale database, it is important to consider both the Azure costs of the service as well as the operational savings of using a managed PaaS service. In this section, we'll focus on the Azure costs of running a Hyperscale database, but it is equally important to assess the operational savings you will make through reducing the time taken to maintain, patch, and operate the underlying infrastructure.

Tip Ensure you're factoring in any cost savings that you might be able to leverage, as discussed in Chapter 16.

The following factors affect the Azure cost of running a Hyperscale database:

- *Region*: The cost of infrastructure does differ across Azure regions, so it is important to investigate the cost of Hyperscale in the regions you're planning to use.
- *Hardware type*: Consider the generation of the hardware used by the database. Premium-series and Premium-series (memory optimized) will cost slightly more than Standard-series (Gen 5) and Premium-series. DC series (confidential compute) costs significantly more than the other types.
- *Number of vCores*: The number of vCores will affect both the primary database and any HA replicas as they always match. Named and geo replicas can have a different number of vCores assigned. These may be able to be scaled dynamically to reduce costs depending on your workload and architecture.
- *Number of replicas*: Consider the number of HA replicas, named replicas, and geo replicas you will be deploying. These may be able to be scaled dynamically to reduce costs depending on your workload and architecture.
- *Database storage*: The amount of data stored in the database affects the cost of the database and the amount of backup storage required. This is likely to grow over time, so it is important to consider how this will change.

- *Backup storage redundancy*: Consider the redundancy required for both the point-in-time-restore (PITR) backups and the long-term retention (LTR) backups. This is impacted by the amount of database storage being used.
- *PITR backup storage*: Consider the amount of storage required for PITR backups.
- *LTR backup storage*: Consider the amount of storage required for LTR backups.
- *LTR retention policy*: Consider the length of time that weekly, monthly, and yearly LTR backups are retained.
- *Reservations*: If you are leveraging a 1- or 3-year reservation, your costs can be significantly reduced. This was discussed in Chapter 16.
- *Azure Hybrid Benefit (AHB)*: Consider whether you are eligible to use Azure Hybrid Benefit to reduce your license costs. These were discussed in Chapter 16.
- *Nonproduction environments*: It is important to factor in the cost of nonproduction environments for testing and other use cases. These may not be the same scale as the production environment or could be removed or scaled down when not needed.
- *Manual or automatic scaling*: If your architecture allows you to be able to scale your Hyperscale databases or replicas up/down/in/out, then you may be able to significantly reduce costs.
- *Microsoft Defender for SQL*: If you are using Microsoft Defender for SQL, this will incur an additional monthly cost.
- *Logging and metrics storage*: If you're sending audit and diagnostic logs or metrics to a Log Analytics workspace or other service, then you should factor this in as well.
- *Related services*: There may be other costs that are incurred by your environment, such as Azure Key Vaults, network bandwidth, virtual networking, and related services. It is important to consider these costs as well.

With these factors you can reasonably accurately forecast the cost of operating a Hyperscale environment.

Tip It is strongly recommended that you leverage the Azure Pricing Calculator to produce a forecast of costs for your Hyperscale database, replicas, and related services (Log Analytics workspaces). You can find the Azure Pricing Calculator at `https://aka.ms/AzurePricingCalculator`.

Summary

In this chapter, we weighed all the benefits of choosing Hyperscale compared to traditional Azure SQL architecture. We discussed the determining factors for that decision such as higher performance, easier scalability, and higher reliability. Some of the key features include the following:

- Rapid constant-time compute scale-up and scale-down
- Large database sizes up to 100TB
- Unlimited log size
- Transaction log throughput of 105MBps regardless of the number of compute node vCores deployed
- Up to four high-availability (HA) secondary replicas
- Numerous read scale-out scenarios utilizing secondary replicas
- Up to 30 named replicas suitable for more complex read scale-out scenarios
- Fast backups and restores regardless of database size
- Multitiered caching storage architecture

In the next chapter, we will review some of the ways to migrate an existing workload to Azure SQL Hyperscale. This will include an overview of the different tools available as well as how to use them.

CHAPTER 18

Migrating to Hyperscale

In the previous chapter, we looked at the types of workloads that would benefit from or require the use of the Azure SQL Database Hyperscale tier. Now that we're familiar with when Hyperscale will be a good fit for you, we need to review the processes required to migrate an existing database into Hyperscale.

This chapter will walk you through the different migration scenarios and identify the migration options available for each. Any database migration should be approached with care, so planning and testing are always recommended. We will run through some approaches that can be used to evaluate Hyperscale with our workload before performing a production migration.

We will cover the various methods and tools available to migration from different source database types as well as review the recommended use cases for each one.

Common Migration Methods

There are many different migration options available to migrate a database to Hyperscale. There are several different factors that you will need to consider when choosing a migration methodology, including the following:

- The source database system you're migrating from
- The volume of writes against your source database
- Whether your database needs to be available for writes during the migration

Regardless of your chosen migration method, there will always be a point that any SQL connections must be terminated and reconnected to the new instance. It is good practice to design your workload to be able to handle transient connection issues with retry logic, as this will make these sorts of migrations simpler.

Z. Barać and D. Scott-Raynsford, *Azure SQL Hyperscale Revealed*,
https://doi.org/10.1007/978-1-4842-9225-9_18

During a migration, the availability of your database will usually be affected. How the availability will be impacted is defined by the following:

- *Online*: The database will remain available for the workloads during the entire process with only a small cutover period at the end.
- *Offline*: The database will be made unavailable during the migration process.

The following are the migration approaches that we are going to cover in the remainder of this chapter:

- *In-place conversion*: We will scale an existing Azure SQL Database to the Hyperscale tier. This is an online mode operation.
- *Microsoft Data Migration Assistant (DMA)*: The DMA generates a schema script and copies data in offline mode from an existing Microsoft SQL database, either on-premises, in an Azure VM, or from a virtual machine hosted in another cloud. This requires you to deploy and manage the compute required to run this process.
- *Azure Data Migration Service (DMS)*: This is a managed service that provides Azure-based compute to perform offline migrations from an existing Microsoft SQL database, either on-premises, in an Azure VM, or from a virtual machine hosted in another cloud. Currently, DMS supports online migrations only if the target database is an Azure SQL MI or SQL Server instance on an Azure VM.
- *Azure Data Studio (ADS) with Azure SQL Migration extension*: This leverages an instance of the Azure Data Migration Service to perform the migration. It effectively provides an alternate interface to create and monitor a migration project in a DMS resource.
- *Data sync and cutover*: This is a more generic migration that uses various tools and technologies to sync or replicate the content of the database and then cut over to the new database once the data has completed syncing.

Choosing the method of migration will depend on a combination of these factors.

Table 18-1. *Available Methods for Migrating to the Azure SQL Database Hyperscale Tier*

Migration Method	From an Azure SQL Database	From an Azure SQL Managed Instance	From Microsoft SQL Server[1]
In-place conversion	Yes, online[2]	No	No
Data Migration Assistant (DMA)	No	No	Yes, offline only
Azure Database Migration Service (DMS)	No	No	Yes, offline only[3]
Azure Data Studio (ADS) with Azure SQL Migration extension	No	No	Yes, offline only[3]
Data sync and cutover	Yes, online and offline	Yes, online and offline	Yes, online and offline

Note It is also possible to migrate from AWS RDS for SQL Server using the Azure Database Migration Service.

In-Place Conversion of an Azure SQL Database to Hyperscale

A common driver to move to Hyperscale is to overcome limitations of an existing Azure SQL Database, for example, if your workload reaches the storage limit of the existing tier or if it will benefit from the enhanced storage architecture. In this case, converting the existing database to Hyperscale may be required and can be performed in-place.

[1] This includes on-premises SQL Server, SQL Server hosted on other cloud providers, and SQL Server on Azure VMs.

[2] In-place conversion to a Hyperscale database with availability zones is not currently supported. Online migration is performed, but a short outage occurs.

[3] Azure DMS can perform only an offline migration to Azure SQL Database.

You can use the in-place conversion method to convert from any of the following Azure SQL Database tiers and configurations:

- Basic
- Standard
- Premium
- General Purpose, Provisioned, or Serverless
- Business Critical, Provisioned, or Serverless

Behind the scenes, when you perform an in-place conversion, Azure will create a new database for you with the desired Hyperscale tier and copy the existing data to the new database. This may take several hours depending on the amount of data in your database and how frequently writes are occurring. Once all the data in the database has been copied, the old database will be taken offline. Any changes to the original database that occurred since the beginning of the copy process will be replayed to the new database. The new database will be brought online in place of the old database.

Tip To minimize the amount of time the database is offline during the final replay process, reduce the number of writes to the database. Consider performing the conversion at a time of lower write frequency.

Because this process creates a new Hyperscale database and copies the existing schema and data into it, we can convert from any Azure SQL Database tier and configuration. However, we cannot perform an in-place conversion from Azure SQL MI or SQL Server on an Azure VM to Hyperscale in this way because of the significant platform differences between the two.

Important Before you can convert an existing Azure SQL Database instance to the Hyperscale tier, you must first stop any geo-replication to a secondary replica. If the database is in a failover group, it must be removed from the group.

Copying an Existing Azure SQL Database

Before performing an in-place conversion to Hyperscale, it is good practice to first perform a test conversion on a copy of the existing Azure SQL Database. This helps us by doing the following:

- Testing our existing workloads against the new Hyperscale architecture
- Determining how long the process is going to take
- Verifying that the conversion process has been completed without any issues

Let's make a copy of an existing Azure SQL Database in either the same logical server or a different one. You can perform this task using the Azure Portal, Azure PowerShell, or the Azure CLI. To copy an Azure SQL Database in the Azure Portal, follow these steps:

1. Select the Overview blade in the existing Azure SQL Database resource.
2. Click Copy.
3. Set the database name to the name of the new database copy.
4. Specify the server to create the new database in.
5. Click "Review + create" and then click Create.

The database will be copied and attached to the chosen logical server. We can then perform an in-place conversion to Hyperscale of the copy and verify the workload performance.

Changing an Existing Azure SQL Database to Hyperscale

To perform an in-place conversion of an existing Azure SQL Database to the Hyperscale tier in the Azure Portal, follow these steps:

1. Select the "Compute + storage" blade in the existing Azure SQL Database resource.
2. Change the service tier to Hyperscale.

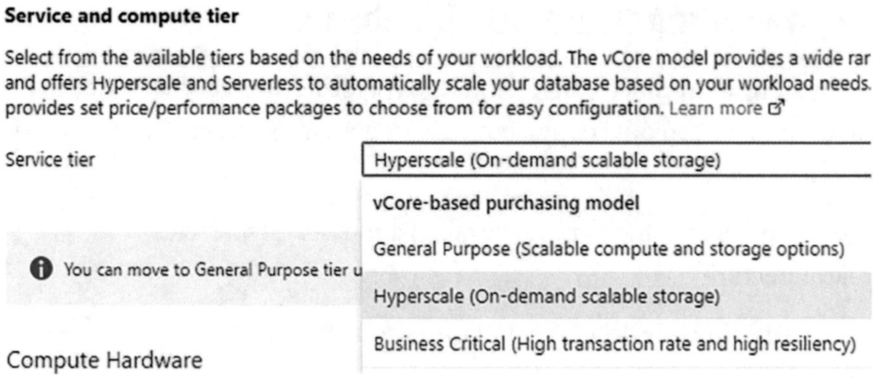

Service and compute tier

Select from the available tiers based on the needs of your workload. The vCore model provides a wide rar
and offers Hyperscale and Serverless to automatically scale your database based on your workload needs.
provides set price/performance packages to choose from for easy configuration. Learn more

Service tier | Hyperscale (On-demand scalable storage)

vCore-based purchasing model

General Purpose (Scalable compute and storage options)

You can move to General Purpose tier u

Hyperscale (On-demand scalable storage)

Business Critical (High transaction rate and high resiliency)

Compute Hardware

3. Configure the number of vCores and high-availability secondary replicas.

Note You cannot currently convert to Hyperscale with availability zones. If you require this, then you will need to use the data sync and cutover method.

---

4. Specify the backup storage redundancy.
5. Click Apply.

Your database will be migrated to Hyperscale. This may take several hours to complete if you have a lot of data in the existing database. It will also go offline at the end for a few minutes while any changed data is replicated.

You can also convert an Azure SQL Database to Hyperscale using Azure PowerShell, Azure CLI, or Transact-SQL, or by updating your declarative infrastructure as code files (Azure Bicep, Terraform).

Tip It is possible to perform a reverse migration back from Hyperscale after you've performed an in-place conversion within 45 days. This process is covered in the next chapter.

Migrating to Hyperscale with Data Migration Assistant

The Data Migration Assistant has been around for several years. It is a straightforward Windows desktop tool that can be used to do the following:

- *Assessment*: It can perform an assessment of a migration source database to identify any unsupported features for the target database system, in our case Azure SQL Database (Hyperscale tier).
- *Migration of schema and data*: It can perform an offline migration, copying both the schema and data from the source Microsoft SQL Server to the destination Azure SQL Database.
- *Migration of schema only*: It can perform the same migration as the previous one, but migrates only the schema.
- *Migration of data only*: It can perform the same migration as the previous one, but only migrates the data.

To perform a migration from a SQL Server instance to an Azure SQL Hyperscale database, the Data Migration Assistant must be installed.

Tip The Data Migration Assistant can be downloaded from `https://www.microsoft.com/download/details.aspx?id=53595`.

You should install the DMS onto a machine that meets the following requirements:

- Has TCP 1433 access to both the source SQL Server and the Hyperscale database
- Is as close to either the source or target database as possible to reduce latency
- Has enough compute and memory assigned to ensure the process performs quickly

If your source SQL Server is on-premises, you'll typically need to create a networking topology that will support your organizing needs, and many organizations will require that the data is not migrated over the public Internet. There are many different network topologies that could be used to enable this connectivity. We won't cover them here as they are beyond the scope of this book.

Migrating with DMS

This process assumes you have both a SQL Server to migrate from and a Hyperscale database in Azure and that both are available and contactable from this machine.

Important Because this is an offline migration process, we should ensure that any workloads using this database have been stopped. Any changes made to this database after beginning the process will not be migrated and will be lost.

The first step of migrating a SQL database using DMS is to perform an assessment to identify any incompatibilities.

1. Start the Data Migration Assistant.
2. Click the + New button to create a new project.

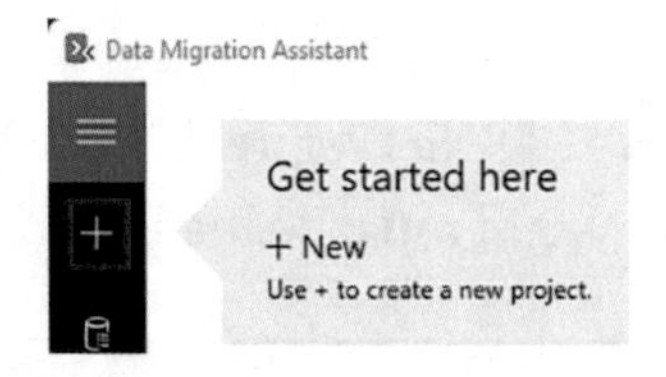

3. Set the project type to Assessment and enter a project name.
4. Set the assessment type to Database Engine.
5. The source server type should be SQL Server, and the target server type should be Azure SQL Database.

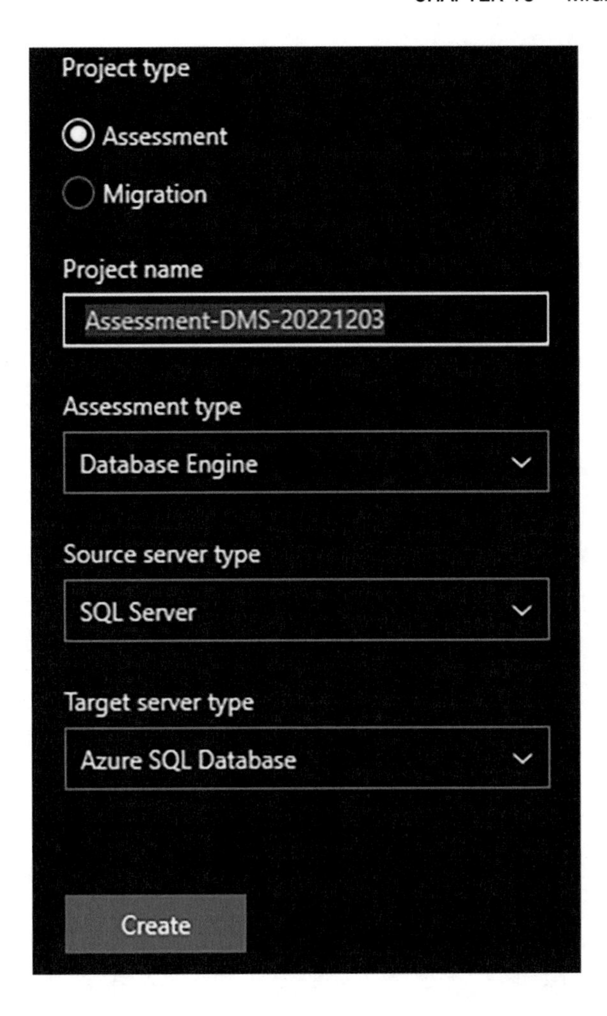

6. Click Create.

7. Select both “Check database compatibility” and “Check feature parity” and click Next.

8. Configure the source SQL Server connection details and click Connect.

9. After a few seconds, if your connection was successful, you will be presented with a list of databases on that server. Select the databases you want to assess and click Add.

10. The list of servers and databases being included in this assessment will be displayed. Click Start Assessment.

11. The results of the assessment will be displayed after a few seconds.

Figure 18-1 shows part of the assessment results screen showing the unsupported features that might impact the workload if migrated to Hyperscale. In our demonstration, these features are not needed, so we can safely ignore this.

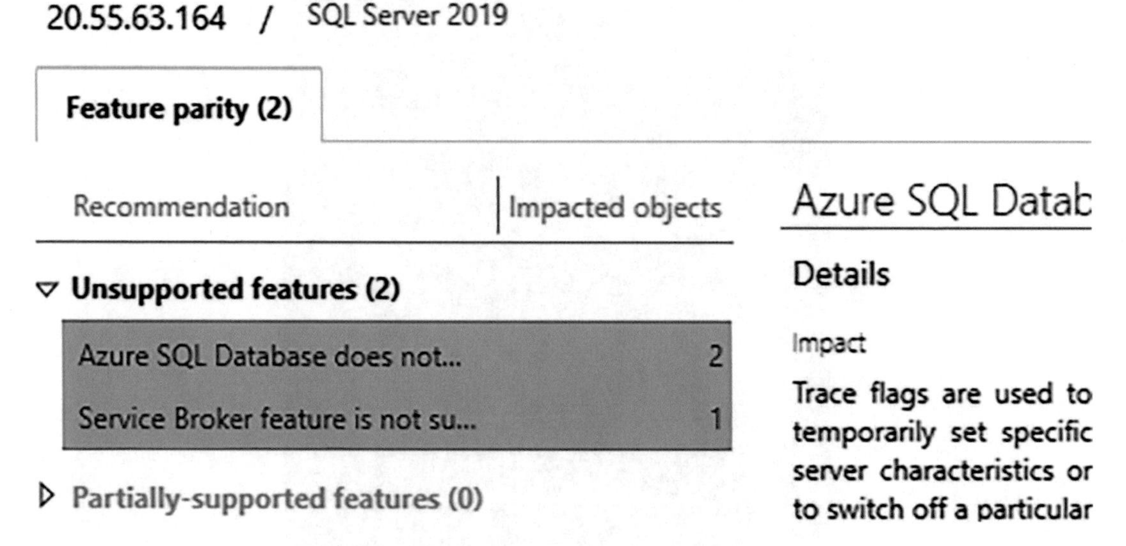

Figure 18-1. *Partial output from an assessment of SQL Server to an Azure SQL Database*

Once we've performed an assessment and determined that our source database and Hyperscale are compatible, then we can perform the migration.

1. In the Data Migration Assistant, click the + New button to create a new project.
2. Set the project type to Migration and specify a name for the project.
3. The source server type should be SQL Server, and the target server type should be Azure SQL Database.
4. The migration scope should be "Schema and data."

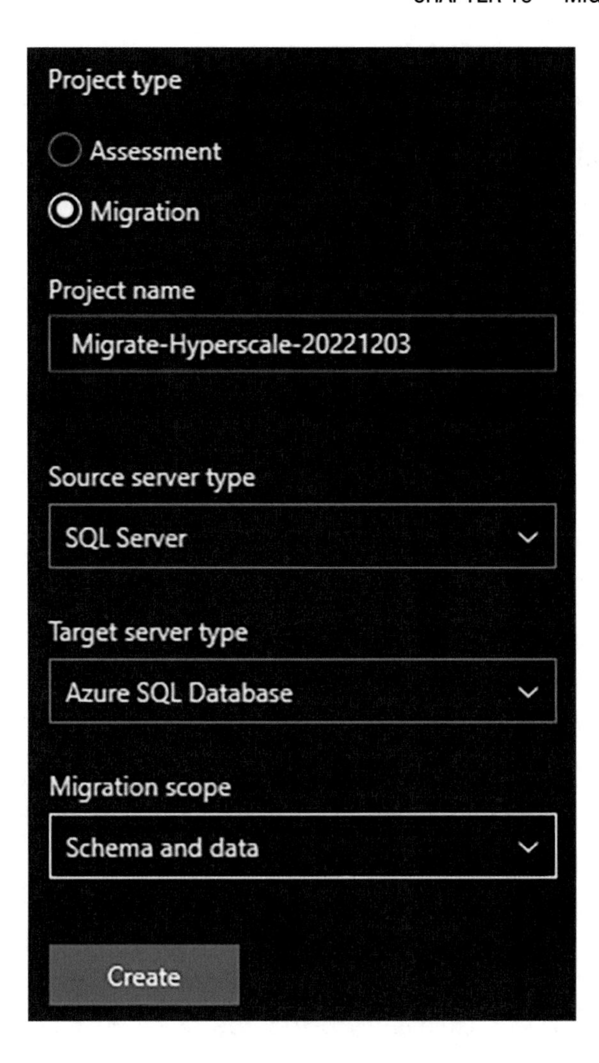

5. Click Create.
6. Configure the SQL Server connection settings in the “Connect to source server” page and click Connect.
7. After a few seconds, the list of databases that can be migrated will be displayed. Select the source database you’d like to migrate from and click Next.
8. Configure the “SQL Server connection settings” in the “Connect to target server” page and click Connect.

9. After a few seconds, the list of target databases will be displayed. Select the target database you'd like to migrate to and click Next.

10. The list of objects that you can migrate will be displayed. You can select each to review any issues with the migration or deselect it to omit it from the migration.

Select the schema objects from your source database that you would like to migrate to Azure SQL Database.

Tables
dbo.customer
dbo.district
dbo.history
dbo.item
dbo.new_order
dbo.order_line

Table: dbo.history
This object is ready to migrate.
No issues found for this object during assessment

11. Click "Generate SQL script" to create the script to create the schema in the target.

12. The schema script will be displayed. You can edit this before deploying and step through any identified issues. Click "Deploy schema" when you're ready.

13. Once the schema has deployed, click the "Migrate data" button.

14. You will see a list of tables available for migration along with the number of rows for each and whether they are ready to migrate. Select the tables you want to migrate.

Selected tables (9/9)

Table name	Row count	Ready to move
[dbo].[customer]	59,480	OK
[dbo].[district]	100	OK
[dbo].[history]	59,494	OK

15. Click "Start data migration."

16. The data will be migrated, and the migration summary will be displayed.

17. You can now close the DMS application.

Figure 18-2 shows an example of the migration summary screen from DMS.

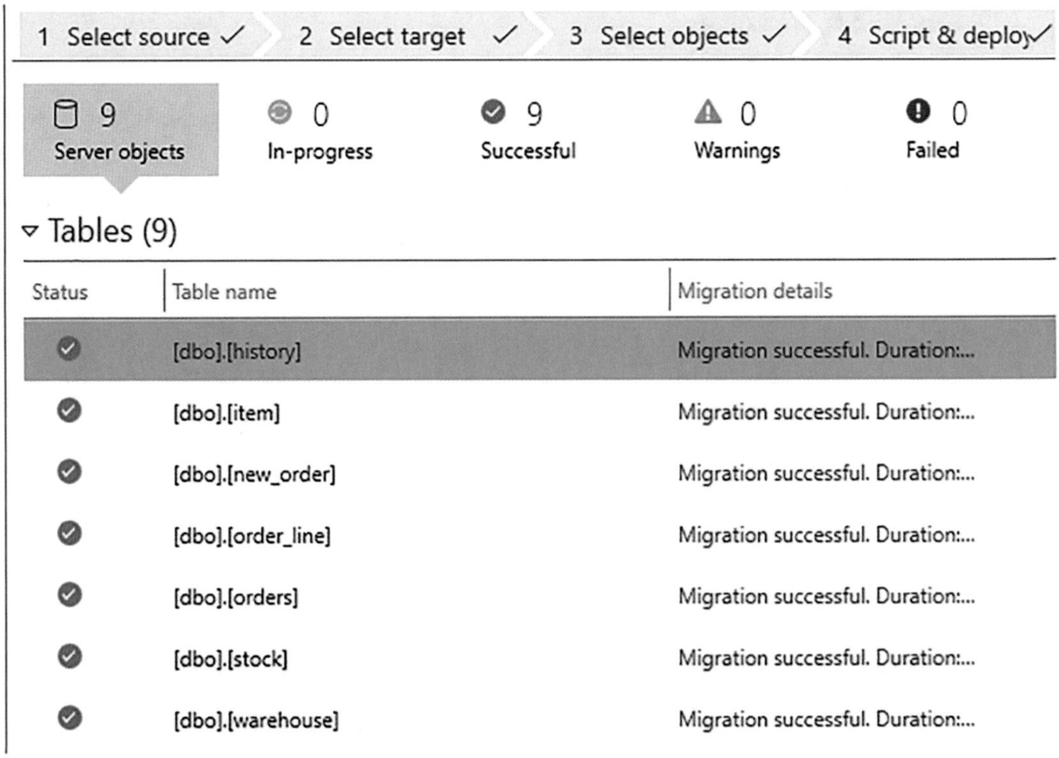

***Figure 18-2.** An example of the DMS migration summary*

The time taken to perform the migration will depend on the volume of data to be migrated as well as a few other factors.

Accelerating a DMS Migration

As this is an offline process, it is usually important to try to reduce the amount of time that the workload is offline. Therefore, it will usually be worth taking temporary steps to accelerate the process.

- Temporarily scale up your source and target database resources, including CPU and storage I/Ops, to reduce the time to copy the data. Once the migration has been completed, you can scale it back down.
- Ensure you've got as much network bandwidth between the source database, the target database, and the Data Migration Assistant machine as possible.

Migrating to Hyperscale with the Azure Database Migration Service

The Azure Database Migration Service is a resource in Azure that provides managed compute to perform offline and online migrations of data between various database management systems. In this case, we'll be talking only about the migration of Microsoft SQL Server to Hyperscale, but the DMS can perform many other data migration types, including AWS RDS for SQL Server to Hyperscale.

There are several benefits to using the Azure DMS service for performing migrations that make it an attractive option when compared with alternatives.

- *Automation*: It is easy to automate, so is excellent for performing migrations at scale.
- *Scale*: It can be used by multiple users or migration tasks at the same time.
- *Monitoring*: It provides centralized monitoring for migration processes.
- *Security*: The migration processes are run from a secure managed environment. When combined with other techniques, such as automation and storing credentials in Azure Key Vault, it can reduce the security risk even further.

When you perform a data migration using the Azure Data Migration Service, it assumes the target database schema already exists. There are multiple ways you might have achieved this. One way is to use the Microsoft Data Migration Assistant schema migration, as documented earlier in this chapter.

Important Online migration is not currently supported when migrating to an Azure SQL Database. When the target database is Azure SQL Database, then only an offline migration is supported. Online can be used only when the target database is an Azure SQL Managed Instance or SQL Server in an Azure VNet.

Unlike other migrations where you are leveraging compute that you deploy and manage for the task, the DMS provisions and manages this for you, providing the supporting infrastructure required to perform the migration. When you deploy the DMS into Azure, it is deployed into an Azure VNet. It will be your responsibility to ensure that the DMS service is able to connect to the source and target databases.

There are two DMS tiers to choose from when you're deploying.

- *Standard*: Use this for large data sizes and with one, two, or four vCores allocated. It supports offline migrations only.
- *Premium*: This provides offline and online migrations with minimal downtime using four vCores. Online is not supported when the target database is an Azure SQL Database (including Hyperscale).

At the time of writing, both the Standard and Premium tiers of DMS are free for the first six months of use.

Database Migration Service Network Connectivity

A DMS resource must be connected to a virtual network. A DMS resource is effectively a virtual appliance that uses an Azure virtual NIC to connect to an Azure virtual network. The DMS resource then connects to both the source database and the target database and copies the data.

For the DMS to operate correctly, you should ensure the following:

- It can connect to the source database on TCP port 1433. If your source database is running on a different port to listen on, you should adjust this accordingly.
- It can connect to the target database on TCP port 1433.
- The network security group attached to the virtual network is configured to allow outbound traffic on TCP port 443 to ServiceTag for ServiceBus, Storage, and Azure Monitor.

- If you are using any firewalls between DMS and your databases, you will need to ensure that TCP port 1433 can traverse them.
- If your source database contains multiple SQL Server instances using dynamic ports, then you may want to enable the SQL Browser Service and allow UDP port 1434.

With these requirements in mind, there are many different architectures and networking technologies that could be used to implement this. Figure 18-3 shows an example.

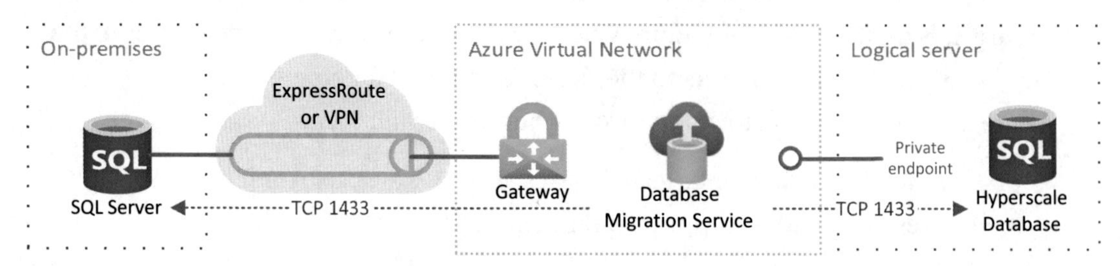

***Figure 18-3.** A typical DMS networking configuration*

The specifics of setting up this networking connectivity will depend on your environment and are beyond the scope of this book.

Registering the DMS Resource Provider

The first task you need to take care of when you choose to use DMS to perform a migration is to register the `Microsoft.DataMigration` resource in the Azure subscription where you're creating the Azure Data Migration Service resource. To do this in the Azure Portal, follow these steps:

1. Select Subscriptions from the portal menu.

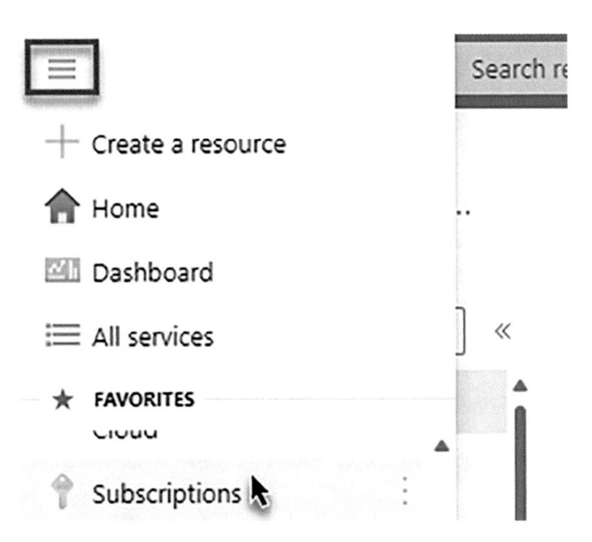

2. Select the subscription you want to install the Data Migration Service resource into.
3. Select the "Resource providers" blade (under the Settings section) in your subscription.
4. Search for *Migration*, select the `Microsoft.DataMigration` resource provider, and click Register if it is not already registered.

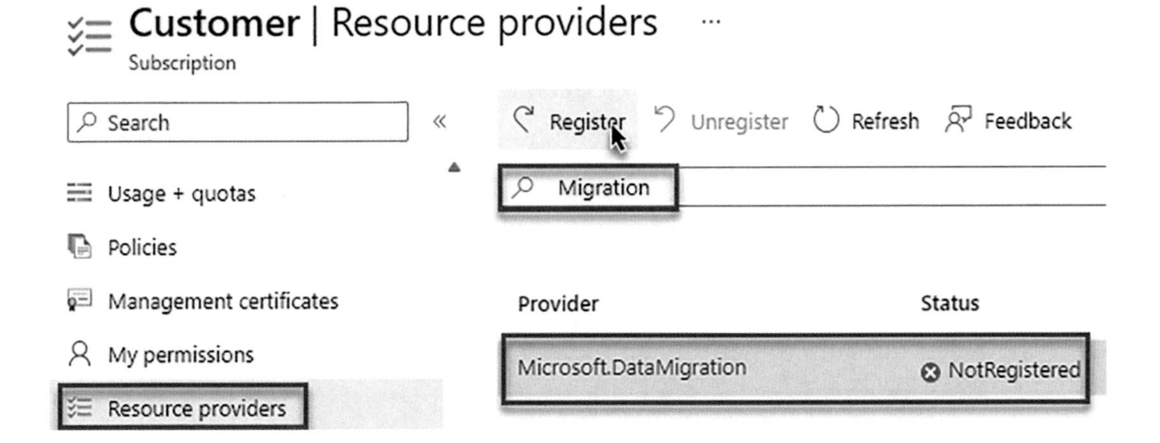

It will take a minute or two to register with the resource provider.

Creating a DMS Resource

The next task is to create the Azure Data Migration Service resource. The resource should be deployed into the same region as the target database.

To deploy the DMS resource in the Azure Portal, follow these steps:

1. Once the resource provider has been registered, enter **Azure Database Migration Service** into the search box at the top of the Azure Portal.

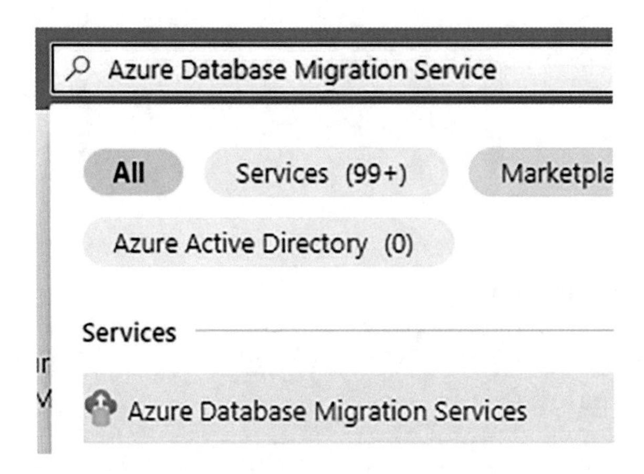

2. Select the Azure Database Migration Services entry.
3. Click + Create.
4. Set the source server type to SQL Server and the target server type to Azure SQL Database.
5. The Database Migration Service will automatically be set to Database Migration Service.
6. Click Select.
7. Select the subscription and resource group you want to create the DMS resource in.
8. Enter a name for the DMS resource into the "Migration service name" field.
9. Set the location to the same region your target database is in.
10. Select "Configure tier" to configure the pricing tier.

11. Select Standard and select the number of vCores allocated to the DMS service. If you were intending on using this DMS resource to also perform online migrations (where supported), then you could select Premium. The more vCores assigned to the DMS service, the greater the performance of any migration projects running on it.
12. Click Apply.
13. Click Next.
14. Select an existing virtual network subnet that is available in the same region as your DMS resource or create a new one by entering a virtual network name.

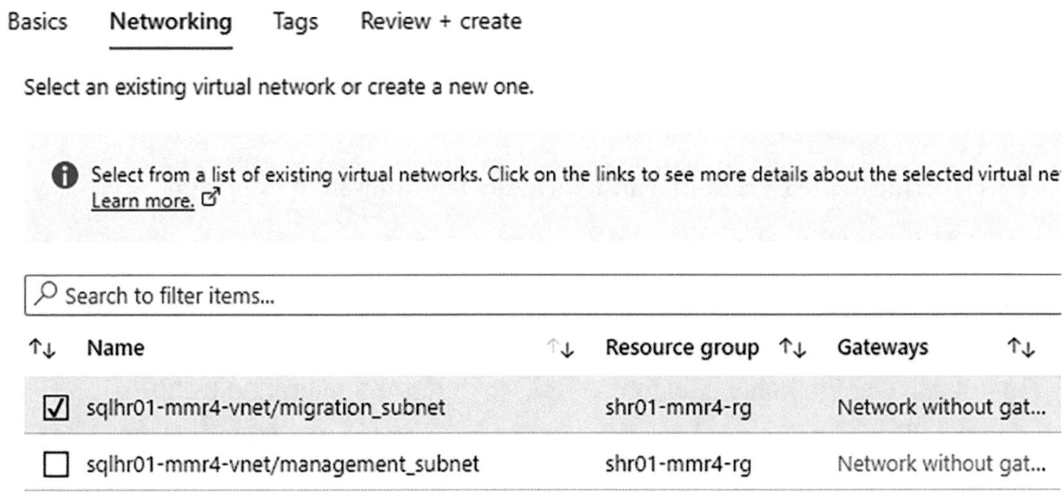

Basics Networking Tags Review + create

Select an existing virtual network or create a new one.

Select from a list of existing virtual networks. Click on the links to see more details about the selected virtual ne
Learn more.

Search to filter items...

Name	Resource group	Gateways
☑ sqlhr01-mmr4-vnet/migration_subnet	shr01-mmr4-rg	Network without gat...
☐ sqlhr01-mmr4-vnet/management_subnet	shr01-mmr4-rg	Network without gat...

15. Click "Review + create."
16. Click Create.

It will take a minute or two to deploy the Azure Data Migration Service as well as the virtual network adapter that is required to connect it to the virtual network.

Performing a Migration with DMS

Now that we have a DMS resource deployed, we can configure a migration project to perform the offline or online migration if we deployed a DMS premium tier instance. A single DMS resource can have multiple migration projects defined.

Important Before using DMS to migrate your database, you should always use the Microsoft Data Migration Assistant to perform an assessment on the source database for any issues that might cause migration problems or prevent your workload from operating correctly in an Azure SQL Database. Using Data Migration Assistant is documented earlier in this chapter.

There are several ways to create and run the DMS migration project.

- Using the Azure Portal
- Using Azure PowerShell
- Using Azure Data Studio and the Azure SQL Migration extension

In this chapter, we will focus on the Azure PowerShell method of using DMS. We will briefly touch on using the Azure Data Studio with Azure SQL Migration.

Tip If you want to know how to use DMS to perform a migration using PowerShell, see `https://learn.microsoft.com/azure/dms/howto-sql-server-to-azure-sql-powershell`.

To create a new migration project in DMS using the Azure Portal, then follow these steps:

1. Open the Azure Database Migration Service resource. Figure 18-4 shows an example DMS resource in the Azure Portal.

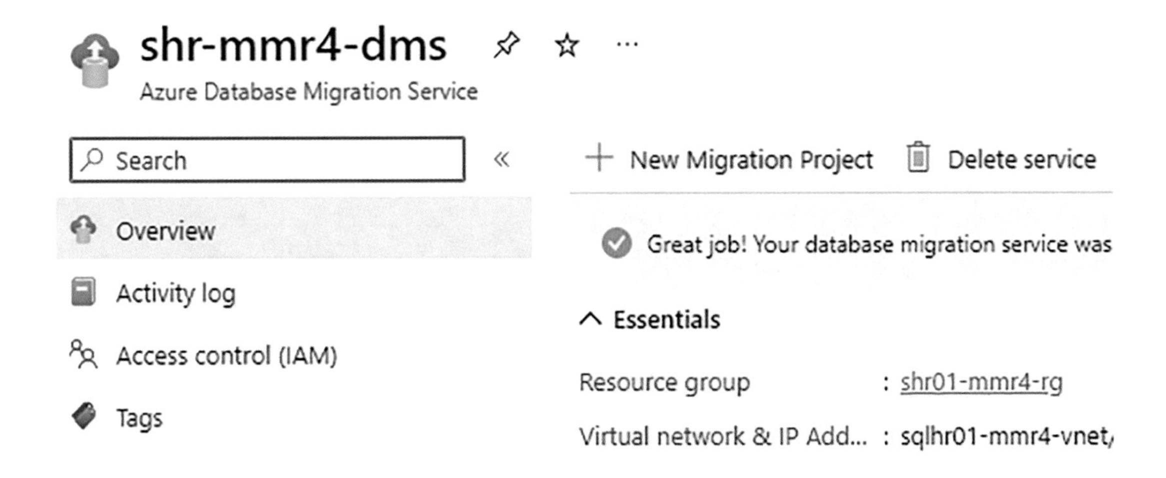

Figure 18-4. *An Azure Database Migration Service resource*

2. Select New Migration Project.
3. Set the project name to a unique name for the migration project.
4. Set the source server type to SQL Server and the target server type to Azure SQL Database.
5. Set the migration activity type to Data Migration.
6. Click "Create and run activity." This will create the migration project and open the SQL Server to Azure SQL Database Migration Wizard, as shown in Figure 18-5.

SQL Server to Azure SQL Database Migration Wizard

Select source | Select databases | Select target | Map to target databases | Configure

Source SQL Server instance name * : Servername.domainname.com
Authentication type : Windows Authentication
User Name * : Enter domain\user name
Password : Enter password

Figure 18-5. *The SQL Server to Azure SQL Database Migration Wizard*

7. Configure the settings to connect to the source SQL Server.

Select source | Select databases | Select target | Map to target c

Source SQL Server instance name * : 20.55.63.164

Authentication type : SQL Authentication

User Name * : shradmin

Password : ••••••••••••••

Connection properties

- [x] Encrypt connection
- [x] Trust server certificate

8. Click Next to select the databases to migrate from the source SQL Server. DMS will estimate the expected downtime to migrate each database based on several factors including target database SKU, source server performance, and network bandwidth.

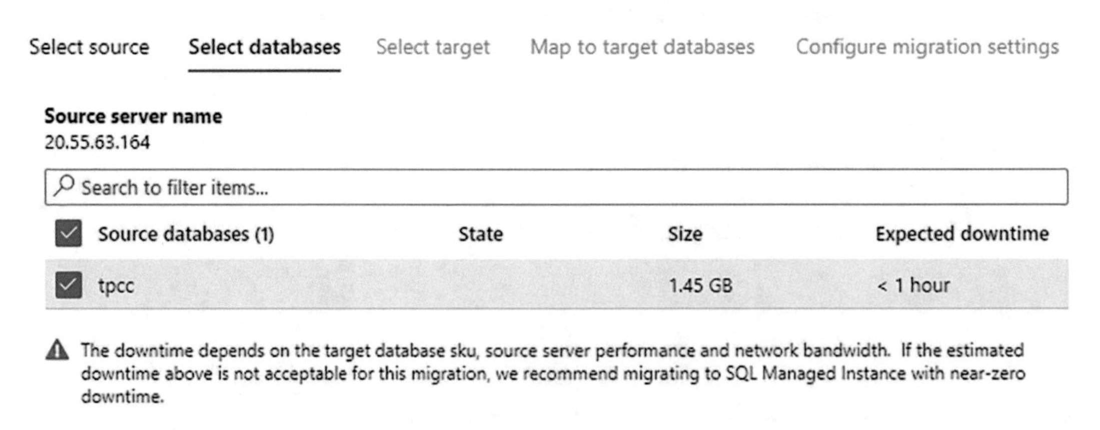

9. Click Next.

10. Configure the settings to connect to the target SQL logical server.

11. Click Next to map the source database to the target database.

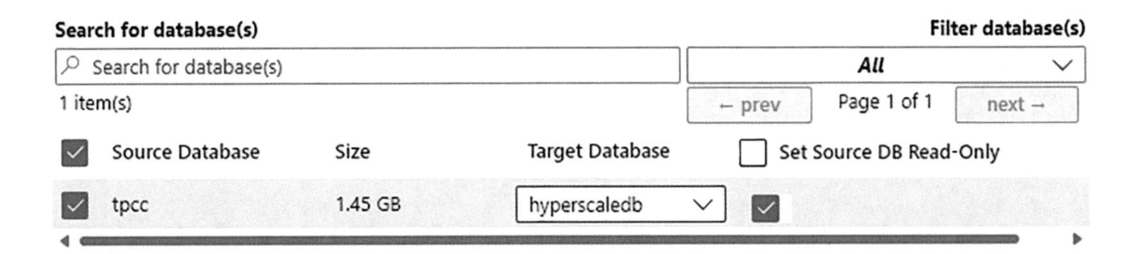

12. Select the target database from the drop-down for each source database. Selecting the check box next to each database mapping will cause the source database to be set to read-only once the migration starts. This is recommended to ensure no data is changed after the migration begins.

13. Click Next to configure the tables to copy from the source to the target. These tables must already exist in the target; otherwise, they will not be able to be selected for migration.

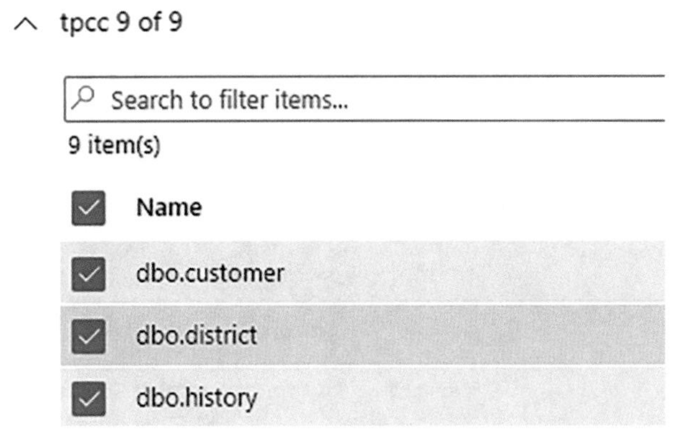

14. Click Next.

15. Set the activity name for this migration activity.

16. Click "Start migration." If you checked the box to make the source database's read-only once migration begins, this will occur now.

17. The migration will begin, and the status of the migration activity will be displayed. Figure 18-6 shows an example of the migration activity status.

Source server
20.55.63.164

Target server
shr01-mmr4.database.windows.net

Source version
SQL Server 2019
15.0.4261.1

Target version
Azure SQL Database
12.0.2000.8

Databases
1

Search to filter items...

Name	Status	size	Migration details	Duration
tpcc	Running	865.06 MB	6 of 9 table(s) completed.	00:00:15

Figure 18-6. *The migration activity status window*

You can review the migration details of each table by clicking the database record in the migration activity status page.

Tables Selected
9

Search to filter items...

Name	Status	Migration details	Duration
dbo.customer	Completed	All rows completed. (~300,000)	00:00:02
dbo.district	Completed	All rows completed. (~100)	00:00:00
dbo.history	Completed	All rows completed. (~396,164)	00:00:00
dbo.item	Completed	All rows completed. (~100,000)	00:00:00

18. Once the migration has been completed, you can then delete any unneeded resources, such as the migration project, migration activity, and the DMS resource itself.

Now that the database migration has been completed, you can reconfigure any workloads to connect to the new Hyperscale database.

Migrating with ADS with the Azure SQL Migration Extension

Azure Data Studio is a recent GUI-based database management solution from Microsoft. It provides somewhat similar functionality to SQL Server Management Studio. It has a broad ecosystem of extensions that provide additional functionality to the product.

ADS can be used to perform an offline migration of data from a SQL Server (wherever it is hosted) into a Hyperscale database. It does this by leveraging both an ADS extension called Azure SQL Migration and an Azure Data Migration Service resource. This is currently in preview at the time of writing. Support for online migration is also planned.

To leverage this method of migration, you must have the following:

- A machine with Azure Data Studio (ADS) installed
- The Azure SQL Migration extension installed into ADS
- An Azure Data Migration Services (DMS) resource deployed, as was demonstrated in the previous section
- A source Microsoft SQL Server and a target Hyperscale database that has network connectivity from the DMS

Tip To learn more about Azure Data Studio, please visit `http://aka.ms/azuredatastudio`.

Using this method provides an alternate interface for creating and monitoring migrations using DMS. Because this feature is currently in preview and the process is like using the DMS service natively, it isn't covered in detail in this book.

Note If you want to learn to use Azure Data Studio with the Azure SQL Migration extension, please see `https://learn.microsoft.com/azure/dms/tutorial-sql-server-azure-sql-database-offline-ads`.

Migrating to Hyperscale Using Import Database

A fairly straightforward method for migrating a database from a SQL Server instance to a Hyperscale database is to simply import a BACPAC file. The BACPAC file needs to be stored in a standard Azure Storage account as a blob that can be accessed by the Azure SQL Database.

To use this migration method, you should have performed the following tasks:

- Created a standard Azure Storage account that contains the BACPAC file. The storage account should be accessible by the logical server that we will import the BACPAC into as a new database.

Note Premium Azure Storage accounts are not supported by the import process.

- Created a BACPAC from your source SQL Server and uploaded it to the Azure Storage account as a blob.

There are several methods and tools you can use to import the BACPAC into the Hyperscale database, including the following:

- The Azure Portal
- Azure PowerShell
- Azure CLI
- The SQLPackage tool
- Azure Data Studio with the SQL Server dacpac extension

For brevity, we'll show the process of using the Azure Portal only.

1. Select the Overview blade in a SQL Server (logical server) resource.
2. Click the "Import database" button.
3. Click the "Select backup" button to configure the location of the BACPAC file.
4. Click the Pricing tier and configure the Hyperscale database.

5. Set the database name, collation, and authentication information for the new database.

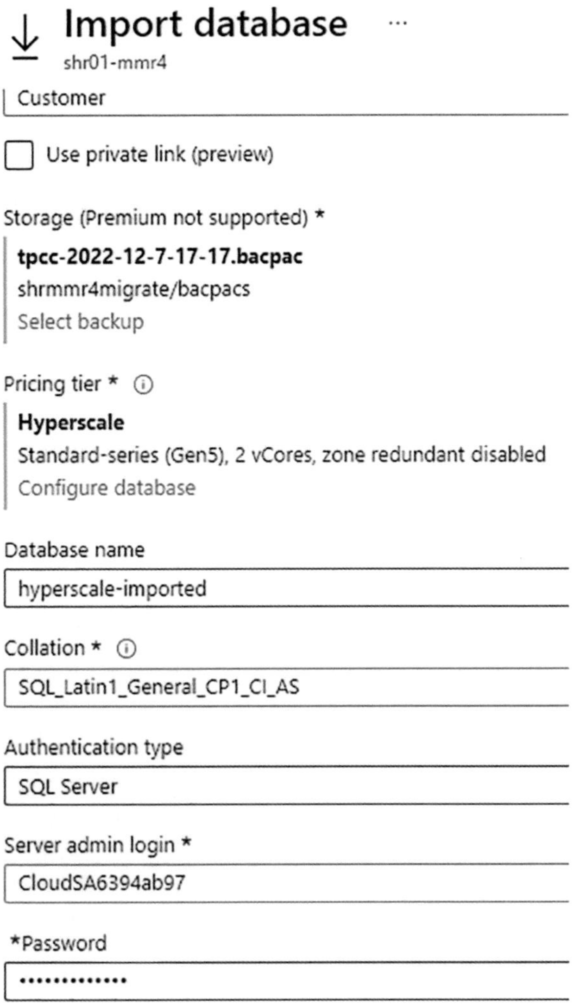

Import database ...
shr01-mmr4
Customer
Use private link (preview)
Storage (Premium not supported) *
tpcc-2022-12-7-17-17.bacpac
shrmmr4migrate/bacpacs
Select backup
Pricing tier *
Hyperscale
Standard-series (Gen5), 2 vCores, zone redundant disabled
Configure database
Database name
hyperscale-imported
Collation *
SQL_Latin1_General_CP1_CI_AS
Authentication type
SQL Server
Server admin login *
CloudSA6394ab97
*Password
•••••••••••••

6. Click OK.

The database import process is performed in a transparent background process and may take some time to complete depending on the size of the BACPAC. Once it has completed, the database will be available in the logical server and can be used like any other Hyperscale database. You can also delete the BACPAC file once the import has been completed.

Migrating to Hyperscale Using a Data Sync and Cutover

The final methodology for migration that we'll discuss in this chapter is the data sync and cutover method. This approach leverages database replication to use Azure SQL Database as a push subscriber with SQL Server as a publisher and distributor.

The process to use this methodology to migrate a SQL Server database to the Azure SQL database is as follows:

1. Enable transactional replication to replicate data from the *source* SQL Server database to the *target* Hyperscale database.
2. Once the target database is in sync with the source database, stop any workloads from using the source database.
3. Wait for the replication to complete the synchronization of the source and target databases.
4. Reconfigure any workloads to use the Hyperscale database.
5. Stop replication and remove the publication.

The benefit of this approach is that the amount of time that the database is offline is minimized, so this could be considered an online migration. However, there are several considerations to be aware of when planning this approach.

- Replication must be configured using Transact-SQL on the publisher or SQL Server Management Studio. Replication cannot be configured using the Azure Portal.
- Azure AD SQL authentication is not supported for the replication. Only SQL Server authentication logins can be used to connect to the Azure SQL Database.
- Replicated tables must have a primary key.
- You must manage and monitor the replication from SQL Server rather than Azure SQL Database.
- Only `@subscriber_type = 0` is supported in `sp_addsubscription` for SQL Database.
- Only Standard Transactional and Snapshot replication is supported by Azure SQL Database. Bidirectional, immediate, updatable, and peer-to-peer replication are not supported by Azure SQL Database.

If you're planning on using this approach for migration to Hyperscale, we recommend examining in detail the various technologies that are required to set this up. Documenting these tools and processes here would take dozens of pages and is beyond the scope of this book.

Tip Because of the many different techniques required to implement replication and the specifics on the limitations, it is recommended that you review this page for more guidance on this process: `https://learn.microsoft.com/azure/azure-sql/database/replication-to-sql-database`.

Summary

In this chapter, we covered some of the many different techniques available to migrate existing SQL databases into the Azure SQL Database Hyperscale tier. These techniques also apply to migrating to any tier in Azure SQL Database.

We also took a brief look at some of the tools available to analyze and migrate your SQL Database instances, including the following:

- *Azure Data Studio (ADS)*: A GUI-based tool for managing SQL databases
- *Microsoft Data Migration Assistant (DMA)*: A GUI-based tool for interactively assessing and migrating schema and data to Azure
- *Azure Database Migration Service (DMS)*: A managed Azure service for performing data migrations in a secure, scalable, and automatable fashion

In the next chapter, we will take a look at the options for reverse migrating away from Hyperscale should you find it necessary.

CHAPTER 19

Reverse Migrating Away from Hyperscale

In the previous chapter, we talked about common migration methods from different database deployments to the Azure SQL Hyperscale service tier. We covered several migration scenarios, various methods, and available migration tools.

In this chapter, we will cover available reverse migration options from the Hyperscale service tier. In the beginning, because of having significantly different underlying architecture, reverse migration from the Hyperscale service tier back to any other Azure SQL PaaS deployment options was not available out of the box. There were no easy ways to go back to any other service tier. Nowadays you do have the option, if you decide that Azure SQL Hyperscale is not the best choice for your workload, to migrate your database back to the General Purpose service tier. Nevertheless, there are some conditions that will need to be fulfilled first.

- Once you migrate your existing database to Hyperscale, you will have a 45-day window to reverse migrate back to the General Purpose service tier.
- You can reverse migrate only to the General Purpose service tier, regardless of your original service tier, before migrating to Hyperscale.
- Only databases that were previously migrated to Hyperscale from another Azure SQL Database service tier can be reverse migrated. If your database was originally created in the Hyperscale service tier, you will not be able to reverse migrate.

Z. Barać and D. Scott-Raynsford, *Azure SQL Hyperscale Revealed*,
https://doi.org/10.1007/978-1-4842-9225-9_19

Tip Bear in mind that migration to Hyperscale and reverse migration back to General Purpose will depend mostly on database size.

Once you have database reverse migrated back to the General Purpose service tier, you can then convert it to an alternative service tier, such as Business Critical (or even back to Hyperscale).

Reverse Migration Methods

There are multiple ways to trigger the reverse migration process; we will cover a few more common ways.

- Azure Portal
- Azure CLI
- Transact SQL (T-SQL)

Azure PowerShell is also supported for performing a reverse migration, but it is omitted for brevity.

Reverse Migration Using the Azure Portal

The most common way to do a reverse migration is via the Azure Portal.

1. Go to the database you want to migrate back to the General Purpose service tier and choose Settings and then the Compute + Storage pane. Figure 19-1 shows the Azure Portal's "Compute + storage" settings and current service tier.

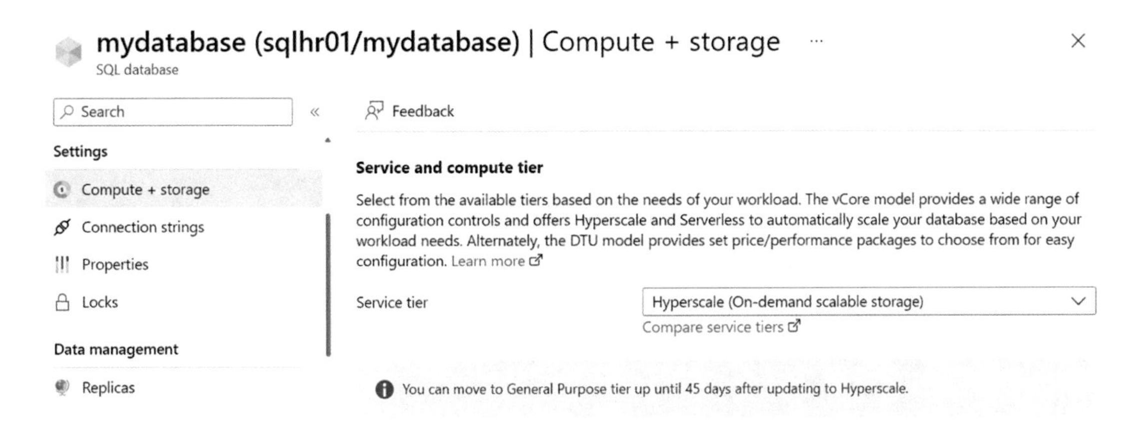

Figure 19-1. *Azure Portal "Service and compute tier" pane of Azure SQL Database*

2. Under the "Service and compute tier" pane, choose the General Purpose service tier from the drop-down menu.

   You will notice that only General Purpose is available for your reverse migration targeted service tier, as shown in Figure 19-2.

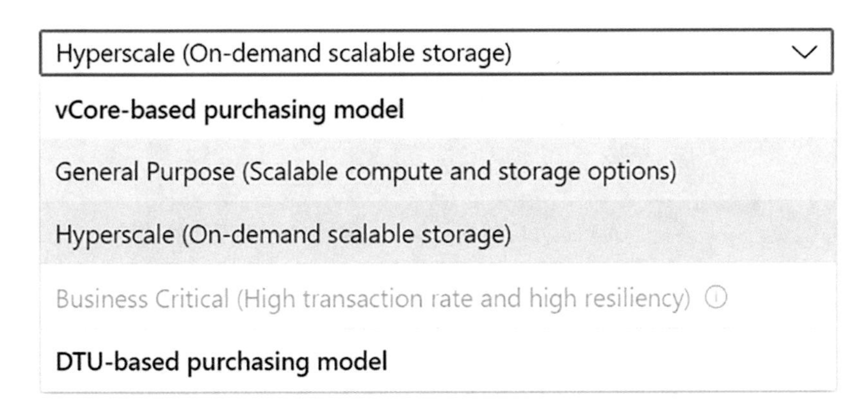

Figure 19-2. *Azure SQL Database service tiers selection*

3. Once you have selected the General Purpose service tier, you can confirm the reverse migration, and the process will begin. The time it takes may vary, depending mostly on database size. Figure 19-3 shows the scaling in progress pane after reverse migration was triggered.

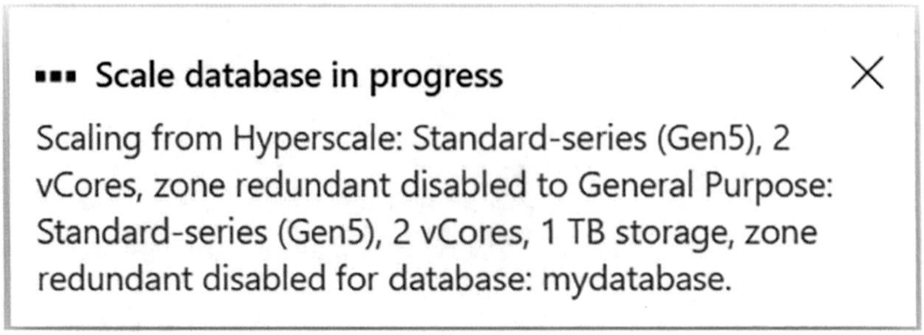

▪▪▪ Scale database in progress

Scaling from Hyperscale: Standard-series (Gen5), 2 vCores, zone redundant disabled to General Purpose: Standard-series (Gen5), 2 vCores, 1 TB storage, zone redundant disabled for database: mydatabase.

Figure 19-3. *The reverse migration in progress*

Monitor Reverse Migration Using the Azure Portal

There are different ways you can monitor the reverse migration progress using the Azure Portal. The simplest way is to perform the following steps:

1. Select the Overview blade on the Hyperscale database resource you are performing a reverse migration of.

2. Scroll down to the bottom of the Overview blade and click the Notifications tab. Figure 19-4 shows the database info message on the Notifications tab.

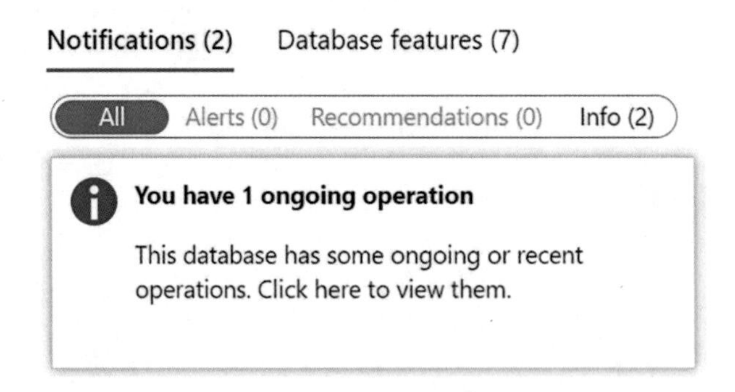

Figure 19-4. *Azure SQL database notification massage*

3. You can review the process of the reverse migration by clicking the operation. Figure 19-5 shows the progress against our Hyperscale deployment.

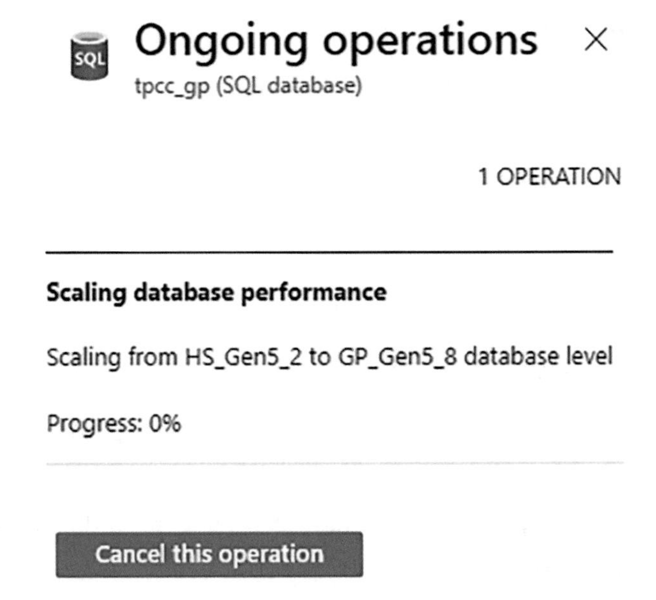

Figure 19-5. *Azure SQL database ongoing operations progress*

You can cancel this scaling operation if it is in progress by clicking the "Cancel this operation" button. You will be asked to click to confirm the cancelation. Figure 19-6 shows canceling the scaling database operation.

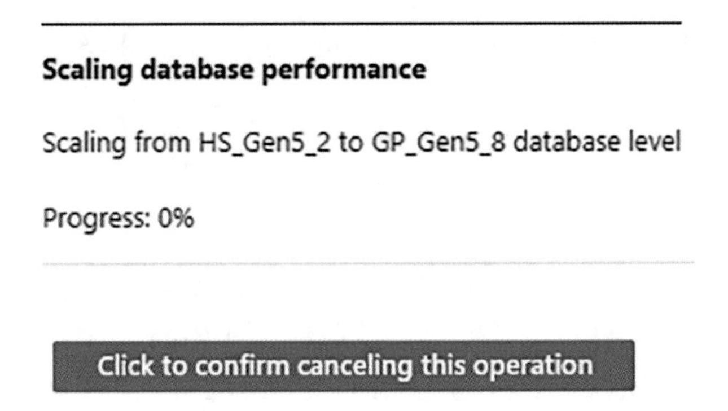

Figure 19-6. *Azure SQL Database canceling ongoing scaling process*

Reverse Migration Using Transact SQL

Reverse migration using T-SQL can be accomplished using different available tools.

- SQL Server Management Studio

- Azure Data Studio
- Azure Portal SQL Database Query Editor

Using any of these mentioned tools, you can run the following T-SQL statement to reverse migrate your Hyperscale database back to the General Purpose tier:

```
ALTER DATABASE [database name] MODIFY (SERVICE_OBJECTIVE = 'GP_Gen5_8',
MAXSIZE = 1024 GB)
```

Monitor Reverse Migration Using T-SQL

To monitor the migration progress using T-SQL, you can query data from the sys.dm_operation_status dynamic management view. This DMV can give you information about progress and the completion percentage as well. To monitor the reverse migration progress, run the following T-SQL statement:

```
SELECT * FROM sys.dm_operation_status
WHERE state_desc ='IN_PROGRESS'
```

Figure 19-7 shows the ongoing scaling database progress including the completion percentage.

session_activity_id	operation	state	state_desc	percent_complete
28872DCC-04CA-4126-B0D4-EE29651B8B53	ALTER DATABASE	1	IN_PROGRESS	0

Figure 19-7. *SQL Server Management Studio query result*

Tip For more information on the sys.dm_operation_status DMV, please see https://learn.microsoft.com/sql/relational-databases/system-dynamic-management-views/sys-dm-operation-status-azure-sql-database.

Reverse Migration Using Azure CLI

One of the most common ways to monitor any activities of your Azure deployments would be by using the Azure CLI from the Azure Portal.

Tip All the Azure CLI commands are easily available at the Microsoft Learn Portal: `https://learn.microsoft.com/cli/azure/reference-index`.

To trigger a reverse migration, you will need a few simple commands.

1. Check the current active subscription with this command:

```
az account show --output table
```

2. Get a list of available subscriptions by running the following:

```
az account list --output table
```

3. Switch to the subscription containing the database that is being reverse migrated.

```
az account set --subscription <subscriptionid>
```

4. Once you are on the right subscription, you can list all the available Azure SQL Database instances hosted on your logical SQL Server.

```
az sql db list --resource-group <resourcegroup> --server
<logicalsqlserver> --output table
```

5. After finding your desired Hyperscale database, you can run this command to perform the Reverse migration process:

```
az sql db update --resource-group <resourcegroup> --server
<logicalsqlserver> --name <databasename> --edition General
Purpose --service-objective GP_Gen5_8
```

Monitor Reverse Migration Using the Azure CLI

Using the Azure CLI, you can monitor any ongoing operation against your Azure resources. This includes your Azure SQL Hyperscale reverse migration. To monitor the reverse migration progress using the Azure CLI, you need to use the `az sql db op list` command. List all operations with the InProgress status for a specific resource group, logical SQL server, and database with this command:

```
az sql db op list --resource-group <resourcegroup> --server
<logicalsqlserver> --database <databasename> --query
"[?state=='InProgress']" --output table
```

To cancel an in-progress operation, you will need the unique name of the operation you want to cancel. You can get the operation name by executing the previous `az sql db op list` Azure CLI command. Once you have the unique name of the operation, you can cancel it using the following statement:

```
az sql db op cancel --resource-group <resourcegroup> --server
<logicalsqlserver> --database <databasename> --name <unique_name_of_the_
operation>
```

Tip Be aware that the cancel operation is not instantaneous, and it might take some time. You can follow the progress using the `az sql db op list` command and looking for the query state `CancelInProgress`.

Common Pitfalls of the Reverse Migration Process

There are a couple of common issues you might run into when performing a reverse migration. In this section, we'll take you through identifying these and mitigating them.

Migrating to an Unsupported Service Tier

The most common mistake is trying to reverse migrate a Hyperscale database back to unsupported service tier, such as Business Critical. Even if you had previously migrated to Hyperscale from the Business Critical service tier, you must still migrate back to General Purpose.

If you try to run the following command against a Hyperscale database that had been previously migrated, you will get an error message:

```
az sql db update --resource-group rg-hyperscale --server hyperscale-
sql-server --name tpcc_gp --edition BusinessCritical --service-objective
BC_Gen5_8
```

Here is the error message:

Update to service objective 'SQLDB_BC_Gen5_8' is not supported for entity '<your database>.'

The solution is to migrate to the General Purpose service tier and then, once that is complete, migrate to the desired service tier.

Database Not Eligible for Reverse Migration

Another common issue is if you attempt to reverse migrate a database originally created as a Hyperscale service tier. Databases that were originally created as Hyperscale are not eligible for reverse migration to a different tier.

If you try to trigger a reverse migration in this situation, you will encounter the ineligibility error shown in Figure 19-8.

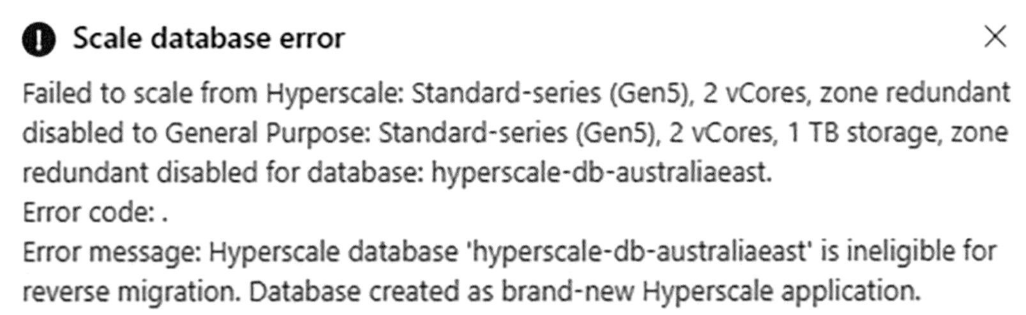

Scale database error

Failed to scale from Hyperscale: Standard-series (Gen5), 2 vCores, zone redundant disabled to General Purpose: Standard-series (Gen5), 2 vCores, 1 TB storage, zone redundant disabled for database: hyperscale-db-australiaeast.
Error code: .
Error message: Hyperscale database 'hyperscale-db-australiaeast' is ineligible for reverse migration. Database created as brand-new Hyperscale application.

Figure 19-8. *Failed to scale database error*

The solution in this case is to back up your Hyperscale database and restore it to a new database with the desired service tier. Alternatively, you could use a different migration methodology, some of which were documented in the previous chapter.

Summary

In this chapter, we showed the process of reverse migration from Hyperscale back to the General Purpose service tier in case you decide that the Hyperscale database is not the right option for your current production workload. We described the different techniques that might use for reverse migration, including the following:

- Azure Portal
- Azure CLI
- Transact SQL

We also showed some of the issues you might encounter during a reverse migration process.

CHAPTER 20

Conclusion

Azure SQL Hyperscale is an exciting evolution of relational database management system architecture. The increasing pace of innovation and change occurring in the public cloud has enabled new, more scalable, decoupled, and distributed design patterns. Access to on-demand computing and scalable object storage has led to a shift toward scale-out architectures. With these fundamental shifts in available technologies and increasing demands for scale, it is the right time for more traditional relational database technologies to leverage these technologies to make the next leap forward. Azure SQL Hyperscale is the most significant jump forward in how SQL Server workloads are implemented since the first implementation of Azure SQL PaaS.

Azure SQL Hyperscale is likely to be the first iteration of this exciting technology. As this technology evolves, supported by the innovations provided by the underlying Azure cloud services, such as storage snapshots, we will see many other enhancements. These innovations will help this approach to become more widely used within the Azure SQL Database family of products. Improvements in speed to scale-outs and scale-ins could lead to greater leverage of on-demand compute and make Hyperscale an attractive option in a greater range of scenarios.

To recap, these are some of the key differences that the shift to Hyperscale architecture enable:

- A significant increase in maximum storage, currently up to 100TB
- Powerful replication configuration options (high-availability replicas and named replicas) share the underlying storage infrastructure, resulting in negligible computing impact on replication
- Streamlined maintenance of extremely high throughput and performance at large capacities due to the decoupled storage and resilient caching architecture

Z. Barać and D. Scott-Raynsford, *Azure SQL Hyperscale Revealed*,
https://doi.org/10.1007/978-1-4842-9225-9_20

- The ability to quickly respond to changing workload requirements by providing rapid horizontal and vertical scaling
- Fast backups and restores of the database without requiring compute on the server, leveraging Azure Blob storage snapshots

These features and many more make Hyperscale a great fit for both simple and complex data scenarios. Understanding why this architecture is so different will help you ensure you're making the most of what it can do. Maximizing the benefit Hyperscale can provide requires knowing its strengths and designing your workloads to make the best use of them. Designing your workloads to make use of scale-out will net you cost savings and increases in scalability and resilience.

The goal of this book has been to provide you with a deeper insight into Hyperscale as well as all the adjoining technologies. With this knowledge, you'll be able to design, deploy, and operate large-scale Hyperscale environments and continue to leverage the dramatic innovations that this enables.

Thank you for taking the journey to learning about this game-changing SQL Server architecture in Azure. We hope the knowledge in these pages helps you maximize your investment in the technology, enabling it to meet your requirements and go far beyond.

Index

A

Z. Barać and D. Scott-Raynsford, *Azure SQL Hyperscale Revealed*,
https://doi.org/10.1007/978-1-4842-9225-9

B

C

D

E

F

G

H

M

N

O

P

Q

R

S

T

U

V

W, X, Y

Z

Printed in the United States
by Baker & Taylor Publisher Services